Smart Innovation, Systems and Technologies

Volume 82

About this Series

The Smart Innovation, Systems and Technologies book series encompasses the topics of knowledge, intelligence, innovation and sustainability. The aim of the series is to make available a platform for the publication of books on all aspects of single and multi-disciplinary research on these themes in order to make the latest results available in a readily-accessible form. Volumes on interdisciplinary research combining two or more of these areas is particularly sought.

The series covers systems and paradigms that employ knowledge and intelligence in a broad sense. Its scope is systems having embedded knowledge and intelligence, which may be applied to the solution of world problems in industry, the environment and the community. It also focusses on the knowledge-transfer methodologies and innovation strategies employed to make this happen effectively. The combination of intelligent systems tools and a broad range of applications introduces a need for a synergy of disciplines from science, technology, business and the humanities. The series will include conference proceedings, edited collections, monographs, handbooks, reference books, and other relevant types of book in areas of science and technology where smart systems and technologies can offer innovative solutions.

High quality content is an essential feature for all book proposals accepted for the series. It is expected that editors of all accepted volumes will ensure that contributions are subjected to an appropriate level of reviewing process and adhere to KES quality principles.

More information about this series at http://www.springer.com/series/8767

Jeng-Shyang Pan · Pei-Wei Tsai
Junzo Watada · Lakhmi C. Jain
Editors

Advances in Intelligent Information Hiding and Multimedia Signal Processing

Proceedings of the Thirteenth International Conference on Intelligent Information Hiding and Multimedia Signal Processing, August, 12–15, 2017, Matsue, Shimane, Japan, Part II

Editors
Jeng-Shyang Pan
Fujian Provincial Key Lab of Big Data Mining and Applications
Fujian University of Technology
Fuzhou, Fujian
China

Pei-Wei Tsai
Swinburne University of Technology
Hawthorn, VIC
Australia

Junzo Watada
Universiti Teknologi Petronas
Teronoh
Malaysia

Lakhmi C. Jain
University of Canberra
Bruce, ACT
Australia

ISSN 2190-3018 ISSN 2190-3026 (electronic)
Smart Innovation, Systems and Technologies
ISBN 978-3-319-87656-6 ISBN 978-3-319-63859-1 (eBook)
DOI 10.1007/978-3-319-63859-1

Printed on acid-free paper

This Springer imprint is published by Springer Nature
The registered company is Springer International Publishing AG
The registered company address is: Gewerbestrasse 11, 6330 Cham, Switzerland

Preface

Welcome to the 13th International Conference on Intelligent Information Hiding and Multimedia Signal Processing (IIH-MSP 2017), which will be held in Matsue, Shimane, Japan, on August 12–15, 2017. IIH-MSP 2017 is hosted by Universiti Teknologi PETRONAS in Malaysia and technically co-sponsored by Fujian University of Technology in China, Taiwan Association for Web Intelligence Consortium in Taiwan, Swinburne University of Technology in Australia, Fujian Provincial Key Laboratory of Big Data Mining and Applications (Fujian University of Technology) in China, and Harbin Institute of Technology Shenzhen Graduate School in China. It aims to bring together researchers, engineers, and policymakers to discuss the related techniques, to exchange research ideas, and to make friends.

We received a total of 321 submissions from Europe, Asia, and Oceania over places including Taiwan, Thailand, Turkey, Korea, Japan, India, China, and Australia. Finally, 103 papers are accepted after the review process. Keynote speeches were kindly provided by Professor Zhiyong Liu (The Institute of Computing Technology, Chinese Academy of Sciences, Beijing, China) on "Cryo-ET Data Processing and Bio-Macromolecule 3-D Reconstruction" and Professor Takashi Nose (Tohoku University, Japan) on "Flexible, Personalized, and Expressive Speech Synthesis Based on Statistical Approaches." All the above speakers are leading experts in related research fields.

We would like to thank the authors for their tremendous contributions. We would also express our sincere appreciation to the reviewers, Program Committee members, and the Local Committee members for making this conference successful. Finally, we would like to express special thanks to the Universiti Tcknologi PETRONAS in Malaysia, Fujian University of Technology in China, Swinburne University of Technology in Australia, Taiwan Association for Web Intelligence Consortium in Taiwan, and Harbin Institute of Technology Shenzhen

Graduate School in China for their generous support in making IIH-MSP 2017 possible.

August 2017

Jeng-Shyang Pan
Pei-Wei Tsai
Junzo Watada
Lakhmi C. Jain

Conference Organization

Conference Founders

Jeng-Shyang Pan	Fujian University of Technology, China
Lakhmi C. Jain	University of Canberra, Australia and Bournemouth University, UK

Honorary Chairs

Lakhmi C. Jain	University of Canberra, Australia and Bournemouth University, UK
Chin-Chen Chang	Feng Chia University, Taiwan

Advisory Committee

Yôiti Suzuki	Tohoku University, Japan
Bin-Yih Liao	National Kaohsiung Univ. of Applied Sciences, Taiwan
Kebin Jia	Beijing University of Technology, China
Yao Zhao	Beijing Jiaotong University, China
Ioannis Pitas	Aristotle University of Thessaloniki, Greece

General Chairs

Junzo Watada	Universiti Teknologi PETRONAS, Malaysia
Jeng-Shyang Pan	Fujian University of Technology, China

Program Chairs

Akinori Ito	Tohoku University, Japan
Pei-Wei Tsai	Swinburne University of Technology, Australia

Invited Session Chairs

Isao Echizen	National Institute of Informatics, Japan
Ching-Yu Yang	National Penghu University of Science and Technology, Taiwan
Hsiang-Cheh Huang	National University of Kaohsiung, Taiwan
Xingsi Xue	University of Birmingham, UK

Publication Chairs

Chin-Feng Lee	Chaoyang University of Technology, Taiwan
Tsu-Yang Wu	Fujian University of Technology, China
Chien-Ming Chen	Harbin Institute of Technology Shenzhen Graduate School, China

Electronic Media Chairs

Tien-Wen Sung	Fujian University of Technology, China
Jerry Chun-Wei Lin	Harbin Institute of Technology Shenzhen Graduate School, China

Finance Chair

Jui-Fang Chang	National Kaohsiung University of Applied Sciences, Taiwan

Program Committee Members

Toshiyuki Amano	Nagoya Institute of Technology, Japan
Supavadee Aramvith	Chulalongkorn University, Thailand
Christoph Busch	Gjøvik University College, Norway
Canhui Cai	Hua-Qiao University, China
Patrizio Campisi	University of Roma TRE, Italy
Turgay Celik	National University of Singapore, Singapore

Thanarat Chalidabhongse	King Mongkut Institute of Technology Larbkrabang, Thailand
Chi-Shiang Chan	Asia University, Taiwan
Kap-Luk Chan	Nanyang Technological University, Singapore
Bao-Rong Chang	National University of Kaohsiung, Taiwan
Feng-Cheng Chang	Tamkang University, Taiwan
Chien-Ming Chen	Harbin Institute of Technology Shenzhen Graduate School, China
Shi-Huang Chen	Shu-Te University, Taiwan
Yueh-Hong Chen	Far East University, Taiwan
L.L. Cheng	City Univ. of Hong Kong, Hong Kong
Shu-Chen Cheng	Southern Taiwan University of Science and Technology, Taiwan
Hung-Yu Chien	Chi Nan University, Taiwan
Jian Cheng	Chinese Academy of Science, China
Hyunseung Choo	Sungkyunkwan University, Korea
Shu-Chuan Chu	Flinders University, Australia
Kuo-Liang Chung	National Taiwan University of Science and Technology, Taiwan
Hui-Fang Deng	South China University of technology, China
Isao Echizen	National Institute of Informatics, Japan
Masaaki Fujiyoshi	Tokyo Metropolitan University, Japan
Pengwei Hao	Queen Mary, University of London, UK
Yutao He	California Institute of Technology, USA
Hirohisa Hioki	Kyoto University, Japan
Anthony T.S. Ho	University of Surrey, UK
Jiun-Huei Ho	Cheng Shiu University, Taiwan
Tzung-Pei Hong	National University of Kaohsiung, Taiwan
Jun-Wei Hsieh	National Taiwan Ocean University, Taiwan
Raymond Hsieh	California University of Pennsylvania, USA
Bo Hu	Fudan University, China
Wu-Chih Hu	National Penghu University, Taiwan
Yongjian Hu	South China University of Technology, China
Hsiang-Cheh Huang	National Kaohsiung University, Taiwan
Du Huynh	University of Western Australia, Australia
Ren-Junn Hwang	Tamkang University, Taiwan
Masatsugu Ichino	University of Electro-Communications, Japan
Akinori Ito	Tohoku University, Japan
Motoi Iwata	Osaka Prefecture University, Japan
Jyh-Horng Jeng	I-Shou University, Taiwan
Kebin Jia	Beijing University of Technology, China
Hyunho Kang	Tokyo University of Science, Japan
Muhammad Khurram Khan	King Saud University, Kingdom of Saudi Arabia
Lei-Da Li	China University of Mining and Technology, China

Contents

Massive Image/Video Compression and Transmission for Emerging Networks

Extraction of EEG Components Based on Time - Frequency Blind Source Separation

Xue-Ying Zhang(✉), Wei-Rong Wang, Cheng-Ye Shen, Ying Sun, and Li-Xia Huang

College of Information Engineering,
Taiyuan University of Technology, Taiyuan 030024, China
tyzhangxy@163.com

Abstract. In order to extract EEG characteristic waves better, this paper adopts the method of combining wavelet transform with time-frequency blind source separation based on smooth pseudo Wigner-Ville distribution. Firstly, the EEG signal is extracted by wavelet transform to reconstruct the β wave band signal and reconstructed as the initial extracted characteristic wave. Then, to remove the other components which are less relevant to get the enhanced beta wave signal, the time-frequency blind source separation technique based on the smooth pseudo-Wigner distribution is used for the initial extracted Target wave. Finally, the features are extracted, and the support vector machine is used to classify and identify the emotional categories. The experimental results show that the recognition rate is improved when the characteristic wave is extracted by using wavelet transform only.

Keywords: EEG · Smoothed pseudo Wigner-Ville distribution · Emotion recognition · β · Wavelet transform · Blind source separation

1 Introduction

Electroencephalography (EEG) as a measure of brain activity and an effective tool, when the human brain in different state of mind and different emotional state, the cerebral cortex position reflects the EEG signal there will be different degrees of difference, Such as EEG in the β wave, if the emotional instability, narrow extreme or by external stimuli, β wave activity will be weakened [1]. Therefore, how to effectively extract the useful information in EEG signals is of great significance for the study of emotional state recognition. Wavelet transform and time-frequency blind source separation (TF-BSS) are very effective signal processing methods. In the literature [2], the multi-path v support vector machine and the time-frequency blind source separation algorithm are combined to extract the ECG signal of the fetus successfully. The literature [3] Using the method of wavelet transform and spatial model to extract the EEG features, the recognition result is improved. Although the wavelet transform can extract the EEG signal of the corresponding frequency band, it ignores the correlation between the electrode signals. While the blind source separation algorithm can obtain the relevant signal between the individual electrode signals, it is inevitable that other irrelevant of the band, so the researchers proposed to combine the two to improve the extraction of the characteristic wave.

J.-S. Pan et al. (eds.), *Advances in Intelligent Information Hiding and Multimedia Signal Processing*, Smart Innovation, Systems and Technologies 82,
DOI 10.1007/978-3-319-63859-1_1

In this paper, wavelet transform is combined with time-frequency blind source separation based on smoothed pseudo Wigner-Ville distribution (TF-SPWVD) based on smooth pseudo-Wigner-Ville distribution. The method of extracting β wave is so as to avoid introducing other interference bands, and the relevant β wave between the electrode signals can be obtained to a great extent.

2 The Correlation Theory

2.1 Emotion Model

There are different definitions of emotion. The following is a description of the dimensional affective model.

The dimension model of emotion is more close to human's natural emotion than discrete emotion model. The theory of "Arousal-Valence" is the most commonly used emotion model. Among them, Arousal indicates the degree of activation of emotion; Valence indicates the positive or negative emotion [4]. In the two-dimensional affective plane, the horizontal axis and the vertical axis express the emotional valence and the dimension of the stimulus. As shown in Fig. 1.

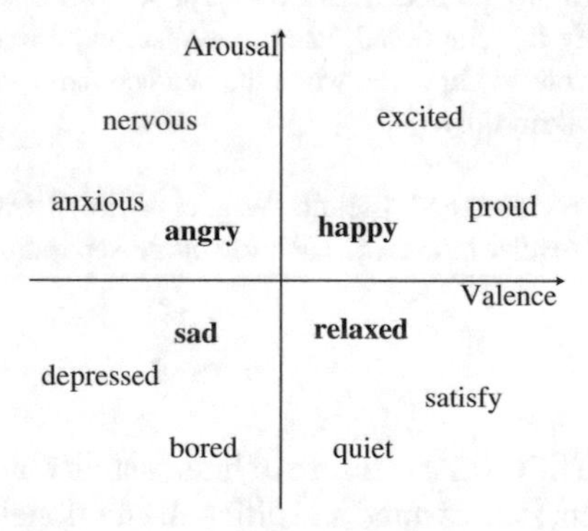

Fig. 1. 2-dimensional emotion model

2.2 Blind Source Separation Based on Smoothed Pseudo Wigner-Ville

In this paper, a blind source separation algorithm (TF-SPWVD) based on a smooth pseudo Wigner-Ville distribution is proposed based on the compact pseudo-Wigner-Ville distribution based on TF-BSS.

2.2.1 Wigner-Ville Distribution

The Wigner-Ville distribution is a kind of quadratic time-frequency representation with energy, with many properties of time-frequency distribution [5]. This method can effectively detect the start and end time and frequency information of the mutation and non-stationary signals.

The Wigner-Ville distribution (WVD) of the signal $s(t)$ is defined as:

$$W_x(t,f) = \int_{-\infty}^{+\infty} x(t + \frac{\tau}{2})x^*(t - \frac{\tau}{2}) \cdot e^{-2j\pi f\tau} d\tau \qquad (1)$$

Where, $x(t)$ is the practical letter, $s(t)$ is the analytical signal. The practical signal $s(t)$ is used as the real part, and the signal $s(t)$ is used as the virtual part of the Hilbert transform, which is the analytic signal $s(t)$ of the practical signal $x(t)$.

The total integration of $W_x(t,f)$ for time and frequency is equal to the energy of the signal $x(t)$.

$$\frac{1}{2\pi}\int_{-\infty}^{+\infty}\int_{-\infty}^{+\infty} W_x(t,f)dtdf = \frac{1}{2\pi}\int_{-\infty}^{+\infty} |X(\omega)|^2 d\omega = \int_{-\infty}^{+\infty} |x(t)|^2 dt = E \qquad (2)$$

According to the Formula (2), WVD is the energy distribution on the plane of t-f.

By Formula (1), WVD has a bilinear form, so it has a cross term, assuming that $x(t)$ is the sum of $x_k(t)$ components:

$$x(t) = \sum_{k=1}^{n} x_k(t) \qquad (3)$$

According to the definition of Wigner-Ville distribution:

$$W_x(t,f) = \sum_{k=1}^{n} W_{x_k}(t,f) + \sum_{k=1}^{n}\sum_{l=1}^{n} 2\mathrm{Re}[W_{x_k,x_l}(t,f)] \qquad (4)$$

Where $\sum_{k=1}^{n}\sum_{l=1}^{n} 2\mathrm{Re}[W_{x_k,x_l}(t,f)]$ is a cross terms. The cross term provides false spectrum distribution, which affects the physical resolution of WVD.

2.2.2 Smooth Pseudo Wigner Ville Distribution

The smoothed pseudo Wigner-Ville distribution (SPWVD) is a time window in which both the time domain and the frequency domain variables are added to reduce the cross term. It is defined as:

$$SPWVD_x(t,f) = \int_{-\infty}^{+\infty}\int_{-\infty}^{+\infty} x(t + \frac{\tau}{2})x^*(t - \frac{\tau}{2})\mathrm{h}(\tau)\mathrm{g}(u)\mathrm{e}^{-j2\pi f\tau} d\tau du \qquad (5)$$

Where $h(\tau)$ and $g(\tau)$ are two symmetric windows, and $h(0) = g(0) = 1$.

The time-frequency blind source separation algorithm based on SPWVD TF-SPWVD steps [6] briefly outline as follows:

(1) Zero meaning the observation signal $x(t)$ to obtain $\bar{x}(t)$, and the time- dependent autocorrelation matrix $R_{\bar{x}}^{\phi}(t,0)$ is calculated.
(2) Decomposing the eigenvalue of the $R_{\bar{x}}^{\phi}(t,0)$, using $W = D_s^{-1/2}(0)V_s^H(0)$ to calculate a whitening array W, and calculated the $z(t) = W \cdot \bar{x}(t)$.
(3) The M SPWVD matrix of $z(t)$ are calculated by the method of Eq. (5), and then $\{V_{ZZ}(t_i, V_i)|i = 1,\ldots, M\}$ is obtained.
(4) Select the points in the time domain [7],$(t_i,f_i), i = 1,\ldots, K$.
(5) The unitary matrix U is obtained by the matrix $\{V_{ZZ}(t_i, V_i)|i = 1,\ldots, K\}$, through the method of joint approximate diagonalization.
(6) Use $\hat{S} = U^H Wx(t)$ to estimate the source signal and then calculate the mixture matrix $A = W^{\#} U \frac{dy}{dx}$.

3 The Experiment

In order to verify the validity of the "TF-SPWVD combined wavelet transform" method, a set of experiments is designed. The time domain characteristics, frequency domain characteristics and power spectrum characteristics are extracted by "TF-SPWVD combined with wavelet transform" and wavelet transform. The validity of "TF-SPWVD combined with wavelet transform" is verified by comparing the recognition results of the two groups.

3.1 The Database

The data set used in the paper is DEAP [8] data. The data set is a public multi-modal emotional data set published by Sanders Koelstra et al. A total of 32 subjects are recorded with 32-channel EEG data (40 channels for the 10-day system) and 8- channel peripheral physiological signals. These videos have been annotated in advance using "Arousal-Valence". These videos are marked with a high awakening (HAHV), low wake-up pleasure (LAHV), low awakening low pleasure (LALV), high awakening low pleasure (HALV), respectively, according to the marked V and A values.

The paper uses the 348 data samples in the DEAP database, and select the four electrode pairs [9, 10] (FP1-FP2, F3-F4, F7-F8, FC5-FC6) in each data sample, as shown in Fig. 2. And intercept the 15 s $\sim$ 45 s data segment of each electrode signal for experimental analysis. The following describes the "TF-SPWVD combined with wavelet transform" the main steps.

Since EEG is a non-stationary signal, it is necessary to perform preprocessing before waveform extraction and signal analysis. The preprocessing mainly includes framing and windowing. Set the frame length 512, frame shift 256 and hamming window for processing.

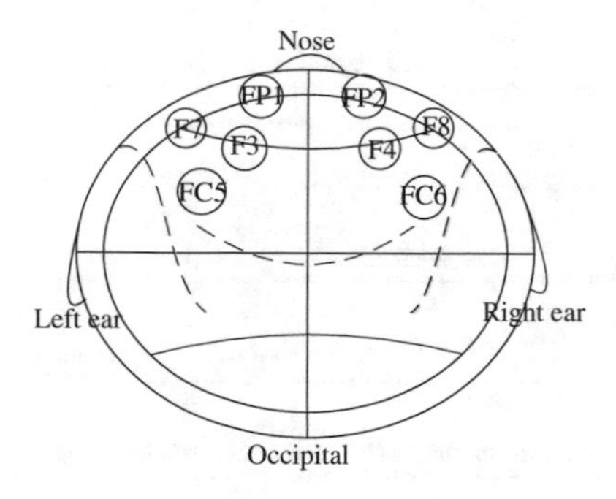

Fig. 2. The electrode distribution

3.2 The Extraction of β Wave by Wavelet Transform

The second layer detail coefficients of the respective electrode signals (for example, FP1, F3, F7, FC5 and FP2, F4, F8, FC6) are extracted and reconstructed to obtain the β wave of each electrode signal, and the target of the preliminary extraction of the β wave is completed.

3.3 Using TF-SPWVD to Enhance the Characteristic of β Wave

Each of the electrodes is extracted as the target signal, and then it is used as the input of the TF-SPWVD with the original signal of the other electrodes on the same side of the brain. For example, the first step is to extract the FP1 electrode channel beta wave, and then with the same side of the F3, F7 and FC5 as the input of TF-SPWVD. The original guide EEG signal itself contains beta waves, and beta waves target signal and original signal has great relevance, after TF-SPWVD, the beta target signal with the original signal in combination as an independent component of the output TF-SPWVD. Figure 3 shows the signal output for the FP1 process.

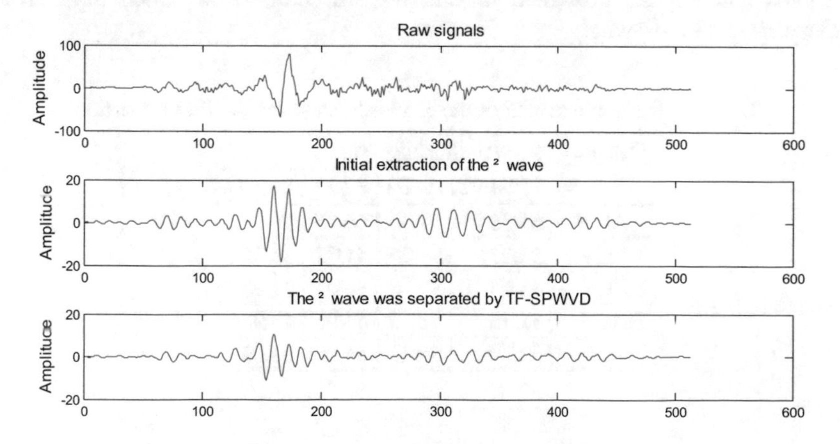

Fig. 3. Signal time domain graphics

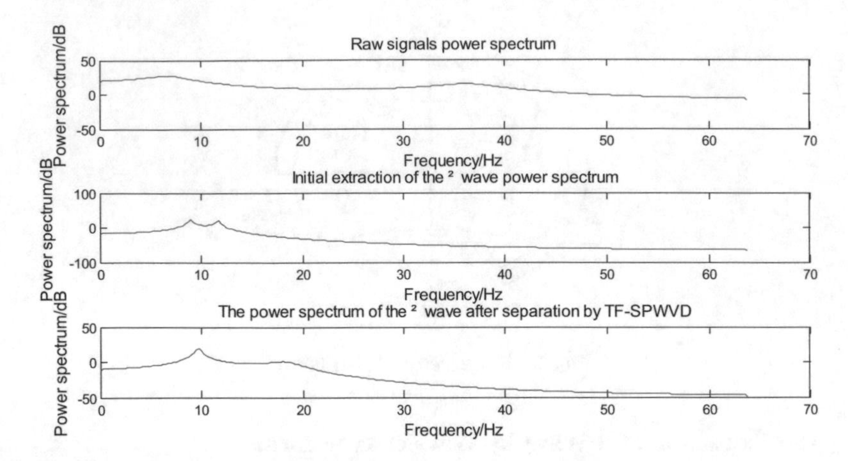

Fig. 4. The signal power spectrum

The power spectrum of each signal in Fig. 3 is estimated and the results are shown in Fig. 4. The β wave extracted by TF-SPWVD is more concentrated in the 16 ~ 32 Hz band.

The time domain characteristics, frequency domain characteristics and power spectrum characteristics of the β wave obtained by TF-SPWVD separation are taken as the input vector of the support vector machine (SVM). In addition, the wavelet transform is used to extract the beta wave, and the same features are extracted.

4 Experimental Results and Analysis

In this paper, support vector machine (SVM) is used as the emotion recognition model, and the electrode pair is used to identify it. The average classification accuracy was obtained by cross validation, the training set was 200, and the test set was 148.

Tables 1 and 2 are presented in a histogram form, from which the following conclusions can be drawn:

Table 1. Recognition rate of the electrode pairs for the first extraction

Category	Recognition rate (%)			
	FP1-FP2	F3-F4	F7-F8	FC5-FC6
HAHV	56.76	59.46	56.76	56.76
LAHV	54.05	54.05	51.35	51.35
LALV	48.65	56.76	54.05	54.05
HALV	51.35	51.35	54.05	45.95
Average	52.70	55.41	54.05	52.03

Table 2. Recognition rate of electrode pairs after TF-SPWVD separation

Category	Recognition rate (%)			
	FP1-FP2	F3-F4	F7-F8	FC5-FC6
HAHV	72.97	81.08	78.38	70.27
LAHV	59.46	70.27	64.86	67.57
LALV	75.68	72.97	59.46	72.97
HALV	64.86	56.76	72.97	75.68
Average	68.24	70.27	68.92	71.62

(a) The β wave extracted from the electrodes F7 and F8 is the target signal, and the extracted β wave is obviously improved in the recognition rate compared with the other electrodes. It is shown that the β wave extracted at the electrodes F7 and F8 is different from the β wave in the other electrode signals. Large correlation, can make the separation of the β wave can be enhanced;

(b) Compared with the average classification result of wavelet transform, the average recognition rate after TF-SPWVD separation has been improved greatly. Combined with Fig. 4, after TF-SPWVD separation, β wave is more concentrated in the 16 ~ 32 Hz, β waveform features protruding, so that the recognition rate is improved;

(c) In addition, the recognition rate of partial electrode pairs was not obvious after TF-SPWVD separation, for example, the recognition rate of FP3-FP4 in LAHV and the recognition rate of F3-F4 in HALV. After the analysis of the results in the literature [10, 11], there are two reasons: firstly, it is likely that the brain activity at the electrode is weaker than other emotions at the electrode; secondly, the extracted β wave is weaker and the band of other adjacent frequency segments is strong at the electrode. And after TF-SPWVD separation, the β wave is weakened due to the small correlation with the wave in the other electrode original signal.

5 The Conclusion

Because of the advantages of wavelet transform and time-frequency blind source separation technology in signal processing, the wavelet transform and TF-SPWVD are combined. Firstly, the EEG is extracted by wavelet transform to obtain the β wave of each electrode, and then SPWVD algorithm is used to enhance the β wave of each electrode, so that the initial β wave is enhanced, and the recognition rate is improved. Finally, the validity of TF-SPWVD combined with wavelet transform is verified by the comparison of recognition rate. However, there are some shortcomings in the text. The degree of enhancement for the β wave has not yet given specific values, and whether the elements that are not related to emotion are also enhanced. In the next step, the degree of enhancement and the inhibition of irrelevant components will be studied more deeply.

Acknowledgments. This work is supported by the National Natural Science Foundation of China (No. 61371193).

References

1. Jenke, R., Peer, A., Buss, M.: Feature extraction and selection for emotion recognition from EEG. IEEE Trans. Affect. Comput. **5**, 327–339 (2014)
2. Liang, H., Xiujuan, P., Qian, L.: FECG signal extraction based on multichannel v-SVR combined with TFBSS. Chin. J. Sci. Instrum. **36**, 1381–1387 (2015)
3. Qingshan, S., Xihao, C., Xugang, X., Qizhong, Z.: Feature extraction of EEG based on DTCWT and CSP. J. Dalian Univ. Technol. **56**, 70–76 (2016)
4. Zeng, Z., Pantic, M., Roisman, G.I.: A Survey of affect recognition methods: audio, visual, and spontaneous expressions. IEEE Trans. Pattern Anal. Mach. Intell. **31**, 39–58 (2008)
5. Xianda, Z.: Modern Signal Processing. Tsinghua University Press, Beijing (2015)
6. Jing, G.: Algorithms of Blind Sources Separation in Time-Frequency Domains. Chongqing University, Chongqing (2012)
7. Joint anti-diagonalization for blind source separation: Belouchrani, A., Abed, Meraim, K., Amin, M.G. IEEE Int. Conf. Acoust. Speech Sig. Process. **5**, 2789–2792 (2001)
8. DEAPDatase. http://www.eecs.qmul.ac.uk/mmv/datasets/deap/download.html
9. Bradley, M.M., Lang, P.J.: Measuring emotion: the self-assessment manikin and the semantic differential. J. Behav. Ther. Exp. Psychiatry **25**, 49 (1994)
10. Jatupaiboon, N., PanNgum, S., Israsena, P.: Emotion classification using minimal EEG channels and frequency bands (2013)

An Algorithm for Asymmetric Clipping Detection Based on Parameter Optimization

Jiwei Zhang(✉), Shaozhang Niu(✉), Yueying Li(✉), and Yuhan Liu(✉)

Beijing Key Lab of Intelligent Telecommunication Software and Multimedia, School of Computer Science, Beijing University of Posts and Telecommunications, Beijing 100876, China
zhangjiwei5510@163.com, szniu@bupt.edu.cn, 3068645478@qq.com, 877250768@qq.com

Abstract. Asymmetric clipping of digital images is a common method of image tampering, and the existing identification techniques of which are relatively meager. Camera calibration technology is an important method to determine the tampering of asymmetric cutting, but the proposed algorithm has made too many assumptions on the internal parameters matrix of the camera, resulting in some error. This paper presents a parameter optimization algorithm based on camera calibration: by keeping the four parameters in the original camera's five internal parameters, after approximate processing, to achieve that a single picture contains no coplanar of the two regular geometric figures can calculate the coordinates of the principal point, and as a basis for the image forensics of the asymmetric cutting tampering. The experimental results show that the proposed algorithm can effectively estimate the camera parameters, the application scope and accuracy can be improved greatly, and can accurately detect the image tampering behavior of asymmetric clipping.

Keywords: Blind forensics · Regular geometric figures · Parameter optimization · Camera calibration · Asymmetric clipping

1 Introduction

With the development of the Internet and the popularity of image processing software, digital image transmission and tampering are becoming more convenient. The means of image tampering–like fuzzy retouching [1]; splicing; copy-move; double JPEG compression; cutting and other means–make the authenticity of the image suffered serious damage, while bringing some political, military, cultural, academic and other aspects of the impact. As a result, our image forensics work is facing challenges.

Image forensics has made some progress in image source authentication [2]; copy-move [3]; splicing [4]; double JPEG compression [5]; etc. Forensics techniques based on computer vision [6] and depth learning [7] have begun to emerge.

Camera calibration technology [8] is an important tool in computer vision. Through the image contains the known geometric shape of the object it can get the camera's internal and external parameters, thereby the image asymmetric cutting tampering

J.-S. Pan et al. (eds.), *Advances in Intelligent Information Hiding and Multimedia Signal Processing*, Smart Innovation, Systems and Technologies 82,
DOI 10.1007/978-3-319-63859-1_2

behavior can be identified. In the past, the algorithm based on camera calibration generally needs to measure the actual size of the calibrated object. However, this greatly limits the scope of the algorithm. The optimization algorithm proposed by us do not need to know the size of the calibration object, it is only necessary to have two or more non-coplanar calibrations in the image to obtain the internal parameters and the external parameters of the camera. By estimating the difference in camera parameters, we can achieve the identification of the authenticity of the image.

The paper is organized as follows: Sect. 2 describes the camera imaging process. Section 3 gives the idea of parameter optimization method to extract evidence of image tampering, while experimental results and conclusion are given in Sects. 4 and 5, respectively.

2 Backgrounds

A basic model of the camera's imaging process is the pinhole model. The basic pinhole model can be understood as a mapping from a three-dimensional continental space to a two-dimensional continental space. The object of the world coordinate system can be transformed into an object of the image coordinate system through a rigid body transformation plus a projection transformation. The transformation from the three-dimensional world coordinate system to the image two-dimensional coordinate system is as follows:

$$x = \lambda \boldsymbol{A} X_c = \lambda \boldsymbol{A} R X_w \tag{1}$$

$$\boldsymbol{A} = \begin{bmatrix} \alpha & \gamma & u \\ 0 & \beta & v \\ 0 & 0 & 1 \end{bmatrix} \tag{2}$$

Where x represents the point in the two-dimensional image coordinate system, λ represents the scale of the transformation, and $\boldsymbol{A}$ is the 3×3 order projection matrix, which is also called the camera internal parameters, with (u, v) the camera's principal point coordinates (for normal images, the coordinates of the principal point are the center of the image), α and β the scale factors in image u and v axes, and γ is the parameter describing the degree of skew of the camera CCD and the corresponding image coordinates. Besides, X_C is a point in the camera coordinate, with R the 3×4 rigid spatial transformation matrix which is also called the extrinsic camera parameters, and X_W represents a world point in the world coordinate.

In order to facilitate the calculation, the model can be further simplified. In the corresponding point selection, we often select the point in the same plane, then the camera's external parameter matrix R can be reduced to 3×3 order, and there:

$$x = \lambda \boldsymbol{A} X_c = \lambda \boldsymbol{A} R X_w = H X_w \tag{3}$$

Where H is a 3×3 homography matrix. If define $H = [h_1, h_2, h_3]$ and $R = [r_1, r_2, t]$, we have (4)

$$[h_1 \quad h_2 \quad h_3] = \lambda \boldsymbol{A}[\boldsymbol{r}_1 \quad \boldsymbol{r}_2 \quad \boldsymbol{t}] \tag{4}$$

Where r_1 and r_2 is the orthogonal vector for the unit, so there are $r_1^T r_1 = r_2^T r_2 = 1$, and $r_1^T r_2 = 0$. Then, two constraints of the camera's internal parameters are obtained:

$$h_1^T \boldsymbol{A}^{-T} \boldsymbol{A}^{-1} h_2 = 0 \tag{5}$$

$$h_1^T \boldsymbol{A}^{-T} \boldsymbol{A}^{-1} h_1 = h_2^T \boldsymbol{A}^{-T} \boldsymbol{A}^{-1} h_2 \tag{6}$$

If there are only four unknown parameters in $\boldsymbol{A}$, it can be seen from the literature [9] that the two sets of calibrators that are not coplanar in the image can obtain two unrelated monocritical matrices, which can be substituted into Eqs. (5) and (6), so the four parameters of $\boldsymbol{A}$ can be obtained.

3 Asymmetric Clipping Detection

The principal point is an important part of the camera's internal parameters, reflecting the important geometric features of the image. If the image is tampered with asymmetric cutting, the center of the image will be offset. According to the relationship between the image center and the principal point, you can determine whether the image has experienced the image asymmetric cutting tampering from the location of the principal point.

3.1 The Estimate of Homography Matrix

The estimate of homography matrix is an important step to solve the problem of computer vision, which determines the accuracy of the internal parameters of the camera. The paper [8] introduced Zhang Zhengyou's calibration method, using the corresponding size of the board to estimate the homography matrix. R-Hardey in the paper [6] described the situation of various camera models and the corresponding method of estimating the homography matrix in detail. In [10], the human eye in the image is modeled as a calibrator, and the homography matrix is obtained without relying on a known size. In the case of the image contains several rules objects, the estimation of the homography matrix can be realized in the paper [9].

However, for a single natural image, there is often no human face and chess board and other fixed objects. Besides, the size of the object in the image is also difficult to obtain, which brought challenges to finding evidence for single image tampering. In the case where the size of the object in the image is unknown, the method of image cropping need to be further optimized.

3.2 Estimating Camera Parameters Using Optimization Methods

The paper [9] implements the method of using the regular objects in the single image to realize the image tamper identification, but makes too many assumptions on the internal parameter matrix. The paper [9] make the assumption that α and β are equivalent, greatly reducing the accuracy of the camera parameters.

This paper solves the problem of too many hypotheses to the internal parameters in the paper [9]. We remove the assumption that α and β are equivalent, after some approximation, the principal points can be obtained by using the regular graphs that are not coplanar in the single image. Determine whether the image has experienced the asymmetric cutting tampering through the offset of principal point. Result in a wider range of the scope of the camera as well as improving the accuracy of the solution of the principal point. The corresponding parameters are optimized as shown in Fig. 1.

$$\boldsymbol{A} = \begin{bmatrix} \alpha & 0 & u \\ 0 & \alpha & v \\ 0 & 0 & 1 \end{bmatrix} \longrightarrow \boldsymbol{A} = \begin{bmatrix} \alpha & 0 & u \\ 0 & \beta & v \\ 0 & 0 & 1 \end{bmatrix}$$

Fig. 1. Parameter optimization

The specific algorithm is as follows:
Define

$$B = \boldsymbol{A}^{-T}\boldsymbol{A}^{-1} = \begin{bmatrix} B_{11} & B_{12} & B_{13} \\ B_{21} & B_{22} & B_{21} \\ B_{31} & B_{32} & B_{33} \end{bmatrix} = \begin{bmatrix} \left[\frac{1}{\alpha^2}\right] & -\frac{\lambda}{\alpha^2\beta} & \frac{v\lambda - u\beta}{\alpha^2\beta} \\ -\frac{\gamma}{\alpha^2\beta} & \frac{\gamma^2}{\alpha^2\beta^2} + \frac{1}{\beta^2} & -\frac{\gamma(v\gamma - u\beta)}{\alpha^2\beta^2} - \frac{v}{\beta^2} \\ \frac{v\gamma - u\beta}{\alpha^2\beta} & -\frac{\gamma(v\gamma - u\beta)}{\alpha^2\beta^2} - \frac{v}{\beta^2} & \frac{(v\gamma - u\beta)^2}{\alpha^2\beta^2} + \frac{v^2}{\beta^2} + 1 \end{bmatrix} \tag{7}$$

B is a symmetric matrix,
Define

$$b = \begin{bmatrix} B_{11} & B_{12} & B_{22} & B_{13} & B_{23} & B_{33} \end{bmatrix}^T \tag{8}$$

Define the i-th column vector of H be expressed as

$$h_i = \begin{bmatrix} h_{i1} & h_{i2} & h_{i3} \end{bmatrix}$$

Then we have:

$$h_i^T Bh_i = V_{ij}^T b \tag{9}$$

$$V_{ij} = \left[h_{i1}h_{j1} \quad h_{i1}h_{j2} + h_{i2}h_{j1} \quad h_{i2}h_{j2} \quad h_{31}h_{j1} + h_{i1}h_{j3} \quad h_{i3}h_{j2} + h_{i2}h_{j3} \quad h_{i3}h_{j3} \right]$$

Thus, Eq. (4) can be written in the form of b:

$$\begin{bmatrix} V_{12}^T \\ V_{11}^T - V_{22}^T \end{bmatrix} b = 0 \tag{10}$$

If there are N images in the case, we have:

$$Vb = 0 \tag{11}$$

Where V is a 2N $\times$ 6 matrix, and if N is greater than or equal to three images, b can be solved (with a scale factor), so that five internal parameters can be obtained. However, we are based on a map of non-coplanar rules of the object to find the coordinates of the principal point, so need a certain approximation. γ will be approximately equal to 0, still retain α and β, then a total of four unknown, theoretically can be solved. Thereby $B_{12} = 0$, $B_{33} = \frac{u^2}{\alpha^2} + \frac{v^2}{\beta^2} + 1$, B_{33} has a quadratic term, then it is difficult to simplify. In the actual calculation, it is found that the value $\frac{u^2}{\alpha^2} + \frac{v^2}{\beta^2}$ is close to zero, so B_{33} will be approximately equal to 1.

Define

$$b = [B_1 \quad 0 \quad B_2 \quad B_3 \quad B_4 \quad 1]^T \tag{12}$$

Then Eq. (10) can be written

$$\begin{bmatrix} v_{11} & v_{12} & v_{13} & v_{14} & v_{15} & v_{16} \\ v_{21} & v_{22} & v_{23} & v_{24} & v_{25} & v_{26} \end{bmatrix} b = 0$$

Further simplification, $\begin{bmatrix} v_{11} & v_{13} & v_{14} & v_{15} \\ v_{21} & v_{23} & v_{24} & v_{25} \end{bmatrix}$ with the D matrix to save, and $\begin{bmatrix} -v_{16} \\ -v_{26} \end{bmatrix}$ With the v matrix to save. So we get the following formula

$$\begin{bmatrix} v_{11} & v_{13} & v_{14} & v_{15} \\ v_{21} & v_{23} & v_{24} & v_{25} \\ \cdots & \cdots & \cdots & \cdots \\ v_{n1} & v_{n3} & v_{n4} & v_{n5} \end{bmatrix} b = \begin{bmatrix} -v_{16} \\ -v_{26} \\ \cdots \\ -v_{n6} \end{bmatrix}$$

Which is $Db = v$. Using the least squares method to calculate the b matrix, then we solve b, and $u = -B_3/B_1$, $\mathrm{v} = -B_4/B_2$ can be obtained from the equation.

4 Experiment and Analysis

4.1 Error Analysis

In order to reduce the error caused by the model, we use a group of (200 photos) with no coplanar regular graphics image as a test sample. Estimating the principal point of the image using the proposed parameter optimization algorithm, and normalize the coordinates of the principal Point. The obtained principal point abscissa can basically fall within 20% of the image width of the radius of the central area.

It can be seen from Fig. 2 that the calculated coordinates of the principal points can fall near the center of the image. The experiments in [9] are also based on 0.2 for the radius, but this method estimates the coordinates of the principal points more accurately, the principal distribution of the region is more concentrated, and with less error.

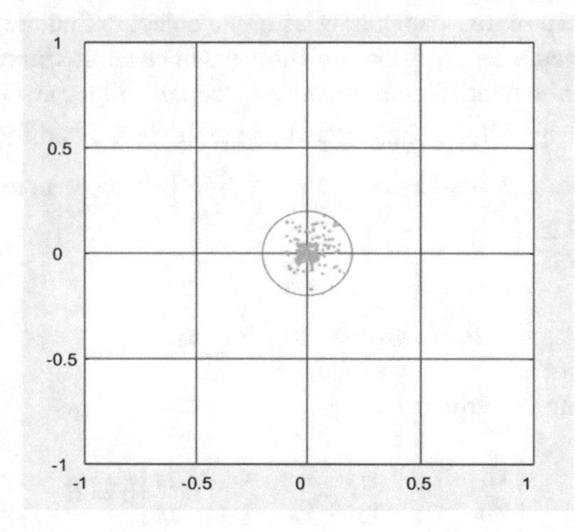

Fig. 2. Estimation of the principal point

4.2 Detection Effect

In Fig. 3 there are two sets of images (a), (c) and (b), (d). (c) is (a) cropped, and (d) is (b) cropped. Each of the images contains a number of regular geometries (text, rules of the decorative patterns, trademarks, etc.) that can be used as a marker for image cutting evidence. For example, the regular walls and the words in Fig. 3, according to the criterion of wall building and the correspondent typeface on the Internet, can be used to identify the worldly coordinate of the object. In addition, Fig. 3(b) has a significant postal logo icon, since the icon rules and the proportion is known, it can be used as an identification of tampering.

Fig. 3. Example image

This method requires two or more non-coplanar regular geometries to determine the single matrix, in addition, through the camera's internal parameter optimization, the accuracy rate has improved greatly. The rectangular box in the image takes 20% of the length of the image as the edge, and the intersection of the red crosses in the rectangle as the position of the principal point. It can be seen from (a) that the principal point of the image falls within the rectangular box, but the principal point of the cropped image (c) falls directly on the outside of the image, the detection accuracy of which is much higher than that of the paper [9]. This result can also be applied to figure (b), the principal point of the original image falls within the square area, after being cropped to image (d), the principal point ran out of the rectangular red box, obviously deviated from the image center, by that we can say the image (d) has experienced asymmetric cutting tampering.

5 Conclusion

It is a prevalent behavior that the tampering of the asymmetric image cutting. In this paper, a method to detect the asymmetric clipping and tampering behavior in digital images is proposed by using the method of camera calibration. Compared to the past method of camera calibration, the algorithm in this paper does not depend on the modeling of objects and the premise of fixed size, the applicable camera range of the algorithm is greatly improved. In addition, the accuracy of the coordinates of the principal points is improved, hence realizing the evidence-obtain process of tampering behavior of the asymmetric cropping image. It can be seen from the experimental results that the proposed method can effectively detect the tampering behavior of asymmetric cropping images. But for the symmetry of cutting behavior, the effectiveness of the method is to be improved for the future focus of the research.

References

1. Bharati, A., Singh, R.: Detecting facial retouching using supervised deep learning. IEEE Trans. Inf. Forensics Secur. **11**(9), 1903–1913 (2016)
2. Ke, Y., Shan, Q., Qin, F., Min, W.: Image recapture detection using multiple features. Int. J. Multimedia Ubiquit. Comput. **8**(5), 71–82 (2013)
3. Ardizzone, E., Bruno, A., Mazzola, G.: Copy-move forgery detection by matching triangles of key points. IEEE Trans. Inf. Forensics Secur. **10**(10), 2084–2094 (2015)
4. El-Alfy, E.S.M., Qureshi, M.A.: Combining spatial and DCT based Markov features for enhanced blind detection of image splicing. Formal Pattern Anal. Appl. **18**(3), 1–11 (2014)
5. Meng, X., Niu, S.: Technology of digital image tampering based on double JPEG compression. In: National Conference on Information Hiding and Multimedia Information Security (2010)
6. Guo, C., Hong, Y.: Automatic camera calibration method using checkerboard target. J. Comput. Eng. Appl. **52**(12), 176–179 (2016)
7. Zhang, Y., Win, L.L., Goh, J., Thing, V.L.L.: image region forgery detection: a deep learning approach. In: Proceedings of the Singapore Cyber-Security Conference (SG-CRC) (2016)
8. Zhang, Z.: A flexible new technique for camera calibration. IEEE Trans. Pattern Anal. Mach. Intell. **22**(11), 1330–1334 (2000)
9. Meng, X., Niu, S.: Asymmetric crop detection algorithm based on camera calibration. J. Electron. Inf. Technol. **34**(10) (2012)
10. Johnson, M., Farid, H.: Detecting photographic composites of people. In: 6th International Workshop on Digital Watermarking, Guangzhou, China, vol. 5041, pp. 19–33 (2007)

Automatic Facial Age Estimate Based on Convolution Neural Network

Jiancheng Zou, Xuan Yang(✉), Honggen Zhang, and Xiaoguang Chen

Institute of Image Processing and Pattern Recognition, North China University of Technology, Beijing 100144, China
XuanBaby_1993@163.com

Abstract. As a new biometric recognition technology, automatic age estimation based on facial image has become an important subject in the field of computer vision and human-computer interaction (HCI). And automatic facial age estimate system has been increasingly used in criminal investigation, image retrieval and intelligent monitoring in recent years. Therefore, the research of facial age estimation has a broad prospect. Convolution neural network (CNN) as a deep learning architecture can extract the essential features of the facial image with a better effect than traditional methods, especially in the case of large changes in imaging shooting conditions. In this paper, an improved method of facial age estimate based on CNN is proposed. By considering the number limitation of the existing age estimation data sets, we adopt the method of fine-tuning the existed network model. The recognition rate can be increased by 3% based on the proposed method. A facial age estimate system has been constructed for applications and the experimental results show that the system can meet the real-time application needs.

Keywords: Facial age estimation · CNN · Deep learning

1 Introduction

With the development of human-computer interaction, intelligent business and social security, the research of human face attributes has attracted people's attention. Age as one of the most important attributes has become a hot topic in the field of image processing. Automatic facial age estimation systems is an important foundation to achieve affective computing and artificial intelligence, which has broad application prospects [1, 2]. The systems can be beneficial in many fields, like human-computer interaction, police investigation, image retrieval and intelligent monitoring etc. With the improvement of face detection algorithm and the innovation of feature extraction technology, the facial age estimation effect was significantly better than before. But there are still many problems. As we all know, the appearance of people changed greatly with the growth of age, and the facial features of different age groups were also different. It is difficult to accurately observe the age of a person. The age characteristics of human face are mainly manifested in the skin texture, skin color, the degree of wrinkles, etc. However, these factors are usually related to personal habits, work environment and so on.

J.-S. Pan et al. (eds.), *Advances in Intelligent Information Hiding and Multimedia Signal Processing*, Smart Innovation, Systems and Technologies 82,
DOI 10.1007/978-3-319-63859-1_3

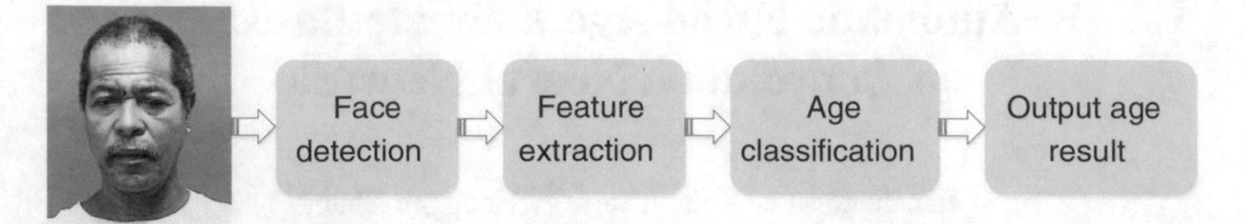

Fig. 1. An overview of the system for age estimation

Therefore, it is still a difficult problem to estimate the age of face images. Age estimates are similar to most classification problems, first extract the model, and then use the classifier for classification. The general process is shown in Fig. 1

Convolution neutral network (CNN) is a new type of neural network. It is a combination of traditional artificial neural network and deep learning technology, and it provides a new idea for classification tasks. Deep learning combines feature extraction and classification, which achieved excellent results in classification problems [3–5]. The method avoids the complicated feature extraction and data reconstruction process in the traditional recognition algorithms [6].

2 Related Work

Since 1970s, a large number of studies on face recognition have emerged. However, due to the complex changes in the age of the face, face age characteristics are affected by a variety of factors, resulting in the age estimation of the relevant research results are not many. However, since twenty-first Century, with the rapid development of pattern recognition and computer vision technology, the age estimation technology of face image has also made great breakthrough and development.

In 1999, Kwon and Lobo [7] were the first to obtain the relevant research results of the age estimation. They are divided into 3 categories according to infants, young adults, and older adults. On the basis of this classification, the classification method is used to estimate the age of face images. Hayashi et al. [8] studied age and gender recognition based on Hough transform for wrinkle texture and face image color analysis. However, the correct rate of the 300 images is only 27%, and the experimental results are not satisfactory. 2004, Nakano et al. [9] mainly studied the texture features of face, and used the edge information of face and neck on the age estimation. They using Sobel operator for edge detection, image edge information on the age of the image classification. Lanitis et al. [10] proposed the use of active appearance model to extract the features of the face of the age estimation firstly. They compared the performance of different classifiers, such as KNN, MLP, and SOM, and thought that the machine could almost estimate the age of a person. In 2013, Wei [11] proposed a new age estimation method, the method takes into account the internal information of age, after the feature extraction, the method of correlation component analysis (RCA) is used to adjust, the LLPP and MFA algorithm are used to reduce the dimension, and the accuracy of age estimation is improved. After several decades of development, the age estimation algorithm has made some scientific achievements. But because of the age change is affected by many factors, involving the age span, lack of training samples and other factors, age estimation problem has faced enormous challenges, in order to achieve high accuracy of age estimation, the majority of scholars also need further research.

3 Proposed Method

The power of CNN models rely on the architecture design with deeper and deeper layers and more and more neurons[a]. But, complexly designed network tend to over-fit the training set, it also needs a huge amount of data to feed in. So, a compact and efficient network is needed to be designed. It should be sufficient to fulfill the task, and easy to train.

Considering that the features of facial age estimate classification task, a new CNN structure is proposed for automatic facial expression recognition in this paper. Two publicly available databases Morph II and FG-Net are used to carry out the experiments. Firstly, we preprocess the face image which includes face detection and face image cropping by using Open CV library. Then, the facial expressions features are extracted using our convolution neural network which under deep learning framework. Finally, the training model is used to classify each facial image as one of the seven age group (young, adult, Middle, middle-aged, quinquagenarian, senile, seniles) considered in this study (Fig. 2).

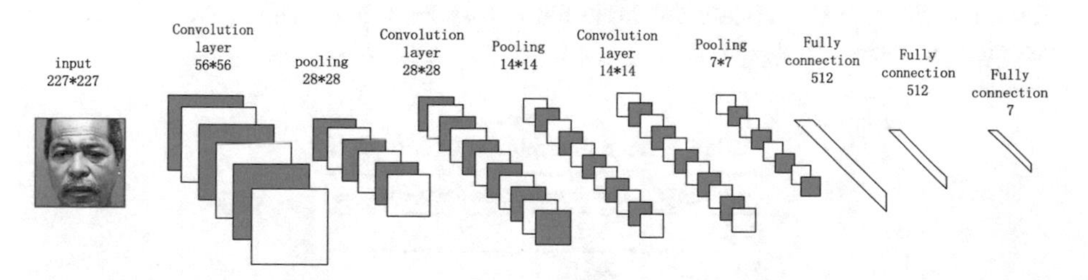

Fig. 2. Facial age estimation network structure

The whole network consists of 3 convolution layers, 3 pooling layers and 3 fully connected layers. The first convolution convolution kernel size is 7 × 7, the step size is 4, no padding. The original image is randomly cropped to get the image area of 227 × 227. Then join the BatchNorm layer, enhance the weight of random initialization to reduce the possibility of overfit. After the first convolution layer, the image size is 56 × 56. Next, to increase the nonlinear properties of network, we use ReLU (Rectified Linear Units) as the activation function. For any given input value x, ReLU is defined by:

$$F(x) = \max(0, x) \tag{1}$$

Where x is the input to the neuron. Using the ReLU activation function allows us to avoid the vanishing gradient problem caused by some other activation functions [12].

With the increasing number of convolution layers, the feature dimension increases rapidly. In order to avoid such a dimension of disaster, we need to use the pooling layer to reduce the dimension. In the experiment, the down-sampling is performed by Max-pooling. After the pooling, the image size becomes 28 × 28. After repeated convolution and pooling, the network is fully connected. This network also uses dropout method to reduce over-fitting, the idea is that the dropout layer after the connection layer, each neuron propagation time of each layer will be fully connected with a certain probability is not involved in the propagation, and then reverse the propagation of these same neurons do not participate, so can effectively to reduce over-fitting, where the probability value is 0.5.

Because of the limited ability of shallow network for feature extraction, the way of training data from scratch is not necessarily suitable for each classification problem. And the age is estimated only more than 50 thousand data sets, the use of deep network training is easy to fall into the over fitting, therefore, we use another way of training deep learning–fine tune. Fine tuning needs to find suitable data sets on the training model of deep convolutional neural networks.

4 Experiments and Results

4.1 Database Description

Morph II: The Morph database contains two databases of album1 and album2, where album2 is the largest publicly available dataset, is Morph II. The age distribution of Morph II is shown in the Table 1.

Table 1. Age distribution of Morph II

Age range	The number of face images
0–9	0
10–19	4971
20–29	14480
30–39	17884
40–49	13668
50–59	4098
60–69	456
70–79	51
Total	55608

4.2 Pre-processing Image

The implementation of the pre-processing was done by using Open CV. It is used to remove the background information, and highlight the age characteristics of the face. Then, all face images are normalize to 227 × 227 pixels (Fig. 3).

Fig. 3. Pre-processing image

4.3 Experimental Results

The experiment is divided into two parts: one is to train the network from scratch, and another is to fine tune the existing network.

During the training process, the super-parameter is set as follows: the initial learning rate is 0.01, momentum is 0.9. The initial weight attenuation is set to 0.0005. lr_policy: set as "step". In each layer, the Gauss distribution with mean 0 and standard deviation of 0.01 is initialized. Every time the learning rate is the same, but in the actual training, will be manually set to adjust the way, in a certain number of iterations, if the accuracy rate does not rise, the super parameter way by multiplying the learning rate reduces to the original 1/10.

The first experiment uses CaffeNet network and the modified network is proposed in this paper. The experimental results are as follows:

Table 2. Comparison of the accuracy rate under different iterations

Network	Number of iteration					
	5000	10000	18000	25000	35000	50000
CaffeNet	0.5320	0.5608	0.6034	0.6429	0.6501	0.6733
Proposed	0.5883	0.6522	0.66162	0.6931	0.7130	0.7391

The experimental results show that the accuracy of the test set increases gradually and tends to be stable with the increase of the number of iterations, its shows that the network has reached the optimal.

Second experiments were conducted by using VGGNet and CaffeNet to fine tune. During the experiment, the initial learning rate is set to the 1/10 of the training time, and the final full join learning rate is set to 10 times of the training time. The experimental results are as follows (Table 3):

Table 3. The accuracy rate under VGGNet and CaffeNet

Network	Number of iteration					
	5000	10000	18000	25000	35000	50000
CaffeNet_finetune	0.5933	0.6188	0.6472	0.6745	0.6983	0.7220
VGG_finetune	0.6016	0.6326	0.6794	0.6928	0.7532	0.7756

Comparison of the Table 2, in the Morph data set, the results of fine-tuning better than training, accuracy rate has been improved. This shows that in the case of insufficient number of samples, fine-tuning is a good choice.

5 Facial Expression Recognition System

The system is mainly concentrated in the test section. A well trained model based on the age data sets on Caffe and the system of age estimation is set up in Windows7 (Figs. 4 and 5).

The system needs to load the pre trained model and the related configuration files. For the input face image, by using Open CV for face detection, we normalize it to 227 × 227, and then estimate the age and output the results in real time. The system implements three data read mode, not only can detect the face image from the camera, but also can detect the face image from the picture file and the video file. The estimation process shows as follows:

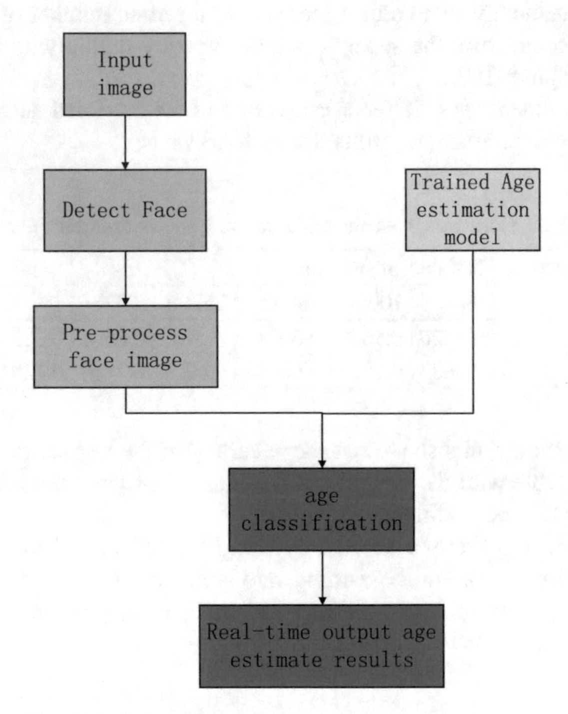

Fig. 4. The system of age estimation

Fig. 5. The results of face detect and face recognition

6 Conclusion

In this paper, a facial age estimation system based on convolution neural network is proposed and the pre-process of image step is added. As we saw the experimental results in both instantaneity and efficiency is improved by adding the adding BN layers to the network. Network fine-tuning is to further improve the identification results. The results show that the fine-tuning in the case of a small number of data sets have a better effect. We plan to investigate the network in the bigger database in order to increase the robustness of facial age estimating algorithm in unknown environments (e.g. with varying light condition, occlusion and others) in the future. And we will add some functions on the recognition system, such as gender identification and other functions. The most important thing is to further improve the recognition accuracy.

References

1. Zhang, D., Tsai, J.J.P.: Machine learning and software engineering. Softw. Quality J. **11**(2), 87–119 (2014)
2. Natarajan, B.K.: Machine Learning, pp. 207–214 (2014)

3. Goodfellow, I.J., et al.: Challenges in representation learning: a report on three machine learning contests. Neural Inf. Process. **64**, 117–124 (2013)
4. Yu, Z., Zhang, C.: Image based static facial expression recognition with multiple deep network learning. In: Proceedings of the 2015 ACM on International Conference on Multimodal Interaction, pp. 435–442 (2015)
5. Kahou, S.E., et al.: Combining modality specific deep neural networks for emotion recognition in video. In: Proceedings of the 15th ACM on International Conference on Multimodal Interaction, pp. 543–550 (2013)
6. Minchul, S., et al.: Baseline CNN structure analysis for facial expression recognition. In: 25th IEEE International Symposium on Robot and Human Interactive Communication (RO-MAN), pp. 26–31 (2014)
7. Kwon, Y.H., Lobo, N.V.: Age classification from facial images. Comput. Vis. Image Understand. 74(1), 1–21 (1999)
8. Hayashi, J.: Age and gender estimation based on wrinkle texture and color of facial images. In: Proceedings of the 16th International Conference, vol. 1(1), pp. 405–408 (2002)
9. Nakano, M., Yasukata, F., Fukumi, M.: Age classification from face images focusing on edge information. In: Negoita, M.G., Howlett, R.J., Jain, L.C. (eds.) KES 2004. LNCS, vol. 3213, pp. 898–904. Springer, Heidelberg (2004). doi:10.1007/978-3-540-30132-5_121
10. Lanitis, A., Draganova, C., Christodoulou, C.: Comparing different classifiers for automatic age estimation. IEEE Trans. Syst. Man Cybern. Part B Cybern. **34**(1), 621–628 (2004)
11. Chao, E.L., Liu, J.Z., Ding, J.J.: Facial age estimation based on label-sensitive learning and age-oriented regression. Pattern Recogn. **46**(3), 628–641 (2013)
12. Krizhevsky, A., et al.: Imagenet classification with deep convolutional neural networks. In: Advances in Neural Information Processing Systems, pp. 1097–1105 (2012)

The Application of Eye Tracking in Education

Yuyang Sun[1(✉)], Qingzhong Li[1], Honggen Zhang[2], and Jiancheng Zou[2]

[1] School of Mathematical Sciences, Capital Normal University, Beijing, China
yyuyy99@163.com
[2] Institute of Image Processing and Pattern Recognition, North China University of Technology, Beijing 100144, China

Abstract. The application of emerging information technologies to traditional teaching methods can not only enhance the value of technologies, but also improve the teaching progress and integrate different fields in education with efficiency. 3D printing, virtual reality and eye tracking technology have been found more and more applications in education recently. In this paper, through the improvement of eye tracking algorithm, we developed an education software package based on eye tracking technology. By analyzing the students' eye movement data, teachers are able to improve the teaching quality by improving the teaching framework. Students can also focus on their own interests more to develop a reasonable learning plan. The application of eye tracking technology in the field of education has a great potential to promote the application of technology and to improve the educational standards.

Keywords: Eye tracking · Education · Software

1 Introduction

Eyes are among the important sensory organs that we receive outside information. The eye is not only an important channel to obtain information, but also send information through the human behavior. Through the location of the pupil we can get the human eye line of sight. As early as the beginning of the 20th century, scholars tried to analyze people's psychological activities through the record of the eye movement trajectory, pupil size, gaze time and other information [1], but the observation method is relatively simple, mainly relying on manual recording. So its accuracy and reliability are difficult to be guaranteed. It is so difficult to accurately reflect the true trajectory of eye movement. The appearance of eye tracker provides an effective means for the psychologist to record the information of eye movements. It can be used to explore the visual signals in a variety of situations, and to observe the relationship between eye movements and human psychological activities [2]. Eye tracker is an instrument based on eye tracking technology to collect eye movement trajectory. With the development of computer, electronic information and infrared technology, eye tracker has been further developed in recent years [3].

With the development of artificial intelligence technology, the research of eye movements can explore the visual information processing and control system, because

J.-S. Pan et al. (eds.), *Advances in Intelligent Information Hiding and Multimedia Signal Processing*, Smart Innovation, Systems and Technologies 82,
DOI 10.1007/978-3-319-63859-1_4

of its wide application value and business prospects in psychology [4], such as medical diagnosis [5], military research [6], commercial advertising evaluation [7] and driving the vehicle [8] and other fields. Eye tracking technology has made rapid development in recent years. The researchers extract and record the user's eye movement data, and then extract the important information by analyzing these data. Eye tracking technology can also be used in education field. We can use the learner's visual psychology to simulate a realistic learning environment, and interact with learners. Eye tracking technology can also be used in an e-learning system by taking appropriate interventions when the user does not focus on the active content on the computer screen for a long time or is concerned with the irrelevant area of the learning task.

2 Eye Tracking Technology

Nowadays, the eye tracking technology is based on the interactive electronic products, which is divided into wearable and non-contact eye movements. The basic principle of eye tracking technology is the use of image processing technology to extract the eye image directly or indirectly using the camera, by recording the position of the infrared light spot reflected by the human cornea and the pupil, so as to achieve the purpose of recording the change of the line of sight. The eye movement system identifies and determines the vector change between the pupil center and the cornea reflex point by analyzing the eye video information collected from the camera. Thereby determining the change of gaze point (Fig. 1).

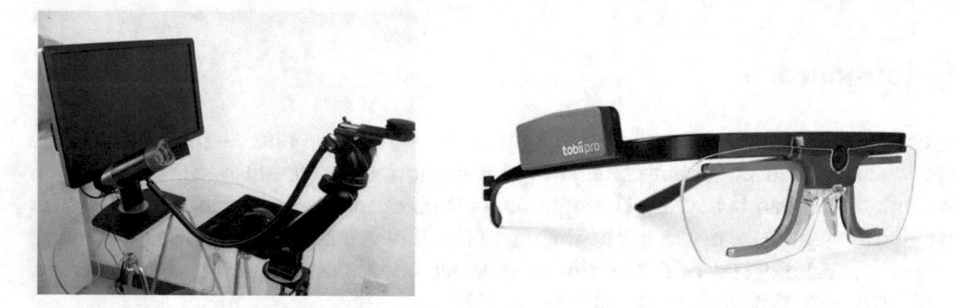

Fig. 1. Non-contact and wearable eye tracker.

2.1 Face Detection and Eye Positioning

At present, face detection classifiers are mostly based on haar features by using Adaboost learning algorithm training. In this paper, the face detection algorithm is also based on Haar feature and AdaBoost classifier, but combined with prior knowledge of facial feature distribution, to locate the eye area, and the human face region and human eye region are positioned in the image (Fig. 2).

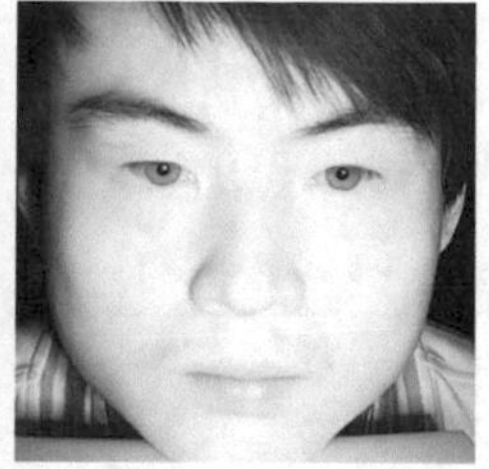
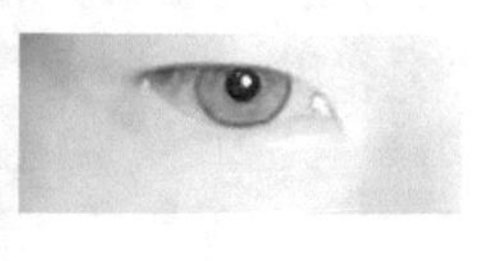

Fig. 2. Face recognition and human eye positioning.

2.2 Pupil Localization and Infrared Spot Detection

Due to eye movement or uneven illumination and other external factors, the pupil area will appear scattered noise information, and these scattered noise on the pupil and cornea reflection high light points will cause errors. Median filtering has some effect on filtering out such noises. So filtering and denoising methods are used as pre-processing of the image.

(1) Infrared spot center
In this paper, the gaze tracking technology is mainly through the near infrared light source in the human cornea to produce the corresponding corneal reflex spot as a reference, combined with the human eye pupil center point, and then through the geometric model for spatial mapping, we get the human eye gaze point.

Step1: Calculate the image gradient. Due to the complexity of the eye image, we estimate the region of interest (ROI) of the eye, since the gray value of the spot in the region of interest is higher than the gray value of thc surrounding pixel, which gives us an idea of locating the infrared spot coordinates. In order to obtain the central position of the spot, it is necessary to divide the spot from the eye images. Before the traditional image segmentation, it is usually necessary to perform edge detection by using the feature image. The traditional edge detection operator has the Sobel operator, Prewitt operator and Roberts operator [9]. The following is a description of the Prewitt operator.

$$\begin{bmatrix} -1 & -1 & -1 \\ 0 & 0 & 0 \\ 1 & 1 & 1 \end{bmatrix}, \begin{bmatrix} -1 & 0 & 1 \\ -1 & 0 & 1 \\ -1 & 0 & 1 \end{bmatrix} \tag{2}$$

In the determination of the center of the spot, more stable edge-based fitting method are based on geometric features, through the circle and other geometric features to determine the center of the spot. There are usually grayscale centroid methods:

$$x_c = \frac{\sum x_n I_n}{\sum I_n}, y_c = \frac{\sum y_n I_n}{\sum I_n} \tag{3}$$

Where (x_c, y_c) is the center of the detected spot and I_n is the pixel gray value of the ROI region. So, if our ROI area is small enough, we can use the centroid method to quickly determine the center of the spot. In order to minimize the ROI, we improved the basis of the Prewitt operator, through the gradient of the image, to determine the approximate edge position and to determine the maximum gradient value of the pixel.

$$g_x = \frac{I(x+1,y) - I(x-1,y)}{2}, g_x = \frac{I(x,y+1) - I(x,y-1)}{2} \tag{4}$$

By using the Eq. (5) to determine the gradient matrix of the ROI region, we can ignore the gradient values at the edges of the image based on our prior knowledge of the pupil position. Finally, we determine the maximum point of the gradient point coordinates p, according to the image pixel size, the rectangular area where the spot position is determined with the p point as the center.

Step2: Threshold processing. After we get the interested area, in order to eliminate the impact of other image areas, we are not interested in for the next image operation, We will get the whole human eye image through the threshold for image segmentation.

$$I_n = \begin{cases} I_n, I_n \in ROI \\ 0, I_n \notin ROI \end{cases} \tag{5}$$

Then, we focus on the segmentation of ROI images. In this paper, we study and improve the traditional maximum interclass variance (OTSU) threshold method.

(2) Pupil center positioning
It is easier to locate the pupil center than the spot. Since the gray value of the pupil and the gray value of the surrounding position are quite different in the human eye region, and the image has been preprocessed before the pupil is positioned, so as to select the appropriate threshold for image segmentation.

2.3 Gaze Tracking

The coordinate extraction of the pupil and the center of the spot are the basis for realizing the gaze tracking. The two-dimensional migration vector can be obtained by calculation, that is, the P-CR vector.

$$x_e = x_p - x_c, y_e = y_p - y_c \tag{6}$$

Where (x_p, y_p) and (x_c, y_c) is the pupil and spot center coordinates respectively, (x_e, y_e) is P-CR vector.

In this paper, we use of 6 parameters of the fitting to ensure that the system has good accuracy and real-time.

$$\begin{aligned} X_{gaze} &= a_0 + a_1 x_e + a_2 y_e + a_3 x_e^2 + a_4 x_e y_e + a_5 y_e^2 \\ Y_{gaze} &= b_0 + b_1 x_e + b_2 y_e + b_3 x_e^2 + b_4 x_e y_e + b_5 y_e^2 \end{aligned} \tag{7}$$

During the calibration process, the user needs to follow the calibration point in order to obtain the corresponding P-CR vector. Using the calibration point and P-CR vector to solve the 12 position parameters, this paper uses the least squares method to solve the model parameters.

$$e^2 = \sum_{i=1}^{9} \left[X_{gaze}^{i} - f(x_e^i, y_e^i)\right]^2 \tag{8}$$

After the parameters are calculated, we calculate the P-CR, the P-CR into the Eq. (7), and to calculate the real scene in the human eye point of view, to achieve the purpose of gaze tracking.

3 The Application of Eye Movement in Mathematics Education

Human eye movement can reflect the human thinking process [9]. The eye movement researcher records the eye movement trajectory based on the eye tracking technique. In the process of mathematics learning, students need to think about the visual mathematical language. In the process of understanding the problem, each different symbol in mathematics represents the different information, the brain for different information symbols, watching, jumping and other different eye movement mode, the response of these eye movement mode of attention, time and pupil diameter and other indicators. In practice, the eyes will be moved to another location once they are gazed at a place after a certain period of time. So if there is a cognitive difficulty in a given area, the fixation time will increase. And, after the students understand the previous content, they can predict the following content. In some areas it will appear the tendency of eye jump. If the students think that part of the content is difficult to understand, there will be short jump distance, so their watching time is longer. At present, the experimenter is the most commonly used eye tracking method with the video image of eye position, that is, to record and analyze the direction of pupil movement by recording video. In this paper, through the real-time eye tracking system designed by ourselves, we can feedback the eye movements in real time. In the aspect of teaching, the type of eye tracker which is often used by teachers is non-contact eye movement. In the experiment, it is necessary to have a host and a camera device. The instrument is used to analyze the subject's cognitive ability by recording the moving characteristics of the eye when reading the equation and the mathematical symbol of the host. The teachers can improve the teaching quality by improving the teaching design through the student's eye movement. In terms of student learning mathematics, we judge the students' acceptance of the teaching content through the eye movement of the students, who can understand the difficult part and the interest part of their own learning. We mainly learn the psycho logical response from the eye movement indicators of the students (Table 1).

Through the two aspects of teaching and learning in mathematics education. Researchers record and analysis the eye movement data of the students, Through the analysis of eye movement data, teachers understand the content of the difficulties

Table 1. Simulated and measured results of the proposed antenna.

Eye movement indicators	Implication
Number of fixations	The number of fixation reflects the learner's proficiency in content
Fixation time	The time of fixation refers to the time when the line of sight stays at a fixed point, reflecting the learner's processing of the content
First fixation point	Reflect the learner's prior knowledge content
Pupil size	Reflect the learners' thinking about the problem
Eye movement track	Reflecting the learner's entire browsing process

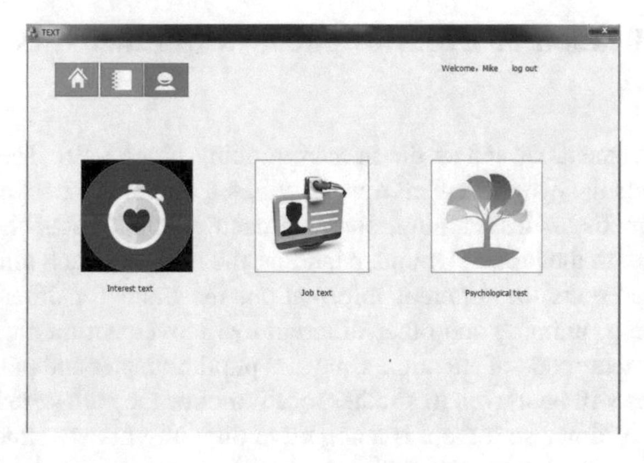

Fig. 3. Eye tracking software interface.

of teaching, so as to modify the entire teaching design. On the other hand, students can understand their own learning interests and optimize their own learning strategies. We use their own in the windows platform to develop real-time eye tracking system, the students of mathematics learning eye tracking experiment, because the system is a beta version, only the number of fixations is provided in the interface, and other indicators will be used in the next version (Fig. 3).

In this software, we developed a multidisciplinary education framework, but in the beta version, we only tested the mathematics disciplines. So this paper take mathematics as an example for application of eye tracking (Fig. 4).

By students' watching four different video or photos, we can analyze the students' eye tracking data. In order to better analyze the fixation point, we add the expression recognition system in the software, while observing the change of the human eye while observing the change of the expression of the student. We can better analyze the learning characteristics of the students. We can make a simple analysis of the students' learning process from the perspective of the students' attention to different videos and the changes of the expression in the whole process.

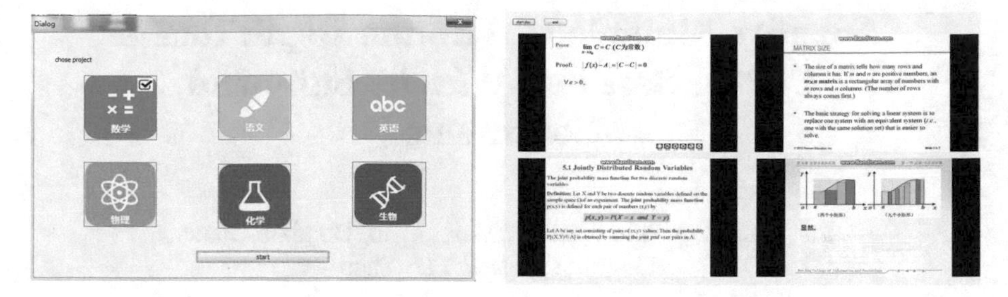

Fig. 4. Mathematics test

4 Conclusions

In this paper, we apply the eye tracking technology to the teaching of mathematics. Teachers can improve the teaching program according to the eye movement data of the students in the course of learning. On the other hand, students can adjust themselves according to the eye movement situation by the human-computer interaction experiences. Due to the abstraction and complexity of mathematics, this paper provides a feasible solution for mathematics teaching. We will expand the software functions and improve eye tracking accuracy in the future study.

References

1. Tao, R.Y., Qian, M.F.: Cognitive meaning of eye movement indicators and the value of polygraph. Psychol. Technol. Appl. **7**, 26–29 (2015)
2. Ling, D.L., Axu, H., Zhi, Y.H.: A study on the text reading based on eye movement signals. J. Northwest Univ. Nationalities (Nat. Sci. Ed.) **35**(2), 43–47 (2014)
3. Morimoto, C.H., Mimica. M.R.M.: Eye gaze tracking techniques for interactive applications. Comput. Vis. Image Underst. **98**, 4–24 (2005)
4. Li, Y.G., Xuejun, B.: Eye movement research and development trend of advertising psychology. Psychol. Sci. **27**(2), 459–461 (2004)
5. Jacob, R.J.K.: The use of eye movements in human-computer interaction techniques: what you look at is what you get. Readings in Intelligent User Interfaces. Morgan Kaufmann Publishers Inc. (1998)
6. Lim, C.J., Kim, D.: Development of gaze tracking interface for controlling 3D contents. Sens. Actuators, A **185**(5), 151–159 (2012)
7. Higgins, E., Leinenger, M., Rayner, K.: Eye movements when viewing advertisements. Front. Psychol. **5**(5), 210 (2014)
8. Hansen, D.W., Ji, Q.: In the eye of the beholder: a survey of models for eyes and gaze. IEEE Trans. Pattern Anal. Mach. Intell. **32**(3), 478–500 (2010)
9. Tatler, B.W., Wade, N.J., Kwan, H., Findlay, J.M., Velichkovsky, B.M.: Yarbus, eye movements, and vision. i-Perception **1**(1), 7 (2010)

Adaptive Multiple Description Depth Image Coding Based on Wavelet Sub-band Coefficients

Jingyuan Ma, Huihui Bai(✉), Meiqin Liu, Dongxia Chang, Rongrong Ni, and Yao Zhao

Institute of Information Science, Beijing Jiaotong University, Bejing 100044, China
{JingyuanMa, 15120341, HuihuiBai, hhbai}@bjtu.edu.cn

Abstract. With the development of multi-view video plus depth technology, the coding algorithm at the depth image has become one of hot research directions. As we know, after wavelet transform, the energy of the image smoothing region is concentrated in low frequency sub-band, while the edge information of the texture region is concentrated in high frequency sub-bands. However, the edge information is very important to the synthetic viewpoint. In order to improve the edge decoding quality and ensure the transmission reliability, we propose an adaptive multiple description depth image coding scheme based on wavelet sub-band coefficients. The low frequency sub-band is encoded by optimized multiple description lattice quantization (OMDLVQ), while the high frequency sub-bands are encoded by embedded block coding with dead-zone. Finally, two streams of the vector quantization are combined with the embedded block coding stream respectively. The experimental results show that this scheme has good performance in transmission reliability and reconstructed image quality.

Keywords: Depth image · Wavelet transform · MDLVQ · SPECK

1 Introduction

In recent years, with the rapid development of digital information technology and high-speed network, the demand on the video has been changed from the traditional 2D video to high frame rate 3D video. Multi-view video plus depth 3D video (MVD) format has been widely used, and the coding algorithm at the depth image has become a hot research object for many experts and scholars. The depth image is used to represent the distance from each point of the scene to camera, and the distance is usually expressed in gray level; the bigger the gray value, the closer from the scene points to camera. Different from the traditional 2D image, the edge of depth image is sharp, and other places are uniform, so depth image has more sparsity. In addition, in MVD encoding, the depth map is not directly used for image or video display, but for view synthesis, the edge information of depth image will directly affect the quality of the synthesized viewpoint. If we adopt the standard texture image encoding method to compress the depth image [1], it will not be able to achieve a good compression and

J.-S. Pan et al. (eds.), *Advances in Intelligent Information Hiding and Multimedia Signal Processing*, Smart Innovation, Systems and Technologies 82,
DOI 10.1007/978-3-319-63859-1_5

high quality of the synthesized viewpoint. Therefore, how to effectively reduce the pressure of network bandwidth, improve the quality of the edge information of depth image and ensure the quality of the synthesized viewpoint have become a hot issue.

At present, the depth image encoding scheme mainly includes three categories. The first kind of depth image encoding scheme is based on the region of interest (ROI) [2, 3]. At first, the method will extract the ROI in depth image, and then encoding respectively for ROI and other areas through different modes, which can obtain a high quality synthesized viewpoint and make good bit allocation. However, this encoding method has the heavy expenses both in hardware and the detection of ROI. The second kind of depth image encoding scheme is based on depth image content. The main principle is to obtain high quality edge information through special edge detection algorithm, and then choose different encoding strategies according to the edge information contained in each block. Wang [4] et al. explained a depth map coding based on adaptive block compressive sensing, where the adaptive sampling rate is used to obtain a good reconstruction image according to edge information contained in each block. However, the edge detection algorithm is obtained by manual intervention, and the result of edge detection has a great influence on the reconstructed depth image. The last kind of depth image encoding scheme is based on up/down sampling. This method compresses and transmits the down-sampled depth image. In the decoder, the depth map is reconstructed by up-sampling according to the relativity. In literature [5], the quality of the synthesized image was improved by modifying the up/down sampling filter. But the two-dimensional median filter has a poor protection for image edge information, which limits the quality of reconstructed image. The coding method above can improve the quality of edge information to a certain extent, but most of them are based on the traditional coding standard, and focus on the rate distortion performance of depth image.

MDC as an error resilient coding method not only takes into account the rate distortion performance, but also guarantees the transmission reliability. In MDC, MDC coding framework based on discrete cosine transform is the mainstream method. But the block operation leads to the block-effect, MDC framework based on wavelet transform is proposed [6]. Bai [7] et al. optimized this framework in the follow-up study. This framework can not only avoid the block-effect, but also have a better rate distortion performance. In this paper, we propose an adaptive multiple description depth map coding scheme based on wavelet sub-band coefficients. Different strategies are adopted to process the different sub-bands coefficients, the transmission reliability is guaranteed, while protecting the high frequency edge information. The experimental results show that there are definite improvements both on depth image and synthesized virtual viewpoint image.

The rest paper is organized as follows. Section 2 delves into the details of the wavelet domain multiple description depth image coding scheme based on the sub-band coefficients. Section 3 highlights the experimental results obtained from applying the proposed scheme. Finally, conclusions are presented in Sect. 4.

2 Proposed Scheme

2.1 Analysis of Wavelet Transform Coefficients

After the wavelet transform, the most of the depth image information is concentrated in low frequency sub-band, while high-frequency sub-bands have less information. Figure 1 shows the four layers wavelet decomposition and coefficients histogram of the Balloons depth image. From the chart we can see that the coefficients of the LL sub-band are large, but only a small number of large coefficients in LH, HL and HH sub-bands, while the rest of coefficients are distributed near the "0" value. The coefficients that near "0" value in high frequency sub-bands are generated by the image smoothing region, while some of the large coefficients are generated by the edge part of the image. Therefore, in order to protect the edge information of depth image, and

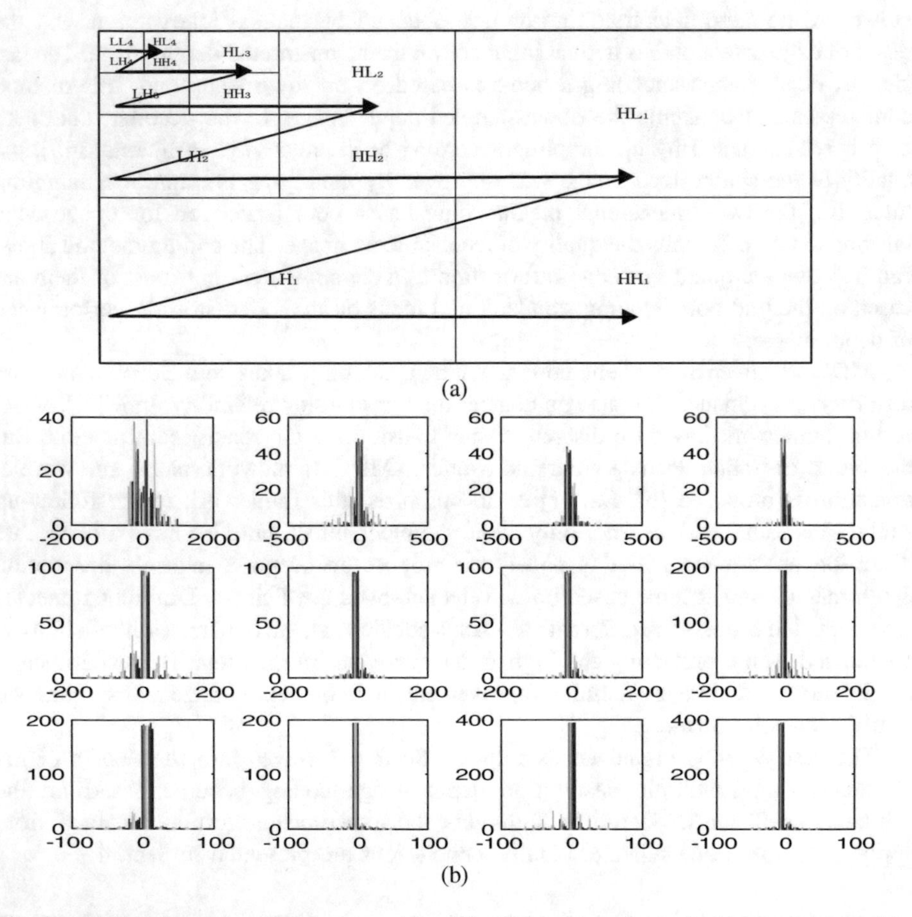

Fig. 1. (a) Four layers wavelet decomposition (13 sub-bands are formed). (b) Distribution of the first 12 wavelet sub-bands coefficients (from LL_4 to LH_1): according to the scanning order of the black arrow in (a).

enhance the quality of the synthesized image, we must strengthen the protection of the large coefficients in high frequency sub-bands.

2.2 Framework of the Proposed Scheme

Figure 2 shows the proposed scheme framework. We use texture plus depth from multiple cameras to synthesize the virtual viewpoint image. In order to prove the compression performance of the proposed scheme to depth image, the texture image we use is not compressed when synthesize the viewpoint. Different from the OMDLVQ coding for 3D depth image proposed by Zhang et al. [8]. In this paper, we use the 10–18 Daubechies wavelet, and a given depth image is decomposed into 13 sub-bands after the decomposition of the four layers. We adopt different strategies to deal with different sub-band coefficients. Specific modules are explained as follows:

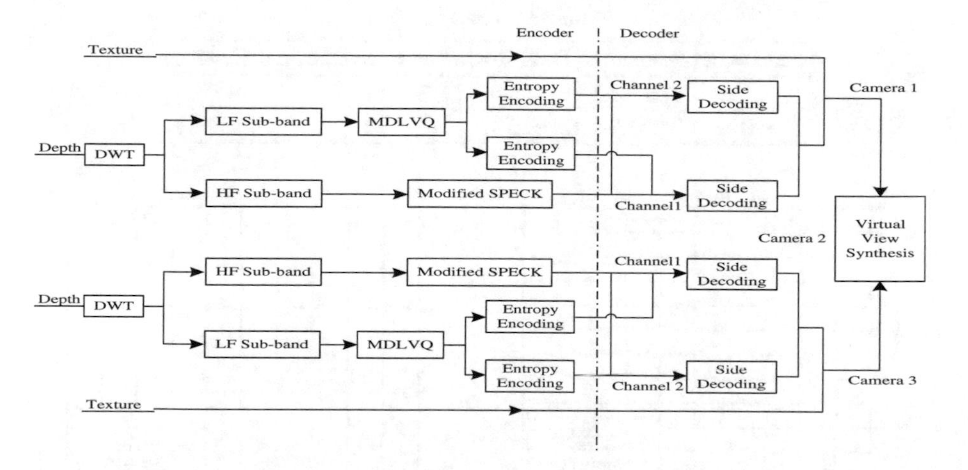

Fig. 2. Block diagram of the proposed scheme.

2.2.1 Low Frequency Sub-band Encoding (LF Sub-band Encoding)

The coefficients of the low frequency component have a great impact on the reconstructed image, and the energy of the image smoothing region is more concentrated in low frequency sub-band. Therefore, the low frequency sub-band coefficients must be quantified by a better quantization model. In the case of two channels, the lattice vector quantization based on A_2 lattice has a better rate distortion performance. So the lattice vector quantizer is applied to 2-dimensional vectors, and finally obtain the quantized lattice point $\lambda, (\lambda \subset A_2)$. Each lattice point $\lambda, (\lambda \subset \Lambda)$ is mapped into two descriptions $(\lambda_1^{'}, \lambda_2^{'}) \in \Lambda^{'} \times \Lambda^{'}$ and transmitted in two channels, subject to bit rate constraints imposed by each individual channel, where Λ is the sub-lattice of $\Lambda^{'}$ with the index N, $\left(\left|\Lambda^{'}/\Lambda\right|\right)$. The index N is used to control the amount of redundancy in a lattice vector quantizer. In OMDLVQ coding method, the value of N is fixed at 7, the quantization steps were taken from 1 to 12. There are two reasons for this, on the one hand, two sub-descriptions can be obtained, on the other hand, the low-frequency information can be retained as much as possible. Finally, we will obtain the rate distortion performance curve of the reconstructed image by modifying the quantization steps.

2.2.2 High Frequency Sub-band Encoding (HF Sub-band Encoding)

The set partitioned embedded block coder (SPECK) algorithm [9] is an improved embedded image coding algorithm base on SPIHT [10] which is proposed by Said and Pearlman. After the wavelet transform for a given depth image, each sub-band is defined as a block, the SPECK algorithm will scan the block with a certain order. If the block is not important, it will be encoded by one symbol, otherwise, it will be divided into four sub-blocks, until the sub-block is decomposed into a block that contains one element. Finally, all the important coefficients in the block are stored and quantified.

Table 1. Test sequences

Sequences	Resolution	Dead-zone	Views	Synthesized viewpoint
Balloons	1024*768	3 ~ 5	1–3	2
Kendo	1024*768	3 ~ 5	1–3	2

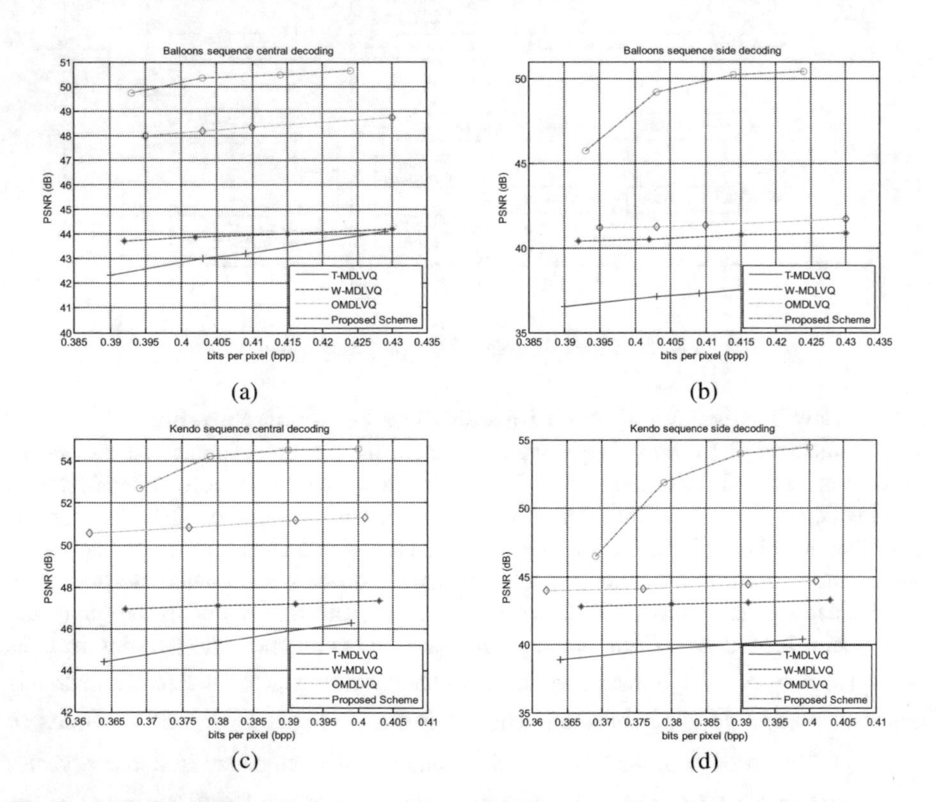

Fig. 3. Objective quality comparison for the depth sequences of *Balloons* and *Kendo*. (a) and (c): central rate distortion performance; (b) and (d): side rate distortion performance.

It is a fully embedded block-based coder which employs progressive transmission by coding bit planes in decreasing order.

As we all know, in high frequency sub-bands the large coefficients can be generated by the edge information of the depth image, while the small coefficients almost have no effect on the reconstructed depth image. Therefore if the large coefficients in the high frequency sub-bands can be well encoded, the quality of the reconstructed image will be greatly improved. Based on SPECK algorithm, we propose a high frequency sub-band quantization encoding scheme with dead-zone. Just like the conventional algorithm SPECK, the modified SPECK coding also has four steps. However, unlike the SPECK algorithm, we deal with the high-frequency sub-bands individually. In the first step, we add two initialization, the first one is all the coefficients located in the dead-zone will be set to 0, and the second item is all the low frequency sub-band coefficients are set to 0. In the second step, quad-tree split method is adopted initially, but the block will be directly divided into twelve sub-blocks when it contains only twelve elements. The redefined block split method and quantization scheme with dead zone greatly increases the number of zero of the wavelet coefficients in the high frequency sub-bands, which can eliminate the useless information. Therefore, there are great improvements both on reconstructed depth image and coding efficiency.

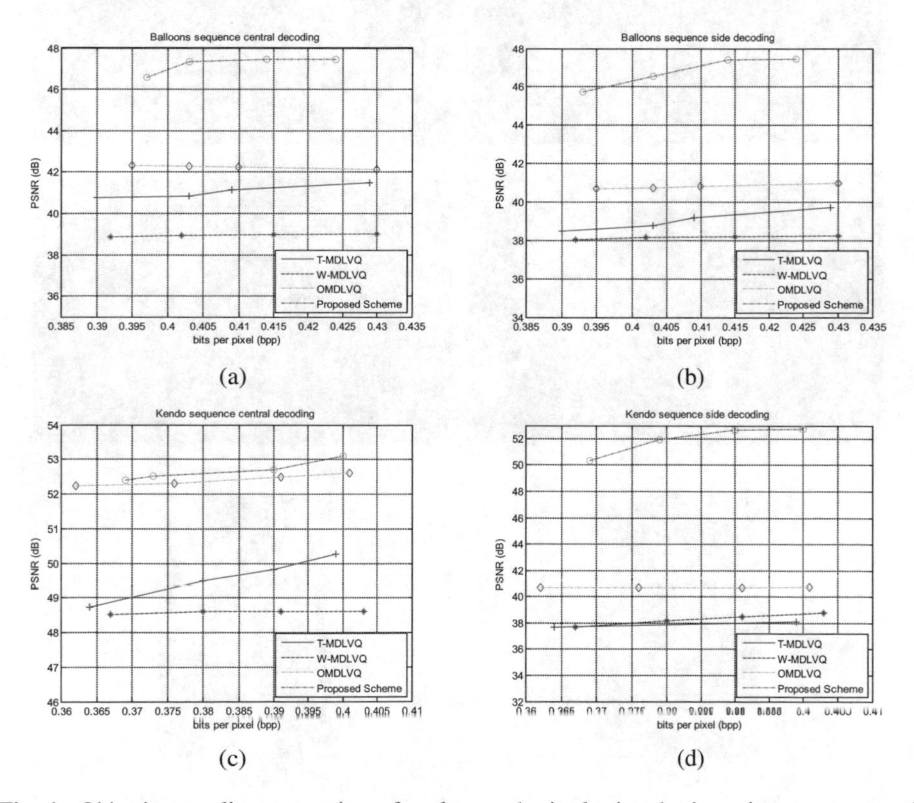

Fig. 4. Objective quality comparison for the synthesized virtual viewpoint sequences of *Balloons* and *Kendo*. (a) and (c): central rate distortion performance; (b) and (d): side rate distortion performance.

3 Experimental Results and Analysis

In this paper, the standard test sequences, shown in Table 1, are selected to verify the effectiveness of the proposed scheme. To highlight the performance of the proposed scheme, there are three schemes selected to make the comparison. The results are presented in Fig. 3, the length of the dead-zone is 5 for *Balloons*, 3 for *Kendo* in this experiment. In the central channel, compared with other schemes, the proposed scheme can achieve 1.8–2.1 dB improvements for *Balloons* and 2.1–3.4 dB for *Kendo* at least. In addition, in the side channel, the proposed scheme can achieve 5.0–8.9 dB improvements for Balloons and 2.7–9.6 dB for *Kendo* at least.

Since the depth image is not directly applied to 3D video display, the comparison of the synthesized images is presented in Fig. 4. In the central channel, the quality of the reconstructed synthetic viewpoint is improved 4.1–5.2 dB for *Balloons* and 0.1–0.4 dB

Fig. 5. Subjective quality comparison for the synthesized virtual viewpoint sequences of Balloons and Kendo. (a), (c) and (e): OMDLVQ method; (b), (d) and (f): the proposed scheme.

for *Kendo*. In the side channel, the quality of the reconstructed synthetic viewpoint is improved 5.2–6.4 dB for Balloons and 9.1–12.0 dB for Kendo. Because of the high frequency sub-band coefficients are well protected, the edge quality of the test sequences has been significantly improved. Figure 5 shows the subjective quality of the synthesized virtual viewpoint of the test sequences, it is clear that the proposed scheme is the most advantageous.

4 Conclusion

In this paper, we fully consider the distribution characteristics of the wavelet coefficients and make a good protection for large coefficients in high frequency sub-bands. Experiments show that the proposed scheme outperforms the three schemes with which we have chosen to compare and has good rate distortion performance both on depth map and synthesized virtual viewpoint, especially the single path reconstruction.

Acknowledgement. This work was supported in part by National Natural Science Foundation of China (No. 61672087, 61402033) and CCF-Tencent Open Fund.

References

1. Lee, J.Y., Wey, H.C., Park, D.S.: A fast and efficient multi-view depth image coding method based on temporal and inter-view correlations of texture images. IEEE Trans. Circ. Syst. Video Technol. **21**, 1859–1868 (2011)
2. Li, G., Qu, W., Huang, Q.: A multiple targets appearance tracker based on object interaction models. IEEE Trans. Circ. Syst. Video Technol. **22**, 450–464 (2012)
3. Wang, M., et al.: Region of interest oriented fast mode decision for depth map coding in DIBR. In: Proceedings of the 2011 IEEE 7th International Colloquium on Signal Processing and its Applications, pp. 177–180 (2011)
4. Wang, T., et al.: Depth map coding based on adaptive block compressive sensing. In: IEEE China Summit and International Conference on Signal and Information Processing, pp. 492–495 (2015)
5. Stark, J.A.: Adaptive image contrast enhancement using generalizations of histogram equalization. IEEE Trans. Image Process. **9**, 889–896 (2000)
6. Servetto, S.D., et al.: Multiple description wavelet based image coding. IEEE Trans. Image Process. **9**, 813–826 (2000)
7. Bai, H., Zhu, C., Zhao, Y.: Optimized multiple description lattice vector quantization for wavelet image coding. IEEE Trans. Circ. Syst. Video Technol. **17**, 912–917 (2007)
8. Zhang, H., Bai, H., Zhao, Y.: Optimized multiple description lattice vector quantization coding for 3D depth image. KSII Trans. Int. Inf. Syst. **9**, 1140–1154 (2015)
9. Said, A., Pearlman, W.A.: A new fast and efficient image codec based on set partitioning in hierarchical trees. IEEE Trans. Circ. Syst. Video Technol. **6**, 243–250 (1996)
10. Islam, A., Pearlman, W.A.: Embedded and efficient low-complexity hierarchical image coder. In: Proceedings of SPIE the International Society for Optical Engineering, pp. 385–388 (2000)

Adaptive Histogram Shifting Based Reversible Data Hiding

Yonggwon Ri[1], Jing Dong[1,2(✉)], Wei Wang[1], and Tieniu Tan[1]

[1] National Laboratory of Pattern Recognition, Institute of Automation, Chinese Academy of Sciences, Beijing 100190, China
yg.ri@cripac.ia.ac.cn, {jdong,wwang,tnt}@nlpr.ia.ac.cn
[2] State Key Laboratory of Information Security, Institute of Information Engineering, Chinese Academy of Sciences, Beijing 100093, China

Abstract. Reversible data hiding (RDH) is a special kind of data hiding technique which can exactly recover the cover image from the stego image after extracting the hidden data. Recently, *Wu et al.* proposed a novel RDH method with contrast enhancement (RDH-CE). RDH-CE achieved a good effect in improving visual quality especially for poorly illustrated images. In *Wu's method*, however, the PSNR of stego image is relatively low and embedding performance is largely influenced by the histogram distribution of cover image. Since PSNR is still considered as one of the most important metrics for evaluating the RDH performance, this paper presents a reliable RDH method based on adaptive histogram shifting for gray-scale images to improve the PSNR of stego image while maintaining the good effect of the contrast enhancement obtained by RDH-CE.

Keywords: Reversible data hiding · Histogram modification · Location map · Contrast enhancement · PSNR

1 Introduction

With the increase of information exchange via internet, protection of the digital products has emerged as an urgent problem in aspects of information technology, such as authentication, security, copyright protection and so on [1]. Data hiding is a useful technique for solving this problem, in which a secret data is embedded within host object. Objects with and without the hidden data are called stego object and cover (original) object respectively [8,10]. Most of works has been focused on digital images.

In most applications, it is usually tolerable that slight degradation is remained within the original image after extracting data. However, in the recent

This work is funded by Beijing Natural Science Foundation (Grant No. 4164102) and the National Natural Science Foundation of China (Grant No. 61502496, No. 61303262 and No. U1536120).

J.-S. Pan et al. (eds.), *Advances in Intelligent Information Hiding and Multimedia Signal Processing*, Smart Innovation, Systems and Technologies 82,
DOI 10.1007/978-3-319-63859-1_6

years, reversible data hiding (RDH) has been intensively investigated as a special kind of data hiding technique which requires the exact restoration of the cover images from the stego images after the hidden data have been extracted [2,9,14]. Hiding ratio and stego image quality are evaluated by bit per pixel(bpp) and Peak Signal to Noise Rate (PSNR) respectively, considering as two important metrics in evaluating the RDH performance [11].

Generally, there are three mechanisms which can be nominated as typical techniques of reversible data hiding: difference expansion (DE), histogram shifting (HS) and prediction error (PE). Although PE [7] and DE [11] methods keep the PSNR of stego images high, the improvement of visual quality is not obvious. Most HS methods [1,3,4,6,13] are performed by modifying the histogram of the cover image. HS algorithms are relatively simple and the location map for HS is small in size for low capacity case, which is an advantage of HS based RDH.

Recently, *Wu et al.* proposed a novel RDH algorithm with contrast enhancement (RDH-CE) based on the principle that image contrast enhancement can be achieved by histogram equalization [13]. Unlike previous works for improving the PSNR value of stego image, they noted fact that data hiding and contrast enhancement could be archived simultaneously. As a result, RDH-CE achieved a good effect in improving visual quality especially for poorly illustrated images while providing considerable embedding capacity. In *Wu*'s method, however, the PSNR of stego image is relatively low and the embedding performance is largely influenced by the histogram distribution of cover image. The PSNR is still considered as one of the most common and important metrics for evaluating the performance of almost all the data hiding methods and express difference between original image and stego image correlated with the security of stego system. Based on this perspective, we present a reliable RDH method based on the adaptive histogram shifting for digital images to improve the PSNR of stego image while maintaining the good effect of the contrast enhancement archived by *Wu*'s work. Experimental results show that our method has improved efficiently the PSNRs of stego images.

The rest of this paper is organized as follows. Section 2 introduces briefly about *Wu*'s algorithm and presents the details of our RDH method based on the adaptive histogram shifting. Our experiment results are presented in Sect. 3. Finally, a conclusion will be drawn in Sect. 4.

2 Adaptive Histogram Shifting Based RDH

In this section, RDH-CE algorithm proposed in [13] is explained briefly at first in Sect. 2.1. Then, the details of our adaptive histogram shifting based RDH method are presented in Sect. 2.2.

2.1 RDH-CE Method

Selecting two highest peaks to be split in histogram of the image and shifting histogram bins for peak split, which is the core technique of RDH-CE.

Consider a gray scale-image C consisting of $j(j \in [0, 255])$ different pixel values. Then the image histogram, denoting G, contains $i(i \leq j)$ nonempty bins, in which the two bins with the highest frequency values are chosen in each time for data embedding. We use G_a and G_b to denote the smaller and bigger bin values corresponding to the two highest peaks in the histogram, respectively. At this step, the embedding procedure is performed in the way that the binary bits are either added to or subtracted from the pixel values i according to Eq. 1,

$$i' = \begin{cases} i - 1 & \text{for } i < G_a \\ G_a - m_k & \text{for } i = G_a \\ i & \text{for } G_a < i < G_b \\ G_b + m_k & \text{for } i = G_b \\ i + 1 & \text{for } i > G_b \end{cases} \tag{1}$$

where i' and m_k are the modified pixel value and the k-th binary bit of the data to be embedded, respectively. The left bins of G_a and the right bins of G_b are shifted to the locations of the next adjacent bins toward the left and right ends (0 and 255) in G, respectively. A location map is generated to record the locations of the pre-shifted (pre-modified) pixels for preventing overflow or underflow. The location map is compressed by JBIG2 [13] to increase the embedding efficiency. A final data to be embedded consists of 3 parts: a location map, a secret data and two peak bin values. Let P_{CE} denotes the amount of peak pairs selected by embedding capacity. The above embedding procedure repeats P_{CE} - 1 times. At the P_{CE}-th embedding, the binary values of the last two peak bins are kept in LSBs of the first 16 pixels in the last row of the image C by LSB replacement, and the stego image C' is produced.

The procedure of data extracting and image recovery starts with extracting the P_{CE}-th peaks, and continues by inverting the procedure of embedding process.

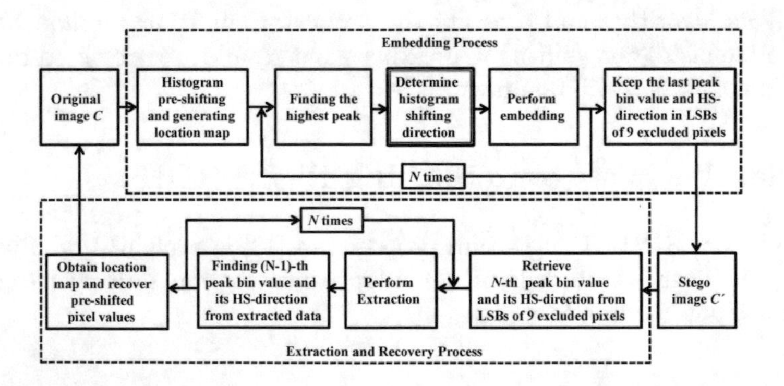

Fig. 1. Block diagram of proposed RDH algorithm

2.2 Procedure of the Proposed Algorithm

The proposed method can be considered as a improved version of RDH-CE. Suppose that N ($N = 2P_{CE}$) peaks are selected for a given 8-bit gray-level image C. To avoid the possible changes of the bounding pixels caused by overflow or underflow, as doing in RDH-CE, $N/2$ histogram bins at the right and left ends in G are pre-shifted simultaneously, and locations of pre-shifted pixels are recorded within a location map. The proposed embedding procedure is performed similar to RDH-CE in which peaks are split for data embedding. In our method, only one highest peak is chosen for data embedding in each time instead of utilizing the two highest peaks, and each of peaks can be split in any direction (right or left). Each time a peak is selected for embedding, its histogram shifting direction (HS-direction) is determined based on the PSNR value of the image to be modified prior to the histogram shifting. The PSNR expresses the visual quality of the modified image as well as the difference between the original image and the modified image. Therefore, each HS-direction is selected adaptively so that the modification of the image caused by data embedding is as minimal as possible in each step. Bin values and HS-directions of previous N-1 peaks are kept together with the location map as side information, meanwhile the value and HS-direction of the last N-th peak are saved in LSBs of the excluded 9 pixels using LSB replacement. The data extracting process and image recovery process are performed by reverse process of the data embedding.

The procedure of the proposed algorithm is illustrated in Fig. 1. The embedding process is performed as follows:

Step 1. Pixels which belong to histogram bins in the range of [0, $N/2$-1] are added by $N/2$ and pixels in the range of [256-$N/2$, 255] are subtracted by $N/2$ excluding the first 9 pixels in the bottom row. Positions of those pixels are recorded in the location map.
Step 2. The image histogram is calculated excluding the first 9 pixels in the bottom row.
Step 3. One highest peak is selected in the histogram, and its HS-direction is calculated as follows:
Step 3.1. Shift histogram bins in the right side of the selected peak toward the right end by one, and the peak is split with binary bits of data.
Step 3.2. Calculate the PSNR value (R-PSNR) of the right shifted image by Eq. 2.

$$PSNR(C, C'_p) = 10log_{10}\frac{MAX_C^2}{(1/mn)\sum_{s,t}\|C(s,t) - C'(s,t)\|} \quad (2)$$

Here, C is the original image; C'_p is the modified image using $p(0 < p < N)$ peaks; MAX_C is the maximum possible pixel value of the original image C.
Step 3.3. Return back the right shifted image into the previous state.
Step 3.4. Shift histogram bins in the left side of the selected peak toward the left end by one, and the peak is split with binary bits of data again.
Step 3.5. Calculate the PSNR value (L-PSNR) of the left shifted image.

Step 3.6. If the R-PSNR is higher than the L-PSNR, then the HS-direction will be set as 1 (right side). Otherwise, it will be set as 0 (left side).
Step 4. To embed data, Eq. 3 is applied to all pixels in the whole image in sequential order excluding those 9 pixels.

$$i' = \begin{cases} i-1 & \text{if } direction = 0(left) \\ G_s - m_k & \text{if } i = G_s \text{ and } direction = 0 \\ G_s + m_k & \text{if } i = G_s \text{ and } direction = 1 \\ i+1 & \text{if } direction = 1(right) \end{cases} \tag{3}$$

Here, G_s indicates the bin value of the selected peak.
Step 5. The value of the peak and its HS-direction are kept as the side information.
Step 6. Then another one peak and its HS-direction in the histogram of the modified image are determined.
Step 7. Step 2 - Step 6 are repeated until all data are embedded.
Step 8. The location map (binary values) is embedded before message bits. The amount of peaks, the length of bitstream in the location map, LSBs of excluded 9 pixels, previous peak values and its HS-directions are embedded when the last peak is split.
Step 9. The last N-th peak value and its HS-direction are embedded into LSBs of excluded 9 pixels of the stego image by LSB replacement.

The extraction process and recovery process are performed as follows:

Step 1. The N-th peak value and its HS-direction are obtained from LSBs of excluded 9 pixels. The amount of peaks, the length of bitstream in the location map, LSBs of excluded 9 pixels are known by using the N-th peak and its HS-direction.
Step 2. The extraction is performed by using Eq. 4.

$$m'_k = \begin{cases} 1 & \text{if } i' = G_s - 1 \text{ and } direction = 0 \\ 0 & \text{if } i' = G_s \\ 1 & \text{if } i' = G_s + 1 \text{ and } direction = 1 \end{cases} \tag{4}$$

where, m'_k is the the k-th binary bit of the data extracted from the stego image.
Step 3. The recovery process is performed by applying Eq. 5 to all pixels excluding the first 9 pixels in the bottom row.

$$i = \begin{cases} i'+1 & \text{for } i' < G_s - 1 \text{ and } direction = 0 \\ G_s & \text{for } i' = G_s - 1 \text{ or } i' = G_s \; (direction = 0) \\ G_s & \text{for } i' = G_s \text{ or } i' = G_s + 1 \; (direction = 1) \\ i'-1 & \text{for } i' > G_s + 1 \text{ and } direction = 1 \end{cases} \tag{5}$$

Step 4. Data are extracted by using the $(N-1)$-th peak value and its HS-direction, and the previous peak value and its HS-direction are determined from the extracted data.

Step 5. Step 2 – Step 5 are repeated until N equals to 1.
Step 6: The location map is reconstructed from extracted data bits, and pre-shifted pixels are retrieved.
Step 7: LSBs of the first 9 pixels in the bottom row are written back, and the original image is fully recovered.

3 Experiments

Our experiments were performed on the USC-SIPI image database in which images were converted into grey-level images with the size of 512*512 [12]. Message bits are pseudo random binary bits generated by the Matlab function rand(). Figures 2 and 3 show 4 typical images and their histograms used in our experiments.

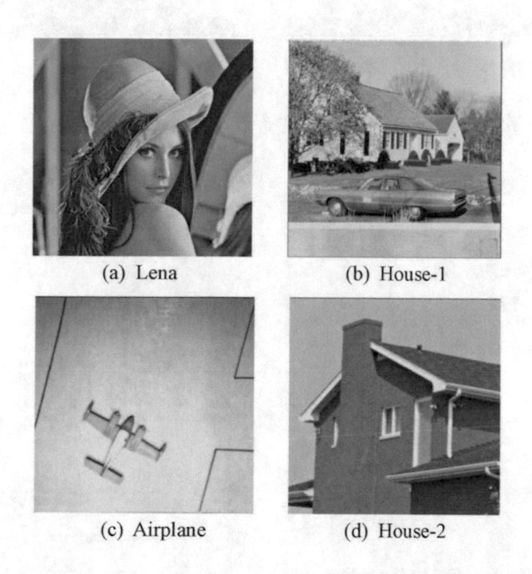

(a) Lena (b) House-1 (c) Airplane (d) House-2

Fig. 2. 4 typical test images used for experiment.

For comparision of our method and RDH-CE method at the maximum hiding ratio based on the number of selected peaks, the embedding performance (Fig. 4) is evaluated with PSNR values of stego images resulted by increasing the number of peaks from 1 to 30. Experiment results vary from image to image, but PSNR values of stego images by the proposed method are generally higher than those by RDH-CE method at the same number of peaks. Especially, as one can see from Fig. 3, the proposed method reveals noticeable results in images with the biased and irregularly distributed histogram, such as House, F-16, Tank, U-2, clock, Jelly beans and so on, rather than in those with the histogram distribution close to uniform or symmetric ones (such as Lena, Baboon, Elaine, Fishing boat and so on). For example, when 20 peaks are used in House (Fig. 2b) for embedding, PSNR values of the proposed method and RDH-CE method are 31.46 dB

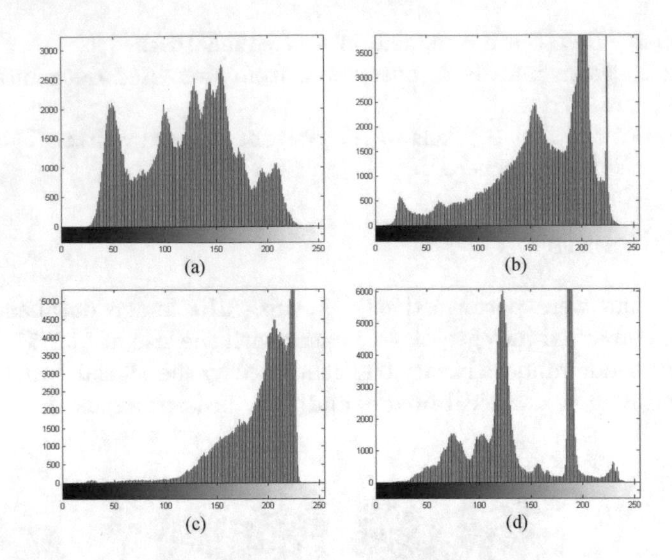

Fig. 3. Histograms of 4 test images in Fig. 2.

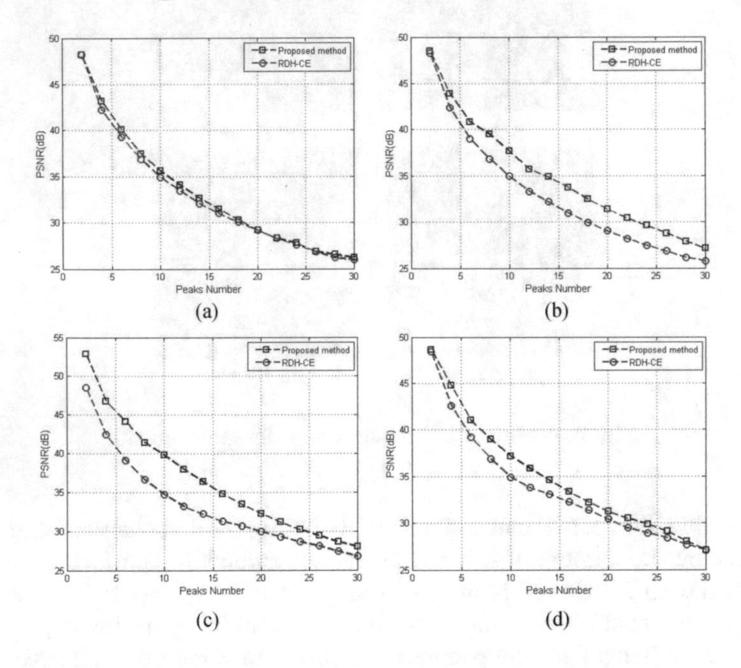

Fig. 4. PSNR comparision of the proposed method and RDH-CE using 4 test images in Fig. 2.

and 29.03 dB at the same amount of embedded data (84,579 bits) respectively. Furthermore, in Airplane (Fig. 2c), although the amount of data (87,764 bits) embedded by using 10 peaks in the proposed method is 2,235 bits more than

Fig. 5. Comparision of visual perception of the proposed method and RDH-CE using 20 peaks. (a), (d) Original images. (b), (e) Stego images by RDH-CE. (c), (f) Stego images by proposed method.

85,529 bits in the RDH-CE method, the PSNR value in the former is 5 dB higher than in the later. Our experiments are also evaluated in terms of the degree of contrast enhancement called the relative contrast error (RCE) [5]. Over all test images, the degree of the contrast enhancement in the proposed method is nearly similar to that in the RDH-CE method. Figure 5 shows visual comparison of the stego images resulted by RDH-CE and proposed method.

Table 1. Comparison of averaged PSNR, RCE and bpp.

	Peaks	PSNR	RCE	bpp
RDH-CE	10	35.35	0.511	0.305
	20	29.87	0.522	0.504
	30	26.75	0.533	0.667
Proposed method	10	36.13	0.511	0.305
	20	30.36	0.522	0.505
	30	27.11	0.533	0.698

Table 1 lists averaged values of PSNR, RCE and hiding ratio resulted by RDH-CE and proposed method over the 32 test images. From average values resulted by using 10, 20 and 30 peaks in the table, it is noted that our method has increased the hiding ratio as well as the PSNR per peak.

4 Conclusions

In this paper, we introduced an adaptive histogram shifting method to *Wu*'s RDH-CE method to improve the PSNR of stego image. One highest peak is selected in each step for data embedding and its histogram shifting direction is determined adaptively based on PSNR of image to be modified. The proposed method can improve efficiently the PSNR of stego image keeping the degree of the contrast enhancement provided by RDH-CE. From the experimental results, it is noted that determining adaptively HS-direction based on distance metrics between original image and the modified image in each step of the embedding process will result in improving PSNR of stego image efficiently.

References

1. Al-Fahoum, A.S., Yaser, M.: Reversible data hiding using contrast enhancement approach. Int. J. Image Process. (IJIP) **7**(3), 248 (2013)
2. Aparna, G.P.K., John, G.: A review on reversible data hiding techniques **9**(3), 44–47 (2014)
3. Chen, H., Ni, J., Hong, W., Chen, T.S.: Reversible data hiding with contrast enhancement using adaptive histogram shifting and pixel value ordering. Signal Process. Image Commun. **46**, 1–16 (2016)
4. Gao, G., Shi, Y.Q.: Reversible data hiding using controlled contrast enhancement and integer wavelet transform. IEEE Signal Process. Lett. **22**(11), 2078–2082 (2015)
5. Gao, M.Z., Wu, Z.G., Wang, L.: Comprehensive evaluation for he based contrast enhancement techniques. In: Advances in Intelligent Systems and Applications, vol. 2, pp. 331–338. Springer (2013)
6. Li, X., Li, B., Yang, B., Zeng, T.: General framework to histogram-shifting-based reversible data hiding. IEEE Trans. Image Process. **22**(6), 2181–2191 (2013)
7. Li, X., Yang, B., Zeng, T.: Efficient reversible watermarking based on adaptive prediction-error expansion and pixel selection. IEEE Trans. Image Process. **20**(12), 3524–3533 (2011)
8. Liu, L., Chen, T., Zhu, S., Hong, W., Si, X.: A reversible data hiding method using improved neighbor mean interpolation and random-block division. Inf. Technol. J. **13**(15), 2374 (2014)
9. Ni, Z., Shi, Y.Q., Ansari, N., Su, W.: Reversible data hiding. IEEE Trans. Circ. Syst. Video Technol. **16**(3), 354–362 (2006)
10. Nosrati, M., Karimi, R., Hariri, M.: Reversible data hiding: principles, techniques, and recent studies. World Appl. Programm. **2**(5), 349–353 (2012)
11. Tian, J.: Reversible data embedding using a difference expansion. IEEE Trans. Circ. Syst. Video Techn. **13**(8), 890–896 (2003)
12. Weber, A.G.: The usc-sipi image database version 5. USC-SIPI Rep. **315**, 1–24 (1997)
13. Wu, H.T., Dugelay, J.L., Shi, Y.Q.: Reversible image data hiding with contrast enhancement. IEEE Sig. Process. Lett. **22**(1), 81–85 (2014)
14. Xuan, G., Yang, C., Zhen, Y., Shi, Y.Q., Ni, Z.: Reversible data hiding using integer wavelet transform and companding technique. In: International Workshop on Digital Watermarking, pp. 115–124. Springer (2004)

Design and Implementation of Network Video Encryption System Based on STM32 and AES Algorithm

Xingyu Tian[1], Chunlei Fan[1], Jia Liu[1], and Qun Ding[1,2(✉)]

[1] College of Electronic Engineering, Heilongjiang University, Harbin, China
{xingyutian1994,qunding}@aliyun.com

[2] Institute of Technology, Harbin, People's Republic of China

Abstract. Due to the rapid development of the Internet, the network video surveillance technology has been widely implemented to overcome the geographical constraints in the network by monitoring the target in any place. There may be some problems when the Internet providing convenience because it's available to global users. For example, video reveal may happen in some situations, which is a threaten to government and national security. So encryption network video surveillance research is imperative. The purpose of this paper is to protect the security of network video surveillance information, and to prevent the useful information from being intercepted and stolen by criminals.

Keywords: Hardware encryption · STM32 · CRYP

1 Introduction

In the 21st century, the Internet has been rapidly developed the camera using analog technology has become the digital network camera. Because the Internet has open, Internet, information sharing and other characteristics, the data collected by the camera should be protected. So the network data encryption is particularly important.

This paper is devoted to the design and implementation of a network video encryption system based on STM32 and AES algorithms. The security system is a new type of equipment that can effectively protect the security of network video data. The innovation of this topic lies in the use of STM32 processor as the main control chip, it has some benefits like low power consumption, high performance, etc., which providing the network data encryption high processing speed [6]. And AES encryption algorithm has the advantages of sufficient key space and high security, When the AES encryption algorithm applying to hardware, it can greatly enhance the anti-attack ability of encryption system and ensure the safe transmission of network video data. The physical map is shown in Fig. 1.

J.-S. Pan et al. (eds.), *Advances in Intelligent Information Hiding and Multimedia Signal Processing*, Smart Innovation, Systems and Technologies 82,
DOI 10.1007/978-3-319-63859-1_7

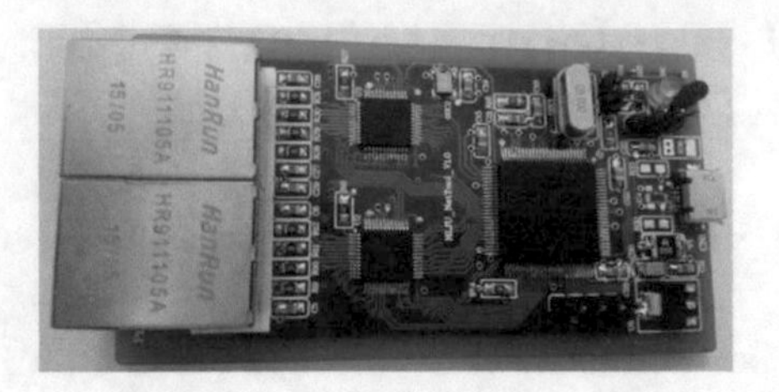

Fig. 1. The physical map.

2 Hardware Design

2.1 Functional Partitioning

The project is based on STM32 and AES algorithm for network video encryption system design and implementation. The overall design is based on STM32 processor chip as the core controller, the standard Ethernet port as the data receiving and sending interface to DM9000 chip as a fast Ethernet controller. In addition, the design of the hardware circuit involves two RJ45 Ethernet ports, which can effectively reduce the equipment installation process. This kind of port can use the network cable to directly connect the computer or camera. In addition, the function of DM9000 is the network data acceptance and transmission, STM32 function is the the network data encryption and decryption processing.

The overall design idea of the network video encryption system is based on STM32F437VIT6 chip of STMicroelectronics, which can encrypt and process network video data through STM32 microprocessor. On this basis, the design of a peripheral circuit, including power supply module, DM9000 Ethernet controller for data frame transceiver module and for program download and debug interface module [1]. The block diagram of the hardware design is shown in Fig. 2.

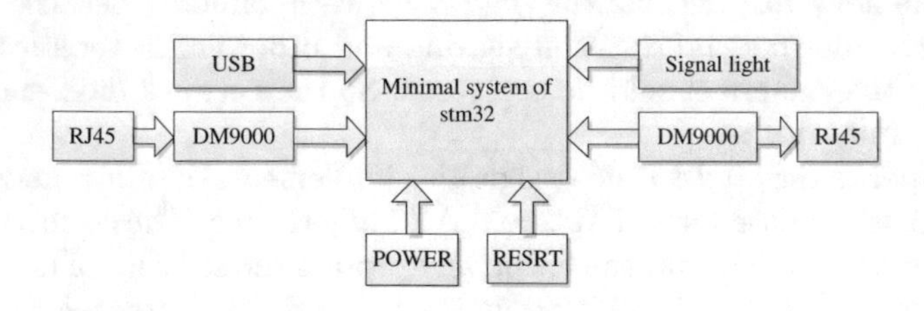

Fig. 2. The block diagram of the hardware design.

2.2 The Design of Peripheral Interface Circuit

The system has the JTAG port and USB interface. JTAG (Joint Test Action Group) is an embedded debugging technology, its function is to debug the system and simulation, and also on the chip within the corresponding test. It is designed to encapsulate the test access port for testing within the chip. The internal node

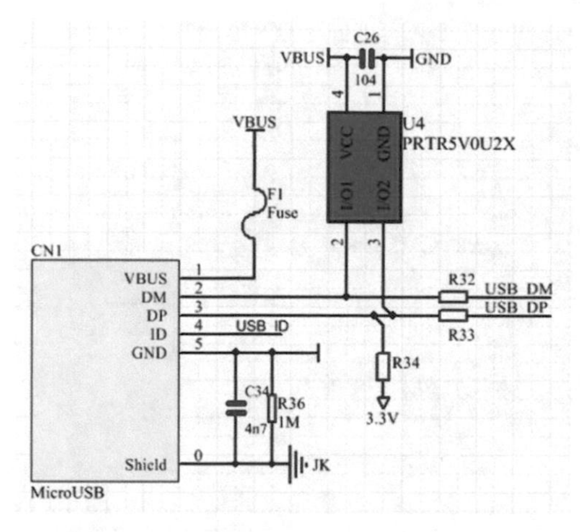

Fig. 3. The design of the USB interface circuit schematic.

can be tested by JTAG test tool. The STM32 processor used in the network video encryption system will support the JTAG protocol [3]. The JTAG interface connection standard for the hardware circuit is 20-pin interface, TDO, TDI, TCK and TMS are four test data output, test data input, test clock and test mode selection. USB is an external bus standard, that is, universal serial bus, USB interface to support the device hot swap and plug and play and other functions, in the circuit design using the most commonly used MicroUSB standard as a USB interface standard, the circuit principle Fig. 3 shows the figure.

2.3 Encryption Processor (CRYP)

The stm32STM32F437VIT6 chip used in the project comes with an encryption processor (cryp). With encryption processor, we can use DES, triple DES or AES (128, 192 or 256) algorithm to encrypt or decrypt the data. This project is mainly used to study AES algorithm for data encryption and decryption. CRYP peripherals are 32-bit AHB2 peripherals. It supports incoming and outgoing data for DMA transfers and has input and output FIFOs (8 words deep) [2].

Because the AES algorithm is very common, so here we do not do a detailed introduction. Let's take a brief look at the AES-ECB encryption algorithm we used. The block diagram of the encryption algorithm is shown in Fig. 4.

However, when using the encryption processor, there are two issues we must pay attention to, because this is the key to the correct execution of the encryption processor.

First, in this project we use the encryption processor to implement the AES encryption kernel. Since the AES algorithm uses a block cipher, the incomplete input data blocks must be padded before encryption (additional bits should be appended to the end of the data string). The padding item needs to be discarded after decryption. Since the hardware does not handle the padding operation, this part of the operation is going to be done by software [5].

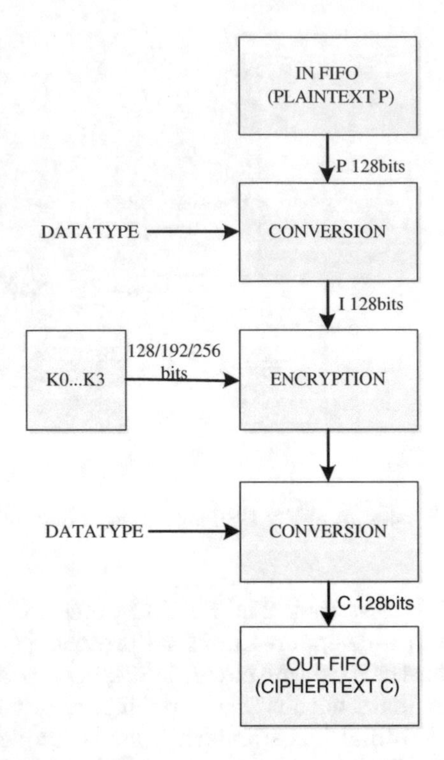

Fig. 4. AES-ECB mode encryption block diagram.

Second, we need to carry out bit exchange operations on the process of data in terms of its data type. When writing data to the CRYP_DIN register, 32 bits (word) data is entered once to the CRYP processor. AES's principle is to process the data stream every 128 bits. For each 128-bit block, bits are numbered from M1 to M128, where M1 is the leftmost bit of the block and M128 is the rightmost bit of the block.

The system memory structure uses the small end mode: Regardless of the data type (bit, byte, 16-bit, half-word, 32-bit word), the least significant data occupies the lowest address location. Therefore, for data reading from the IN FIFO, it is necessary to perform bit, byte, or halfword swapping operations (depending on the type of data to be encrypted) before entering the CRYP processor. Before CRYP data is written to the OUT FIFO, it needs to perform the same exchange operation.

3 The Software Design

3.1 The Overall Framework of Network Video Encryption Based on UIP - UDP Technology

Network communication programming to use the network socket (Socket). Socket is a descriptor, it is also the basic operation of network communication unit.

Data transmission is a special I/O, providing different hosts to communicate with each other between the end of the process, these processes of the communication before the establishment of a Socket, and through the Socket read and write operations to achieve network communication functions [4].

In this paper, the communication between the encryption and decryption board is based on UDP network data communication technology. In addition, the system is going to improve the speed of encryption by replacing the TCP/IP protocol stack with UIP protocol stack, UIP widely used with STM32 chip.

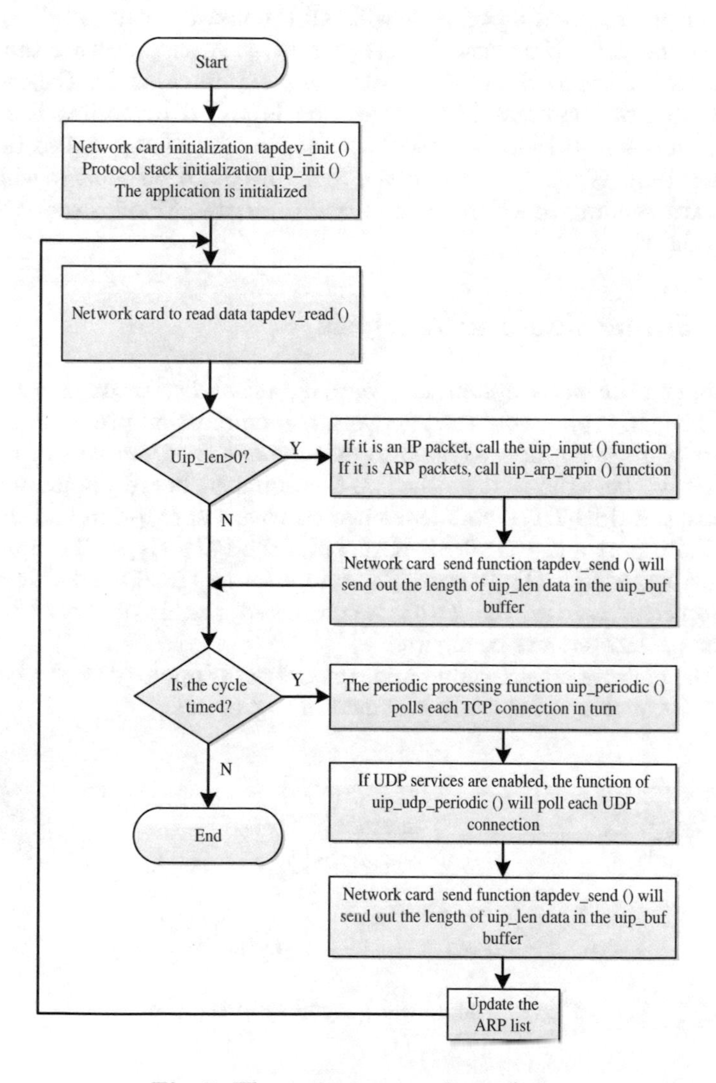

Fig. 5. The main program loop of uip.

3.2 The Main Program Loop of UIP

When all the initialization, configuration and other work is completed, uip constantly in the main loop is going to run. The basic processing flow chart shown in Fig. 5.

When the device starts running, the first part is the network card, protocol stack and application initialization. When the initialization is complete, the card will check the device to ensure there is no incoming data [7]. If there is a new data, it will call a different function to deal with the packet in terms of their types. After processing the packet, it will call the network card sending function to send out the data. If no new data is present, it checks whether the periodic timer expires. If the periodic timer does not expire, it ends. Otherwise, the timer is reset first, and the TCP connection is polled by calling the periodic processing function. If there is a UDP service, the function is called to poll the UDP connection. When the data in the cache needs to be sent, it will call the network card sending function to send out the data. Finally, the ARP cache table is updated.

4 The Simulation and Analysis

After designing the entire system hardware circuit and software encryption program and the UDP protocol-based network communication program, the circuit and the software program are tested to determine the correctness of the circuit design. Firstly, the serial port on the two development board parameters is used to configure the USB-TTL module connected to the encryption and decryption board COM10 port and PC's USB port, open the PC's HyperTerminal and set the parameters, select COM10 port, The rate is set to 115200, data bit 8, no parity bit, stop bit is 1. After the setting is completed, the circuit board is powered on and the parameters are configured [8].

After the parameters are configured, the network topology can be connected. The network topology of the test is shown in Fig. 6.

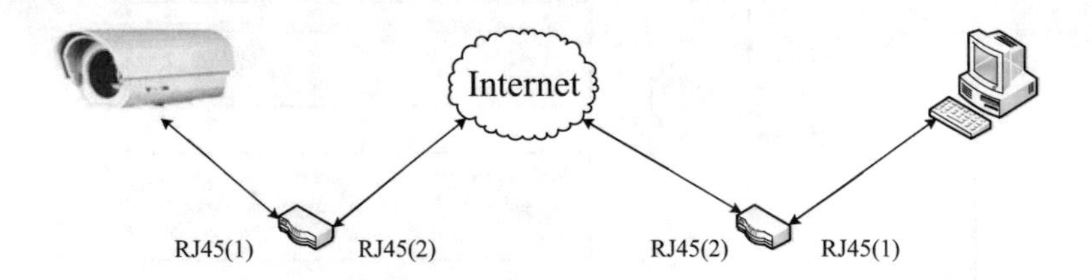

Fig. 6. The network topology of the test.

The RJ45 (1) port of the encryption board is connected to the HD webcam (model DS-2CD1203D). After receiving the data frame from the webcam, the encrypted packet is processed by the STM32 processor. The encrypted data is passed to the two ports, and the two ports pass the UIP-UDP network communication program to the encrypted data frame through the network to the decryption board. The decryption board decrypts the ciphertext data frame according to the AES decryption program. To the PC side, and the PC side shows the video.

On the PC side, we need to search the network device by the device network search software. The search result is shown in Fig. 7.

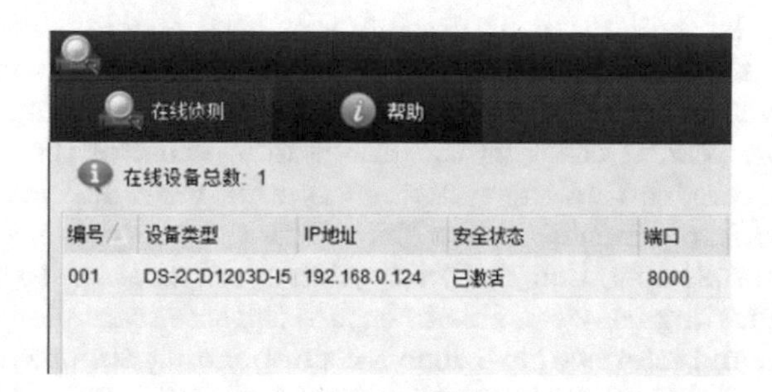

Fig. 7. The results of the network device search.

From the figure you can clearly see the searching result of this section of the web camera, model DS-2CD1203D and the IP address of this device is 192.168.0.124. After getting this information the access of this web camera is obtained to get the front of the network video data, in the IE browser address bar enter 192.168.0.124 IP address and enter the network video data can be obtained, the figure shown in Fig. 8.

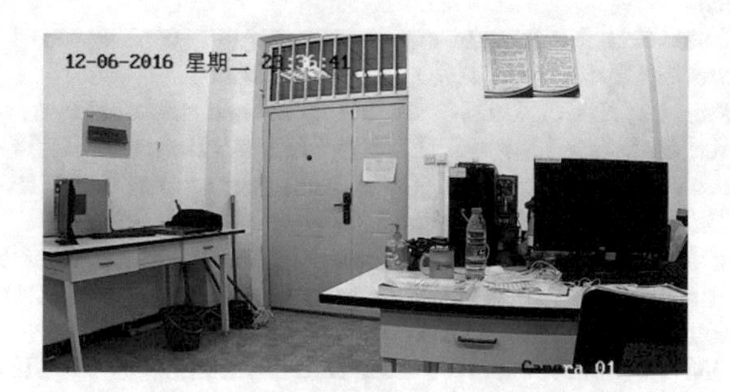

Fig. 8. The display of the Video data.

5 Conclusion

This project uses AES encryption theory for network video data encryption and application STM32 processor block encryption in the field of information security is an important research topic. The project design and implementation of a new type of network video encryption system, can effectively protect the network video surveillance data security. The innovation of this project is: Firstly, the use of STM32 processor as a video encryption system, the main control chip, it has low power consumption, high performance, etc., to ensure that the network video data encryption high-speed, high performance. Secondly, the encryption system is no longer using the PCI standard, but the use of external devices and switches between the PC and a device can simplify the install operation, only need Powering the system, it is possible to complete the data encryption on the application layer, it makes it possible for those people who's not familiar with computer to use the encryption system. Thirdly, the system uses a high-performance encryption technology, so in the follow-up system upgrade process, you can achieve without changing the circuit in the case of the replacement of the encryption core, so that the system encryption reset and upgrade are very convenient The Experimental results show that the encryption device has the advantages of simple operation, fast encryption speed and easy to be deciphered. After repeated testing has been able to ensure the safe transmission of information, and speed and interface have done a corresponding optimization design, to meet the needs of the public, the application prospects.

Acknowledgment. Project supported by the National Natural Science Foundation of China (Grant Nos. 61471158), and Project supported by the Modern sensor technology of Universities in Heilongjiang (Grant No. 2012TD007).

References

1. Ferri, C., Viesas, A., Moreshet, T.: Energy efficient synchronization techniques for embedded architectures. In: Proceedings of the 18th ACM Great Lakes Symposium on VLSI (2008)
2. Barkan, E., Bihametal, E.: Instant ciphertext-only cryptanalysis of GSM encrypted communication. J. Cryptology **21**(3), 392–429 (2008)
3. Zhong, Y., Jiang, C., Xu, J., Lu, X., Li, Y.: Research and design of hardware debugging method for embedded system. Electr. Measur. Instrum. **4**, 94–107 (2006)
4. DAVICOM Semiconductor Inc. DM9000A Ethernet Controller with General Processor Interface Data Sheet (2005)
5. Wang, Q.: An interrupt management scheme based on application in embedded system. In: Proceedings of 2008 International Conference on MultiMedia and Information Technology, pp. 449–452 (2008)
6. Bergeron, J.: Writing Testbenches Functional Verification of HDL Models, pp. 36–40. Kluwer Academic Publishers (2000)
7. Wang, H., Liang, J., Jay Kuo, C.-C.: Overview of robust video streaming with network coding. J. Inf. Hiding Multimedia Signal Process. **1**(1), 36–50 (2010)
8. Liu, P., Wu, Y., Gao, Y., Jia, K.: Hierarchical coding algorithm for medical video based on H.265/HEVC. J. Inf. Hiding Multimedia Signal Process. **7**(5), 906–915 (2016)

Fast Intra Mode Decision Algorithm for 3D-HEVC Transcoding

XiuJun Feng[1,2(✉)], PengYu Liu[1,2(✉)], and KeBin Jia[1,2(✉)]

[1] Beijing Laboratory of Advanced Information Networks, Beijing, China
[2] Faculty of Information Technology,
Beijing University of Technology, Beijing, China
{liupengyu, kebinj}@bjut.edu.cn

Abstract. To satisfy the popularity of 3D application, 3D-HEVC has been developed as a new video encoding standard for multiviews. What's more, as the development of internet and mobile services, the use of 3D video application from the big screen to mobile devices become an inevitable trend. So an efficient transcoding for 3D videos is necessary. In 3D-HEVC, 3D video is comprised by multiview video and corresponding depth maps. The computational complexity is further increased caused by depth maps. In order to reduce the computational complexity caused by the mode decision for depth maps, an efficient fast intra mode decision for 3D-HEVC depth maps down-sizing video transcoding based on SVM (support vector machines, SVM) is proposed in this paper. Compared with the HTM16.0 which employs the original intra mode decision, the proposed algorithm can save about 16% of transcoding time without distortion of synthesized intermediate views.

Keywords: 3D-HEVC · Depth Modeling Mode (DMM) · Down-sizing · SVM · Transcoding

1 Introduction

With the popularity of 3D application, many 3D representation devices have been produced and 3D video services have been investigated in recent years. Meanwhile as the development of Internet and increasing diversity of services, especially mobile terminal, the use of 3D video application from the big screen gradually shift to mobile handheld devices become an inevitable trend because of the advantage of handheld mobile devices. In order to satisfy the increasing demand for better visual quality and cater to this trend, an efficient down-sizing video transcoding about 3D video is definitely need.

In order to cater for the popularity of 3D application, 3D-HEVC has been developed. In 3D-HEVC, 3D video is represented using multiview video plus depth format. As the depth maps introduced, the computational complexity of 3D-HEVC is increased than traditional 2D video encoding. Studies have shown that it can be find that intra mode decision of depth maps accounts for a large part of the computing time in the whole encoding progress [6]. Meanwhile, in last decades, there are many researches have been studied traditional 2D video transcoding algorithms, but there are few papers on 3D video transcoding. So combined the transcoding research status, it is worthy of

J.-S. Pan et al. (eds.), *Advances in Intelligent Information Hiding and Multimedia Signal Processing*, Smart Innovation, Systems and Technologies 82,
DOI 10.1007/978-3-319-63859-1_8

researching how to reduce computation of intra mode decision for depth maps. On the purpose of speed up the intra mode decision of depth maps process, a fast intra mode decision algorithm about 3D-HEVC down-sizing video transcoding based on SVM is proposed in this paper. Experimental results show that the proposed algorithm can get good performances without distortion of synthesized intermediate views.

The reminder of this paper is organized as follows. In Sect. 2 the principles of intra prediction about depth maps in 3D-HEVC is briefly introduced. Our proposed mode decision algorithm for depth maps based on SVM is fully expounded in Sect. 3. Then our experimental results are shown in Sect. 4. Finally this paper is briefly concluded.

2 3D-HEVC Depth Maps Intra Coding

As the latest encoding standard about multiview videos, 3D-HEVC is a high compression video coding standard due to the contribution of several modified and additional techniques. Intra prediction in 3D-HEVC exploits the directional spatial correlation to reduce the spatial redundancy within a single frame, by means of checking the similarity among pixels in the previous coded blocks with the pixels in the current block. In 3D-HEVC, there are two more types of intra mode prediction for depth maps coding besides 35 types of original intra mode prediction. Traditional intra prediction modes cannot handle the depth maps because of its characteristics. It has sharp edges and large regions with nearly constant values in depth maps. The process of intra prediction about

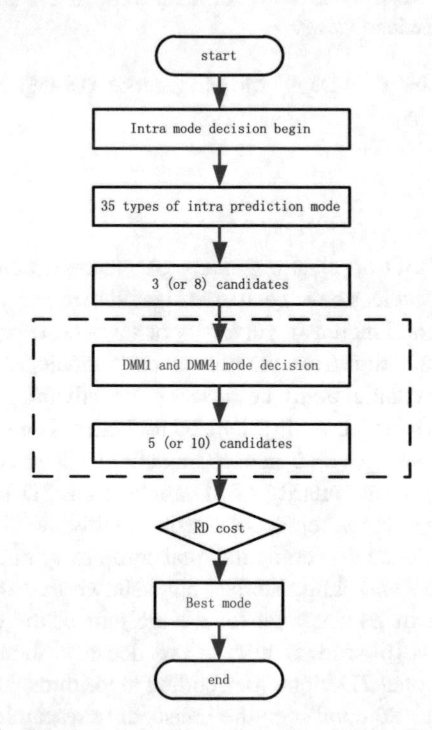

Fig. 1. Intra mode decision of depth maps in 3D-HEVC

depth maps is shown in Fig. 1. In order to eliminate ringing artifacts caused by unique characteristics of depth maps, Depth Modelling Modes (DMM) is introduced.

DMM is a new intra prediction mode for depth map coding. It's used on the purpose of a better representation of edges in depth maps, where each region is represented by a constant value. After the 6th meeting of JCT-3V, the additional intra mode prediction contains DMM1 (above) and DMM4 (below) (as shown in Fig. 2). DMM1 is Explicit Wedgelet signaling and DMM4 is Inter-component prediction of Contour partitions. The information required for such a model consists of two elements. One is the partition information which specifying the region each sample belongs to, the other is region value information which specifying a constant value for a samples of the corresponding region [1]. For the partition of DMM1, a search over a set of default Wedgelet partitions is carried out and the Wedgelet partition with the minimum distortion between the original signal and the Wedgelet approximation is selected. The prediction of Contour partition is determined by a thresholding method. The mean value of the four corner luma samples of the texture reference block is set as the thresholding and depending on whether the value of a sample is above or below the sample position is marked as part of region P_1 or P_2 in the resulting Contour partition pattern.

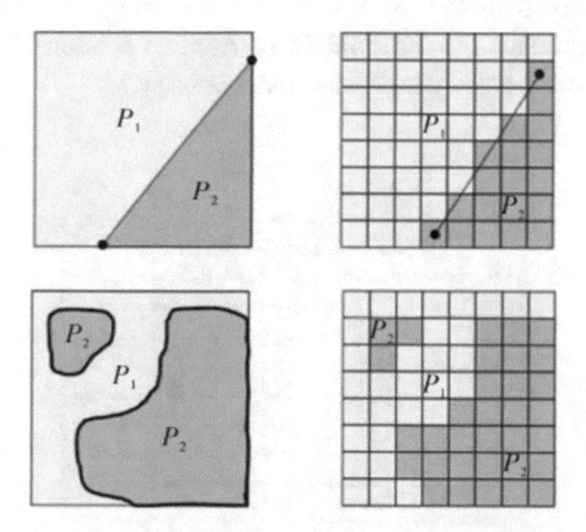

Fig. 2. DMM in 3D-HEVC for depth maps

And referring to the experiment we've done, it can be found that DMM is a time-consuming process compared to the remaining encoding process. A fast intra mode decision about depth maps for transcoding is necessary on the purpose of reducing the computational complexity of 3D-HEVC transcoding.

3 Proposed Algorithm

The high complexity of 3D-HEVC video transcoding creates an opportunity for applying machine learning algorithms to reduce the complexity of transcoder. From the existing classification models available in the literature, referring to the existing transcoding algorithms, we choose Support Vector Machines (SVMs) due to its excellent performance in pattern recognition. In this section we are going to describe the process of using SVMs to build a mode decision classifier for very low complexity transcoding.

3.1 Complexity Analysis of Intra Mode in 3D-HEVC for Depth Maps

In this paper we focus on the new introduced characteristics, depth maps. Same as texture pictures, pictures are divided into a sequence of coding tree units (CTUs), all being the same size, and each covering a square pixel region of the picture. From the experiment we've done, it can be found that the DMM isn't always selected as the best mode in CTUs as one of intra mode prediction for depth maps calculated by (1).

$$\mathrm{J(m)} = \mathrm{D(m)} + \lambda * R(m) \tag{1}$$

Taking one of test video sequences, balloons, as an example, after encoding by HTM16.0, it turns out that 41% CTUs in test video sequences whose best mode do not contain DMM (as shown in Fig. 3) and time-consuming of DMM mode decision is shown in Fig. 4. Compared to original intra mode decision, DMM mode decision accounts for more time on the whole encoding process.

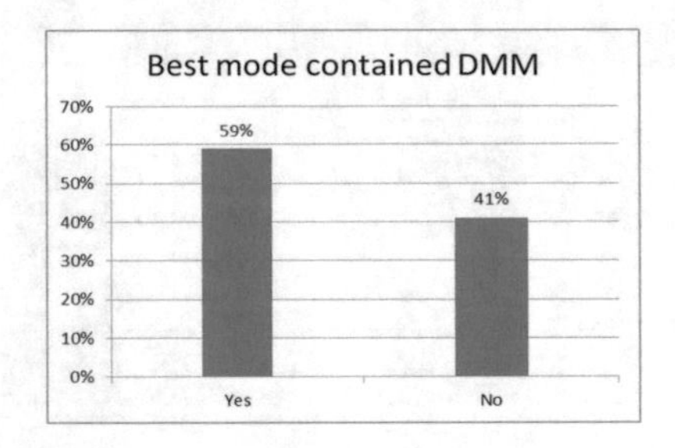

Fig. 3. DMM selected as best mode percentage among CTUs

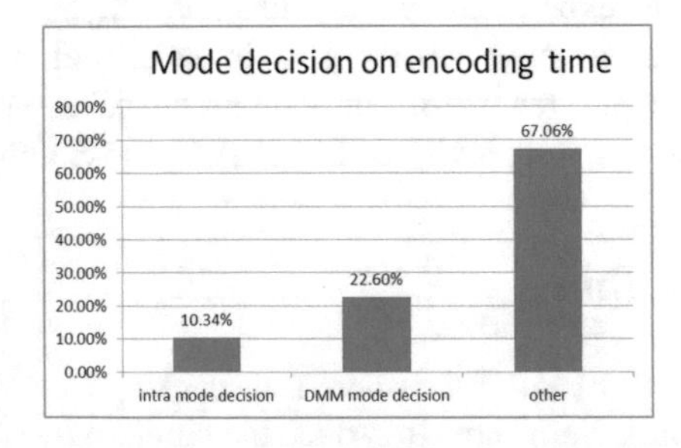

Fig. 4. Time-consuming of mode decision on encoding time

In order to further analyze the use of DMM, the proportion of whether best modes in each CTU contain DMM or not for the remaining test video sequences is statistically analyzed. 100 frames of test video sequences have been chosen respectively. The experiment is conducted in HTM16.0 configured by All Intra, and (25, 34) is selected as QP pair for texture video and depth map in 3D-HEVC. And the result is as below, Fig. 5. From the result, it is obviously found that more large regions with nearly constant value of the corresponding depth of video, less proportion of best modes in each CTU containing DMM.

Although good quality and high compression efficiency can be achieved by the mode decision optimization, it can be still found that best modes in a large number of CTUs do not contain DMM mode. Compared to original intra mode decision, DMM mode decision is a more time-consuming process. So a fast intra mode decision algorithm about depth maps for down-sizing transcoding is very desirable.

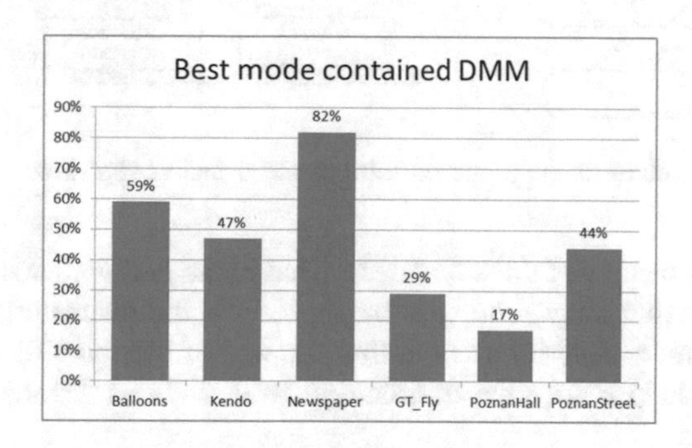

Fig. 5. Best mode contained DMM in CTU

3.2 Feature Vectors for Mode Decision Classification

Based on the analysis in the previous section, the complexity of 3D-HEVC video transcoding creates an opportunity for applying machine learning algorithms to reduce the complexity of the transcoder. From the existing different classification models available in the literature and the previous researchers about video transcoding [2, 3, 5, 7, 8], SVMs is chosen due to their excellent performance in pattern recognition. The detailed intra mode classification scheme using SVMs is illustrated in Fig. 6. By discarding the improbable DMM modes using SVMs classifier, only the remaining modes of intra modes is used and the process of calculating RD cost can be early terminated compared with the original mode decision.

The SVMs classifier is chosen to be learnt as pattern recognition. In view of own characteristics represented by depth map, it only has gray value and without chroma information. And combined the experiment have done, a group of feature vectors FV_p, are applied in our proposed algorithm.

FVp = [mean value of depth maps of each CTU, variance value of depth maps of each CTU, entropy of depth maps of each CTU].

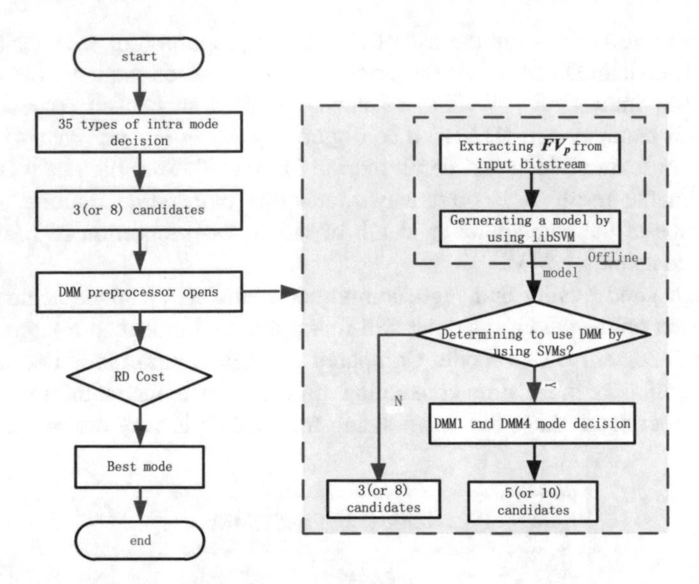

Fig. 6. Flowchart of the proposed intra mode decision algorithm using SVMs

The reason choosing that three features is that mean and variance can well describe the properties of depth maps, and entropy represents the characteristics of the image gray-scale distribution. And it turns out that the use of the three features as a training set of SVMs can help reduce the transcoding time without distortion of synthesized intermediate views.

4 Experimental Results

In this section, two kinds of resolution of test video sequences: 1024 × 768; 1920 × 1088, have been tested to prove the effectiveness of proposed algorithm. The proposed algorithm is implemented in the 3D-HEVC reference software HTM16.0. A high quality, easy to use and free libSVM [4] software package was used in our experiment for SVMs training and prediction.

In the training process, an extension set of experiments with test video sequences and performed offline have been conducted. Then we sample test videos for a quarter. Three kinds of videos, balloons, Kendo, Newspaper, are sampled from 1024 × 768 to 512 × 384. The left videos, GT_Fly, PoznanHall, and PoznanStreet, are sampled from 1920 × 1088 to 960 × 544. In this paper, the experiment is conduct in HTM16.0 configured by All Intra, and (25, 34) is selected as QP pair for texture video and depth map.

Then the experimental platform loads the mode file calculated by training process when transcoding. The performance compared to the original encoder is shown in Table 1. Peak Signal to Noise Ratio (PSNR) about texture video (Δvideo PSNR) and

Table 1. Experimental results comparison between proposed algorithm and trivial transcoder

Test sequences	Δvideo PSNR (dB)	Δsynth PSNR (dB)	ΔEnc Time (%)
Balloons	0.00	0.00	11.00%
Kendo	0.00	0.00	11.45%
Newspaper	0.00	0.00	23.92%
GT_Fly	0.00	0.00	34.11%
PoznanHall	0.00	0.00	23.15%
PoznanStreet	0.00	0.00	6.76%
Average	0.00	0.00	15.89%

synthesized intermediate view (Δsynth PSNR) evaluate coding property jointly, and running time (ΔEnc Time) measures computational complexity, and calculated as follows.

$$\Delta \text{PSNR} = \text{PSNR}_{\text{original}} - \text{PSNR}_{\text{proposed}} \tag{2}$$

$$\Delta \text{Enc Time} = \frac{\text{time}_{\text{original}} - \text{time}_{\text{proposed}}}{\text{time}_{\text{original}}} \times 100\% \tag{3}$$

From the result, it is obvious that the proposed fast intra mode decision about 3D-HEVC depth map algorithm save about 16% coding time without distortion of synthesized intermediate views. The corresponding depth maps of video having more constant region, more encoding time saved. And taking video sequences, balloons, as an example, one frame synthesized using depth-image-based rendering (DIBR) techniques is shown as below (in Fig. 7). From the viewing point, there is no difference.

Fig. 7. Left generated by original algorithm, right by proposed algorithm

5 Conclusion

In this paper, a fast intra mode decision for 3D-HEVC depth maps transcoding algorithm is proposed. The proposed algorithm exploits the correlation between the coding information of original high resolution video and the coding modes of down-sized video using SVMs. The proposed scheme was implemented in 3D-HEVC reference software HTM16.0. Experiments were conducted to evaluate the performance of proposed algorithm and the results shows that the fast intra mode decision scheme about depth maps based on SVMs can reduce about 16% of the computational complexity without distortion of synthesized intermediate views.

Acknowledgments. This paper is supported by the Project for the National Natural Science Foundation of China under Grant No. 61672064, the National Key Research and Development Program of China under Grant No. 2016YFB0901200, the Beijing Natural Science Foundation under Grant No. KZ201610005007, 4172001, the China Postdoctoral Science Foundation under Grants No. 2015M580029, 2016T90022, the Beijing Postdoctoral Research Foundation under Grant No. 2015ZZ-23 and Project supported by Chaoyang District of Beijing Postdoctoral Research Foundation under Grant No. 2016ZZ- 01-15.

References

1. Chen, Y., Tech, G., Wegner, K., Yea, S.: Test model 11 of 3D-HEVC and MV-HEVC. In: Joint Collaborative Team on 3D Video Coding Extensions (JCT-3 V) Document JCT3 V-K1003, 11th Meeting, Geneva, FR, February 2015
2. Lu, Z.Y., Siu, W.C., Jia, K.B.: Efficient mode decision for H. 264 down-sizing video transcoding using support vector machines. In: APSIPA ASC 2010-Asia-Pacific Signal and Information Processing Association Annual Summit and Conference (2010)
3. Lu, Z.Y., Jia, K.B., Siu, W.C.: Low-complexity intra prediction algorithm for video down-sizing transcoder. In: 2011 Visual Communications and Image Processing (VCIP), Tainan, pp. 1–4 (2011)
4. Chang, C.-C., Lin, C.-J.: A Library for Support Vector Machines. http://www.csie.ntu.edu.tw/~cjlin/libsvm/
5. Wang, J., Li, L., Zhi, G., et al.: Efficient algorithms for HEVC bitrate transcoding. Multimedia Tools Appl., 1–21 (2016)
6. Zhao, J., Zhao, X., Zhang, W., et al.: An efficient depth modeling mode decision algorithm for 3D-HEVC depth map coding. Optik Int. J. Light Electron. Opt. **127**(24), 12048–12055 (2016)
7. Zhang, J., Dai, F., Zhang, Y., et al.: Efficient HEVC to H.264/AVC transcoding with fast intra mode decision. In: Advances in Multimedia Modeling, pp. 295–306. Springer, Heidelberg (2013)
8. Guang, D., Tao, P., Song, S., et al.: Improved prediction estimation based H.264 to HEVC intra transcoding. In: Advances in Multimedia Information Processing – PCM 2014, pp. 64–73 (2014)

Implementation of a Drone-Based Video Streamer

Zhifei Fan, Baolong Guo(✉), and Jie Hou

School of Aerospace Science and Technology, Xidian University, Xi'an 710071, China
zffan029@163.com, blguo@xidian.edu.cn

Abstract. Image transmission technology is one of the key technologies in the field of UAVs. In this paper we present a video streamer application for UAVs. It enables a drone to publish a live video to streaming media server directly. The Qualcomm Snapdragon Flight, a highly integrated board that targets consumer drones applications, is the experimental platform we use. The camera on the board is used to capture real time video images. And these video sequence images are encoded to H.264 video stream. Then the stream is published over wireless network follows the RTMP specification.

Keywords: UAV · Live video · H.264 · RTMP

1 Introduction

In recent years, UAVs have been widely used in various fields, especially in the field of consumption. Consumer interest in drones has increased dramatically with more than one million new drones expected to be sold to U.S. consumers alone by the end of 2015 [1]. People have a great interest in taking video from a drone. Image transmission and flight control are the main technologies in the field of UAVs.

Currently, the most common way of image transmission is to send the video captured by a drone to the smart device such as mobile phone, which is connected to the remote control device. The disadvantage of this way is that we have to use a smartphone or tablet which is paired with the drone to watch the video it captures. Today, the live video technology has been very mature. If the drone can publish the live video to the streaming media server directly, anyone who wants to watch the video just need to know the website of the streaming media server. Therefore, it is very meaningful to develop a video streamer application in a drone.

The prerequisite for publishing a live video is to encode the video sequence images in real time. H.264 is a video codec standard which can achieve high quality video in relatively low bitrates. It can be seen as the successor of the

B. Guo—This work was supported by the National Natural Science Foundation of China under Grants No. 61571346.

J.-S. Pan et al. (eds.), *Advances in Intelligent Information Hiding and Multimedia Signal Processing*, Smart Innovation, Systems and Technologies 82,
DOI 10.1007/978-3-319-63859-1_9

existing formats (MPEG2, MPEG4, DivX, etc.) as it aims in offering similar video quality in half the size of the formats mentioned before [2]. It was developed by the ITU-T Video Coding Experts Group (VCEG) together with the ISO/IEC JTC1 Moving Picture Experts Group (MPEG). Cisco has taken their H.264 implementation, and open sourced it under BSD license terms [3]. And it is named OpenH264.

The Real-Time Messaging Protocol (RTMP) was designed for high-performance transmission of audio, video and data between Adobe Flash Platform technologies, including Adobe Flash Player and Adobe AIR. RTMP is now available as an open specification to create products and technologies that enable delivery of video, audio and data in open AMF, SWF, FLV and F4V formats compatible with Adobe Flash Player [4].

The main objective of this paper is to develop an application that enables a drone to publish a live video to streaming media server directly. The hardware development platform is Qualcomm Snapdragon Flight. It features advanced processing power, real-time flight control on the Qualcomm Hexagon DSP, built-in 2×2 Wi-Fi and Bluetooth connectivity, and global navigation satellite system optimized to support highly accurate location positioning.

The remainder of this paper is organized as follows. Section 2 gives a brief introduction of H.264 video compression standard and RTMP protocol. Section 3 describes the implementation of the video streamer application and presents the experimental results. Finally, Sect. 4 concludes the paper.

2 Related Protocols

2.1 H.264 Video Compression Standard

H.264 is a block-oriented motion-compensation-based video compression standard and there are three profiles: baseline, main and extended. The baseline has lower capability and error resilience that is usually used in video conferencing and mobile video. The main has high compression quality that is usually used in broadcast. And the extended added some features for efficient streaming.

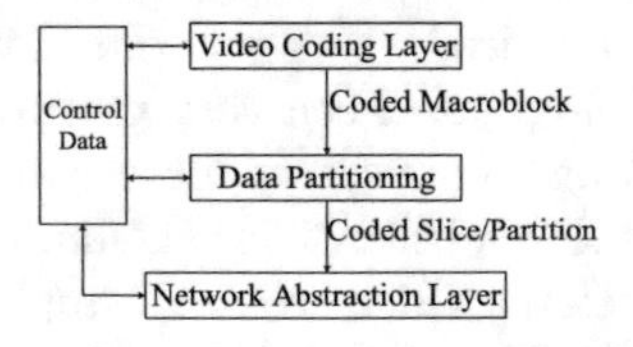

Fig. 1. Structure of H.264 video encoder

There are a video coding layer (VCL) and a network abstraction layer (NAL) in H.264 video coding standards as shown in Fig. 1 [5]. The VCL contains coded video data and these data will be wrapped under NAL for transmission.

The main goal of the NAL is the provision of a "network-friendly" video representation addressing conversational and "non conversational" applications. The coded video data is organized into NAL units, each of which is effectively a packet that contains an integer number of bytes. Each NAL unit is divided into two parts: header and payload. Header is the first byte of NAL unit, and the remaining bytes are the NAL unit payload.

The structure of NAL unit is shown in Fig. 2. The NAL Unit header consists of three parts: forbidden_zero_bit, nal_ref_idc and nal_unit_type [6]. Where the nal_ref_idc represents the importance of the current NAL unit.

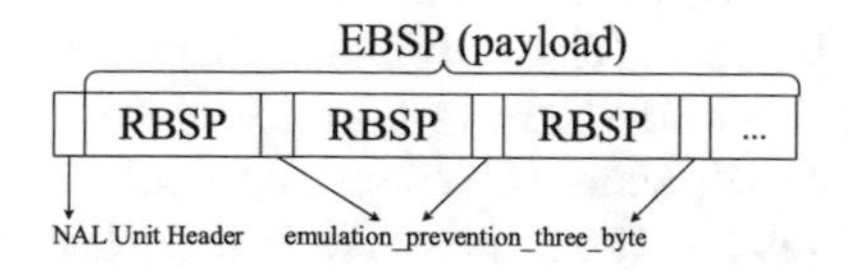

Fig. 2. The structure of NAL unit

The EBSP (Encapsulate Byte Sequence Payload) is composed of several RBSPs (Raw Byte Sequence Payload) and emulation_prevention_three_bytes (0x03). Each RBSP contains a SODB (String of Data Bits), a stop bit which equals to one and some zero bits as shown in Fig. 3. The SODB is the real raw H.264 stream and it's a bit stream. So the number of bits for a SODB may not be byte-aligned and therefore difficult to process. By adding a stop bit and several zero bits, the byte-aligned is finally achieved.

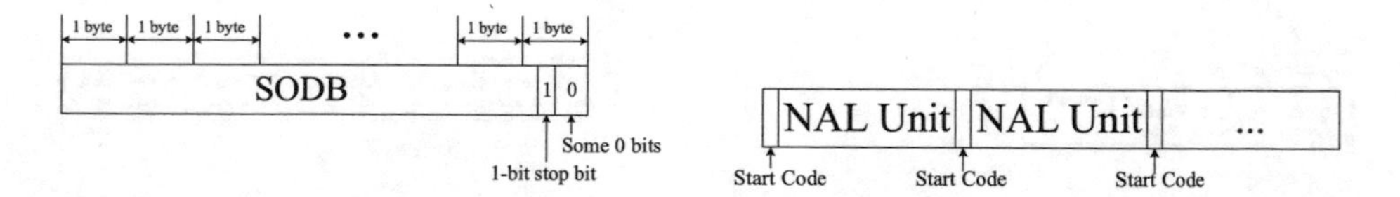

Fig. 3. Message format

Fig. 4. Chunk format

There are two ways to pack a NAL unit for different systems, Packet-Transport System and Byte-Stream Format. The byte stream format is shown in Fig. 4. In byte stream format the NAL units is prefixed with the start code (0x000001 or 0x00000001) in order to identify their boundaries.

2.2 RTMP Specification

Adobe's Real Time Messaging Protocol (RTMP) provides a bidirectional message multiplex service over a reliable stream transport.

An RTMP connection begins with a handshake [7]. In the handshake process, the client and the server each send the same three chunks. RTMP protocol

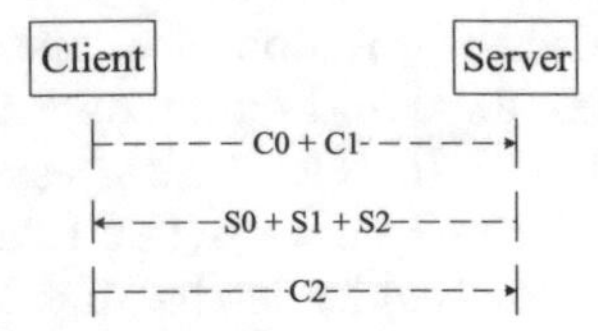

Fig. 5. Handshake

doesn't specify the order in which these six chunks are sent, just to meet some requirements. In order to reduce the communicate times, the transmission order shown in Fig. 5 could be used. The handshake begins with the client sends C0 and C1 chunks. When C0 and C1 have been received, the server sends S0, S1, and S2. And the client will send C2 when received S0, S1 and S2. After the server received the C2 chunk, the handshake is complete.

Netconnection and Netstream are used to send various commands between the client and the server. The NetConnection is a higher-lever representation of connection between the server and the client, and NetStream represents the channel over which audio streams, video streams and other related data are sent [8].

The server and the client send RTMP messages over the network to communicate with each other. There are two parts in RTMP message, a header and its payload. The Message Header consists of Message Type, Length, Timestamp and Message Stream Id (see Fig. 6). Where the Length field represents the size of the payload in bytes and the Stream Id field identifies the stream of the message [9].

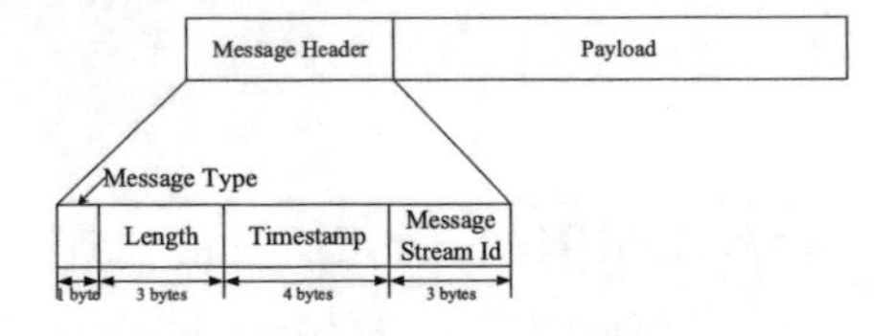

Fig. 6. Message format

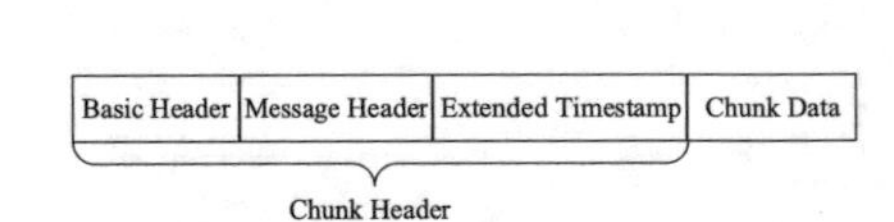

Fig. 7. Chunk format

The message needs to be split into several chunks to transmit in the network. Each chunk consists of a header and its data. The header contains Basic Header, Message Header, and Extended Timestamp (see Fig. 7). Where the Basic Header encodes the chunk stream ID and the chunk type and the Message Header encodes information about the message being sent.

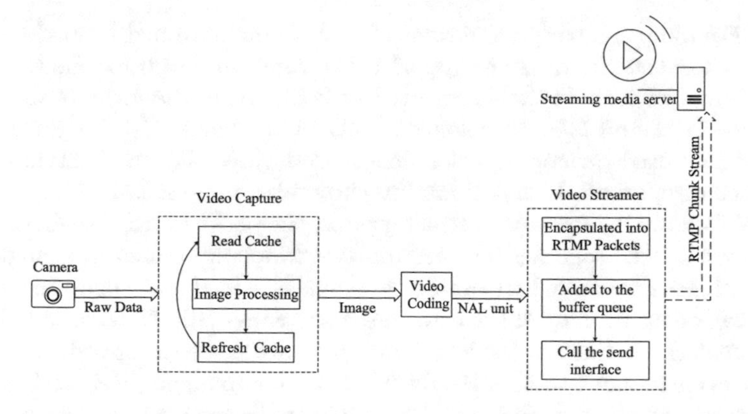

Fig. 8. Module structure

3 Implementation of the Software

3.1 Module Structure

The module structure is shown in Fig. 8. The raw data obtained from the camera on the board is the input of the video capture module, and the output is the data of each frame. Then the video coding module encodes the video sequence images into NAL units. These NAL units are encapsulated into RTMP packets and added to the buffer queue in video streamer module. Finally, RTMP packets are sent to the streaming media server by calling the send interface. Each module is described in detail in Table 1.

Table 1. Description of each module

Modules	Description	
Video capture	Input:	Parameters of the camera
	Output:	Video sequence images
	Function:	1. Get raw data from the camera
		2. Extract the data of each frame (NV21)
		3. Convert the format of each frame to I420
Video coding	Input:	1. Parameters of The Encoder
		2. Video Sequence Images
	Output:	NAL Units
	Function:	Encode the video sequence images
Video streamer	Input:	NAL Units
	Output:	RTMP Chunks
	Function:	1. Encapsulate each NAL unit into RTMP packet
		2. Added each RTMP packet to the buffer queue
		3. Call RTMP send interface to send RTMP packets

Image Format. The original format of each frame obtained from the buffer is NV21 but the encoder doesn't support this format very well for the time being. So the format of each frame is converted to I420, one of the formats the encoder supports. NV21 and I420 all belongs to YUV420 format. YUV is a color space which widely used for coding color images and video. Where Y stands for the luminance component, U and V are the chromatic components.

YUV formats fall into two distinct groups, the packed formats where Y, U, V samples are packed together into macropixels which are stored in a single array, and the planar formats where each component is stored as a separate array, the final image being a fusing of the three separate planes [10]. NV21 and I420 are in the planar format. Each of the four Y components shares a set of U and V components. So there are a total of Height*Width Y components, Height*Width/4 U components and Height*Width/4 V components in each frame. The position in byte stream of NV21 and I420 is different (see Fig. 9). The Height and Width in Fig. 9 refer to the height and width of each frame.

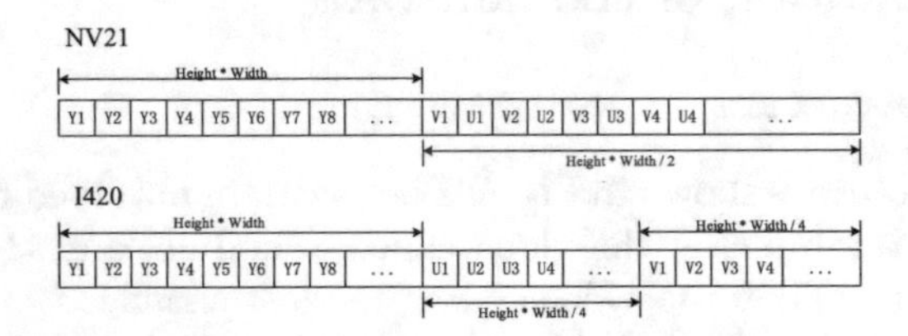

Fig. 9. Position in byte stream of NV21 and I420 format

NAL Unit. OpenH264 is an open source H.264 codec and it is suitable for using in real time applications. It provides a library named libopenh264. Once an image is generated, the encoder will be called and the coded video data is organized into NAL units. NAL units are classified into VCL and non-VCL NAL units. The VCL units contain the data that represents the values of the samples in the video pictures, and the non-VCL units contain any associated additional information such as parameter sets [11]. A parameter set contains information that is expected to rarely change and offers the decoding of a large number of VCL NAL units. There are two types of parameter sets: picture parameter set and sequence parameter set. The picture parameter set (PPS) contains data that is common to the entire picture and the sequence parameter set (SPS) contains data that is common to all the pictures in a sequence of pictures.

RTMP Chunk Stream. The chunks are transmitted over the network. While transmitting, each chunk must be sent in full before the next chunk. And the chunk size is configurable, it can be set using a control message named Set Chunk Size. The maximum chunk size defaults to 128 bytes. Larger chunk sizes reduce CPU usage but commit to larger writes that can delay other content on lower bandwidth connections. Smaller chunks are not good for high bit rate streaming [12].

3.2 The Process of the Software

When the program is running, the drone attempts to access to the streaming media server via Wi-Fi. Then the encoder and camera are initialized according to the supplied parameters. When a preview stream is started, a callback function is used to obtain the raw data of each frame. After processing the raw data, the encoder is called to compress the video sequence images. The output of the encoder is NAL unit (byte stream format) which is the input of the streamer. The streamer processes each NAL unit and eventually send the data to the streaming media server in the form of RTMP Chunk. The video can also be saved in the board.

At present, the application enables a drone to publish a live video over Wi-Fi network. But the application areas of this research is not limited to Wi-Fi transmission. Compared with the Wi-Fi network, the transmission delay will be reduced if the drone publishes a live video over 4G network.

3.3 Experimental Results

As show in Fig. 10, the Snapdragon Flight board is publishing a live video to the streaming media server directly, and the video can be watched from the website. We conducted some simple delay tests in Wi-Fi network and the video size is 720*480 and the frame rate is 30 fps. At present, the delay is usually around 1100 ms.

Fig. 10. Publishing a live video

We have learned that the transmission delay of DJI Matrice100 UAV is usually from 800 ms to 1200 ms. So the delay of this application is indeed greater. The delay is related to the current network environment, and more importantly, the parameters of encoder and streamer modules need to be further optimized.

4 Conclusion and Future Work

This paper gives a brief introduction of the H.264 video codec standard and RTMP protocol, and analyzes the structure of the NAL unit, RTMP Message and RTMP Chunk. Then develops a video streamer application by using the Qualcomm Snapdragon Flight as the experimental platform. The application enables drones to publish a live video to the streaming media server directly follows the RTMP protocol. Anyone who wants to watch the video just need to know the website of the streaming media server. Compared with the common ways of image transmission, it is more convenient.

In the future, we need to do further research on H.264 video codec standard and RTMP protocol. It's essential to optimize some parameters of the encoder and improve the quality of the live video that is published in actual flight. In addition, we will develop the video streamer application in other open source UAV platforms.

References

1. Qualcomm Developer Network. https://developer.qualcomm.com/hardware/snapdragon-flight
2. Digiarty Home. https://www.winxdvd.com/resource/h264.htm
3. OpenH264. https://www.openh264.org/
4. Real-Time Messaging Protocol (RTMP) specification. http://www.adobe.com/cn/devnet/rtmp.html
5. Wiegand, T., Sullivan, G.J., Bjontegaard, G., Luthra, A.: Overview of the H.264/AVC video coding standard. IEEE Trans. Circ. Syst. Video Technol. **13**(7), 560–576 (2003). IEEE Press, New York
6. Recommendation ITU-T H.264: Advanced video coding for generic audiovisual services
7. Vun, N., Ooi, Y.H.: Implementation of an android phone based video streamer. In: 2010 IEEE/ACM International Conference on Green Computing and Communications & 2010 IEEE/ACM International Conference on Cyber, Physical and Social Computing, pp. 912–915. IEEE Press, New York (2010)
8. Adobe Systems Inc.: Adobes Real Time Messaging Protocol
9. Lei, X., Jiang, X., Wang, C.: Design and implementation of streaming media processing software based on RTMP. In: 5th International Congress on Image and Signal Processing, pp. 192–196. IEEE Press, New York (2012)
10. YUV pixel formats. https://www.fourcc.org/yuv.php
11. Wikipedia, Network Abstraction Layer. https://en.wikipedia.org/wiki/Network_Abstraction_Layer
12. Zhao, P., Li, J., Xi, J., Gou, X.: A mobile real-time video system using RTMP. In: 4th International Conference on Computational Intelligence and Communication Networks, pp. 61–64. IEEE Press, New York (2012)

Advances in Speech and Language Processing

Dialog-Based Interactive Movie Recommendation: Comparison of Dialog Strategies

Hayato Mori[1], Yuya Chiba[2], Takashi Nose[2], and Akinori Ito[2](✉)

[1] Faculty of Engineering, Tohoku University, Sendai, Japan
hayato.mori.r8@dc.tohoku.ac.jp
[2] Graduate School of Engineering, Tohoku University, Sendai, Japan
{yuya,aito}@spcom.ecei.tohoku.ac.jp,
tnose@m.tohoku.ac.jp

Abstract. The user interface based on natural language dialog has been gathering attention. In this paper, we focus on the dialog-based user interface of movie recommendation system. We compared two kinds of dialog systems: the system-initiative system presented all the information about the recommended item at a time, and the user-initiative system provided information of the recommended item based on a dialog between the system and the user. As a result of dialog experiment, the users preferred to the user-initiative system for availability of obtaining required information, while the system-initiative system was chosen for the simplicity of obtaining the information. In addition, it was found that the appropriateness of the system's replies in the dialog affected the user's preference to the user-initiative system.

Keywords: Recommendation system · Dialog · Movie database

1 Introduction

Recommendation systems are systems to present information about various items (such as movies, songs, books, etc.) to a user based on the user's preference. Numerous recommendation systems have been developed so far [1], and those systems have also been deployed in the real use such as amazon.com. Usually, the recommendation system gathers the user's preference either explicitly or implicitly, and selects items that match the preference from the database. Then the selected items are presented to the user as the results of the recommendation. Most research works focus on the recommendation engine, or how to select items that match the preference. Besides, presentation of the selected items is another issue to solve, but there are only a few works that focus on this part because basically presentation of the items is task-dependent.

In this paper, we focus on the presentation of the selected items through natural language dialog. Most recommendation systems are web-based [2–4], and several systems employ user interfaces based on natural language [5, 6]. The MadFilm system [6] is a recommendation system based on natural language dialog, which gathers the user's preference explicitly through dialog and makes recommendation of movies

J.-S. Pan et al. (eds.), *Advances in Intelligent Information Hiding and Multimedia Signal Processing*, Smart Innovation, Systems and Technologies 82,
DOI 10.1007/978-3-319-63859-1_10

based on the user's request and interest. However, it has not been studied what kind of dialog strategy is preferred from the recommendation system aspect. Thus we compared two kinds of dialog-based recommendation systems: a system with user-initiative dialog and that with system-initiative dialog.

2 The Dialog-Based Movie Recommendation Interface

2.1 The Dialog System Based on QA Database

First, we developed a simple movie recommendation interface. The system is based on the example-based dialog system [7], which exploits the example-response database. Figure 1 shows an overview of an example-based spoken dialog system. When a user inputs a sentence, the system determines an example sentence in the database that is most similar to the input sentence. We used word-based cosine similarity as the similarity. When an example sentence is determined, the corresponding reply sentence is sent to the speech synthesizer, and the reply speech is generated. When the similarity between the input sentence and the most similar example sentence is less than a pre-determined threshold, the system randomly makes a reply among the set of prepared sentences, such as "Could you please say it again?" or "Please explain it with different words."

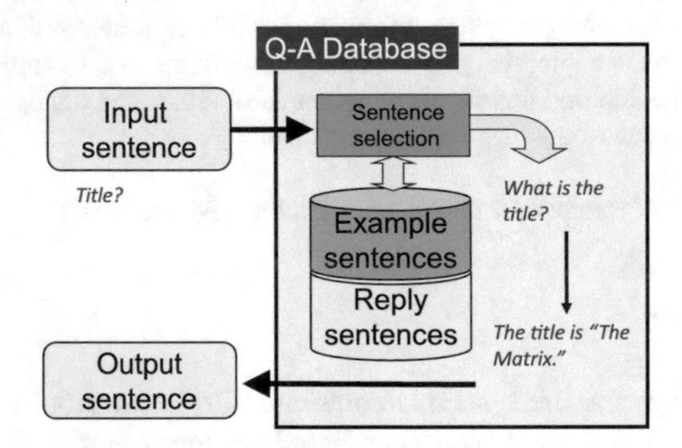

Fig. 1. The dialog system based on QA database

2.2 Movie Database

We exploited IMDb database [8] as the source of the information for the dialog. The IMDb is the shared database of movies, containing various information. In this work, we exploited the title, genre, plot, director, writer, cast, rating, income, country and year. In addition, tags were imported from MovieLens [9].

2.3 Automatic QA Database Generation

The QA database should be prepared movie by movie. To expedite QA database creation, we exploited the automatic QA database generation method [10]. For generating the QA database, we prepared the database description table and the example sentence templates. Figure 2 shows the example of the automatic QA generation. The database description table provides the templates of the reply sentences and the query type for each keyword. Example sentence templates are used for generating the example sentence when a keyword and the query type is given.

Movie Database

Keyword	Content
Title	The Matrix
Cast	Kianu Reeves
Rating	8.3

Database description table (movie independent)

Keyword	Type	Response
Title	What	is <Title>
Cast	Who	is <Cast>
Rating	How	is <Rating> out of 10

Example sentence templates (movie independent)

Type	Example
What	What is the <Keyword> / What about the <Keyword>
Who	Who is the <Keyword> / What about the <Keyword>
How	What is the <Keyword> / How about the <Keyword>

Generated QA database

Example	Response
What is the title	Title is The Matrix
Who is the cast	Cast is Kianu Reeves
How about the rating	Rating is 8.3 out of 10

Fig. 2. Automatic generation of QA database

Using these information sources, the QA database is generated as follows. First, one entry is chosen from the movie database, such as (Title, The Matrix). Then we search the keyword "Title" in the database description table, obtaining the entry (Title, What, is <Title>). We know here that the query type of this keyword is <What>, and generates the reply sentence by concatenating the keyword and the response template and substituting the placeholder <Title> with the content of the keyword "Title" in the movie database, obtaining a sentence "Title is The Matrix." After that, we search the entries of the example templates with the specified query type, "Title" in this case, and find the entry "What is the <Keyword>". Then we obtain the example sentence by substituting the <Keyword> part with the keyword "Title", obtaining "What is the title."

3 Experiment

3.1 Experimental Paradigms

We made an experiment to compare two strategies for presenting movie information. The first one, the system-initiative, first asked the genre the user was interested in, then the system chose one title and presents all information at a time. The four kinds of information, "Genre", "Story", "Income" and "Rating" were presented in this order. The second one, the user-initiative, also first asked the genre, and presented only the title of the chosen movie. After that, the user could ask information he/she was interested in, such as "Who is the cast?" or "How about the rating?" Here, we expected that a user might feel better for the user-initiative system because the user could choose the question he/she was interested in. Figure 3 shows the examples of dialogs conducted by the two systems.

Speaker	Contents
SYS	Input the genre you would like to be recommended
USER	SF
SYS	I recommend "Terminator 2: Judgment Day"
SYS	The genre of this movie is action, SF and thriller.
SYS	A cyborg, identical to the one who failed to kill Sarah Connor, …
SYS	The gross was 520 million dollars
SYS	The user rating of this movie was 8.0 out of 10
SYS	Input the genre you would like to be recommended

(a) The system-initiative dialog

Speaker	Contents
SYS	Input the genre you would like to be recommended
USER	SF
SYS	I recommend "Terminator 2: Judgment Day"
SYS	What would you like to know about this movie?
USER	How is the story?
SYS	A cyborg, identical to the one who failed to kill Sarah Connor, …
USER	Who is the cast
SYS	Cast is Arnold Schwarzenegger
USER	Finish
SYS	Input the genre you would like to be recommended

(b) The user-initiative dialog

Fig. 3. Examples of the system-initiative and the user-initiative dialogs

One session consisted of two dialogs with different systems and different genres. After making two dialogs, the user was asked to fill a questionnaire for subjective evaluation to choose which system was felt to be better. One participant conducted six sessions, where the orders of the systems were different in the first and second half of the six sessions.

3.2 Experimental Conditions

Eight titles corresponding eight genres (adventure, mystery, comedy, drama, SF, fantasy, thriller, romance) were prepared. In the experiment, when a user inputted the genre, one title was presented deterministically.

Eight persons participated in the experiment. Half of the participants evaluated the system-initiative system first, and the rests evaluated the user-initiative system first. A user inputted the sentence using the keyboard, and the system presented the reply using both texts on the screen and synthesized speech.

After finishing one session (two dialogs with different systems), a participant was asked the following five questions:

- Which system did you like?
- Which system provided you with the required information?
- Which system was simpler for obtaining the required information?
- Which system did you want to use again?
- Which system made you want to watch the movie?

Since one participant conducted six sessions, one participant filled the questionnaire six times.

3.3 Result of the Subjective Evaluation

Figure 4 shows the experimental results. This graph shows the ratio of the participants who chose one of two systems for each question. The item "Total" shows the result of the majority vote by all the five questions. More than 60% participants chose the user-initiative system for questions "Which system provided you the required information?" "Which system did you want to use again?" and "Which system made you want to watch the movie?" However, 63% participants chose the system-initiative system for "Which system was simpler for obtaining the required information?" because the system-initiative system automatically provided the information of the movie and the participant had no need to input further request. The preference question "Which system did you like?" was 50-50 for the two systems.

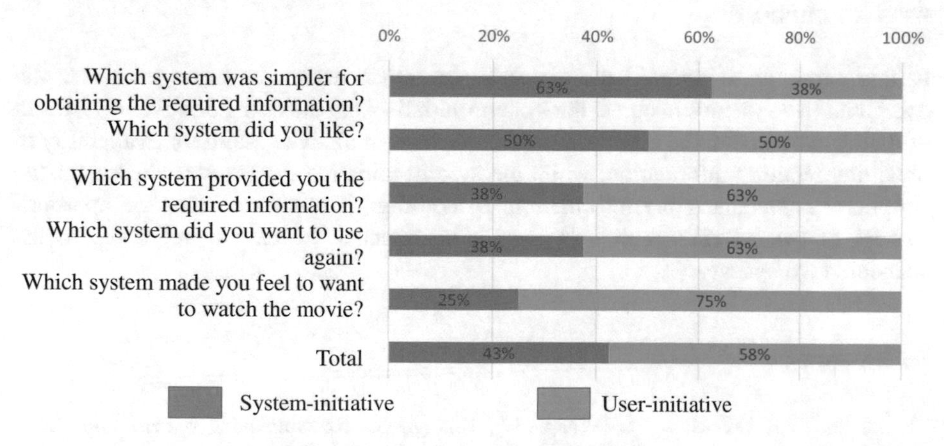

Fig. 4. Results of the subjective evaluation

3.4 Result of the Objective Evaluation

Next, we analyzed the subjective evaluation by considering objective evaluation result. We observed the "answer rate," which is a ratio of the reply of the user-initiative system that are not inappropriate. The answer rate differed from participant to participant, possibly because of the tendency of input sentences of the participant. The relation between the answer rate and the selection rate (the result of "Total" in Fig. 3) for each participant is shown in Fig. 5. Here we can observe the positive correlation between the answer rate and the selection rate, meaning that the system's appropriate answer makes the system to be preferred to.

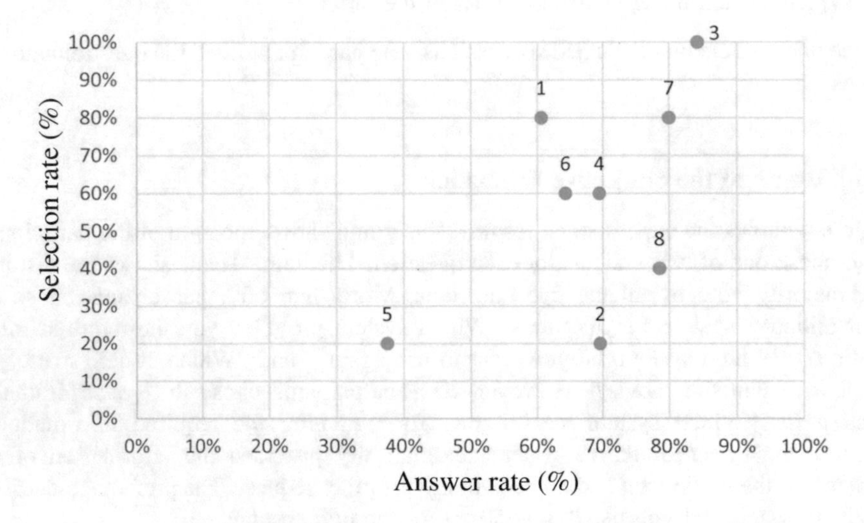

Fig. 5. The answer rate and the selection rate

4 Conclusion

In this paper, we introduced the dialog-based movie recommendation interface, and compared two system strategies: the system-initiative and the user-initiative. As a result of dialog experiment, the users preferred to the user-initiative system for availability of obtaining required information, while the system-initiative system was chosen for the simplicity of obtaining the information. In addition, it was found that the appropriateness of the system's replies in the dialog affected the user's preference to the user-initiative system.

References

1. Lu, J., Wu, D., Mao, M., Wang, W., Zhang, G.: Recommender system application developments: a survey. Decis. Support Syst. **74**, 12–32 (2015)
2. Yuan, X., Lee, J.H., Kim, S.J., Kim, Y.H.: Toward a user-oriented recommendation system for real estate websites. Inf. Syst. **38**(2), 231–243 (2013)

3. Choi, W., Choi, C.H., Kim, Y.R., Kim, S.J., Na, C.S., Lee, H.: HerDing: herb recommendation system to treat diseases using genes and chemicals. Database (2016). baw011
4. Meehan, K., Lunney, T., Curran, K., McCaughey, A.: Context-aware intelligent recommendation system for tourism. In: 2013 IEEE International Conference on Pervasive Computing and Communications Workshops (PERCOM Workshops), pp. 328–331 (2013)
5. Thompson, C.A., Goker, M.H., Langley, P.: A personalized system for conversational recommendations. J. Artif. Intell. Res. **21**, 393–428 (2004)
6. Johansson, P.: Madfilm-a multimodal approach to handle search and organization in a movie recommendation system. In: Proceedings of the 1st Nordic Symposium on Multimodal Communication, pp. 53–65, September 2003
7. Nisimura, R., Lee, A., Saruwatari, H., Shikano, K.: Public speech-oriented guidance system with adult and child discrimination capability. In: Proceedings of International Conference on Acoustics, Speech, and Signal Processing, (ICASSP 2004), p. I-433 (2004)
8. Internet Movie Database http://www.imdb.com. Accessed 1 Feb 2017
9. MovieLens. http://movielens.org. Accessed 1 Feb 2017
10. Ito, A., Morimoto, T., Makino, S., Ito, M.: A spoken dialog system based on automatically-generated example database. In: Proceedings of International Conference on Audio, Language and Image Processing (ICALIP), pp. 732–736 (2010)

Response Selection of Interview-Based Dialog System Using User Focus and Semantic Orientation

Shunsuke Tada(✉), Yuya Chiba, Takashi Nose, and Akinori Ito

Graduate School of Engineering, Tohoku University, Sendai, Japan
shunsuke.tada.q4@dc.tohoku.ac.jp,
{yuya,aito}@spcom.ecei.tohoku.ac.jp,
tnose@m.tohoku.ac.jp

Abstract. This research examined the response selection method of an interview-based dialog system that obtains the user's information by the chat-like conversation. In the interview dialog, the system should ask about the subject that the user is interested in to obtain the user's information efficiently. In this paper, we proposed the method to select the system's utterance based on the user's emotion to a focus detected from the user's utterance. We prepared the question types corresponding to the semantic orientation, such as the positive, neutral, and negative. The focus was detected by the CRF, and the question type was estimated from the user's utterance and the system's previous utterance.

Keywords: Spoken dialog system · User focus · Interview dialog · Conditional random field · Support vector machine

1 Introduction

Recently, non-task-oriented spoken dialog systems [1, 2] have been focused on in contrast to the traditional task-oriented dialog systems. It is known that the user's information, such as preference or emotion is efficient to improve the task success or user satisfaction of the dialog system. The many research works studied to obtain the user's information (e.g., [3]). Many of these studies constructed the dialog system for a task-specific dialog. On the other hand, we usually get to know the other person from the chat-like conversation considering the human-human conversation. Therefore, it is useful to examine the dialog system obtaining the user's information from chat-like conversation.

The target dialog of this paper is interview-formed conversation, where the system actively poses questions to collect the user's information. In the interview dialog, the system should dig into the subject that the user is interested in to collect the user's information efficiently. The conventional interview systems [4, 5] tend to repeat the questions for collecting the user's information, but the dialog with such systems may be stressful for the user. We used the user focus [6] as the subject that the user's interested in. Table 1 shows an example of the interaction between the user and a target system. Here, S1 and S2 represent the system's utterance and U1 represents the user's utterance. In this example, the system detects the user focus (i.e., *Godzilla*), and asks

J.-S. Pan et al. (eds.), *Advances in Intelligent Information Hiding and Multimedia Signal Processing*, Smart Innovation, Systems and Technologies 82,
DOI 10.1007/978-3-319-63859-1_11

Table 1. Example of interview dialog of target system

S1:	Have you seen any movies lately?
U1:	I saw *Godzilla*
S2:	What do you like about *Godzilla*?

further questions on the assumption that the user has the positive emotion to the detected focus. Generating such question, the system has to not only detect the focus, but also consider the semantic information of the user's utterance.

In this paper, we proposed the response selection method based on the user's semantic orientation. The system detects the focus of the user's utterance, then estimates the question type corresponding the semantic orientation. A definitive response is selected based on the user focus and the estimated question type. We prepared four question types to represent the user's emotion. The question type is estimated from the user's utterance and system's previous utterance.

2 Proposed Response Selection Method

2.1 Overview of Procedure

A procedure of selecting the system's utterance is summarized as follows:

Step 1: Focus detection from the user's utterance
Step 2: Feature extraction from the user's utterance and system's previous utterance
Step 3: Question type classification of the next system's utterance
Step 4: Utterance selection based on the user focus and estimated question type

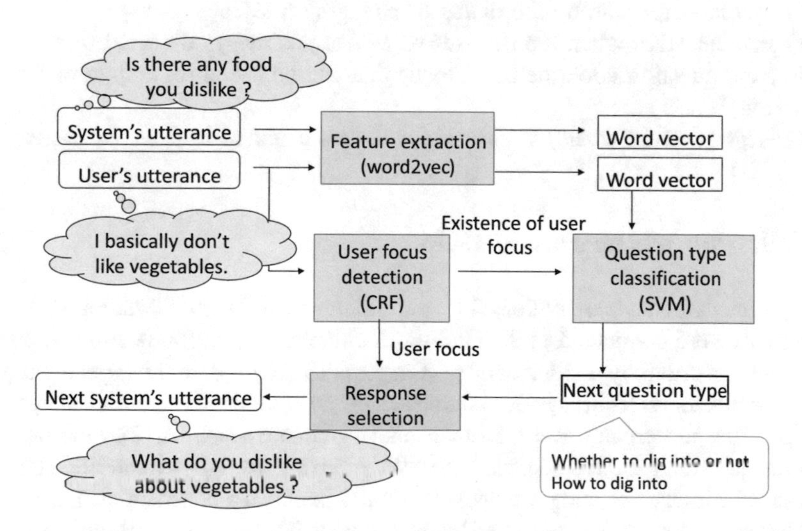

Fig. 1. Proposed system

Figure 1 shows an overview of the proposed system. The target language of this system is Japanese. In this paper, the noun and noun phrase that the system can ask the question about them were defined as the user focus. Figure 2 shows an example of the focus. The user focus was annotated based on IOB2 format. B-F and I-F represent the beginning of the user focus and inside of the user focus, respectively.

Utterance	好き	な	動画	は	ゲーム	の	プレイ	動画	など	です	ね
	(My favorite videos are *game play videos.*)										
Label	O	O	O	O	B-F	I-F	I-F	I-F	O	O	O

Fig. 2. Example of user focus annotation based on IOB2 format

Question type of the system's next utterance is estimated from the user's utterance and system's previous utterance. We employed the embedding vector of the utterances, existence of the user focus, and question type of the system's previous utterance as the features. Finally, the system generates the next utterance based on the templates corresponding to the question type.

2.2 Labels of Question Type

We prepared four labels to annotate the system's utterance. Here, the experimental dialog data of this paper only include the questions of the system and the user's response to them. The question types excluding OQ correspond to the user's emotion. The question types were summarized as follows:

OQ: question for starting the dialog or changing a topic
DQ Pos: question when the user seems to feel positive to detected focus
DQ Neu: question about detected focus that can be asked regardless of the user's emotion
DQ Neg: question when the user seems to feel negative to detected focus

3 Collection of the Dialog Data

The experimental data were collected by a simulator of an interview dialog system. The experiments were conducted in a sound-proof chamber. The sounds and videos of the dialog were recorded by a video camera (Canon iVIS HF G10) and a lapel microphone. A speaker talked with the system controlled by an operator who was outside of the chamber. The participants were 10 male undergraduate students. The dialog system posed the questions about a specific topic. We prepared 10 topics including cooking or movies for the interview, and presented to the participants in random order. The questions posed to the speaker were determined by the operator. The operator selected the question using a touch panel, and the system read it by a synthesized voice. The system could generate the questions based on the template such as "What do you like about W?"

in addition to the prepared sentences. Moreover, the system could generate the backchannels by pressing the software buttons. The backchannels seem to be important for the interview dialog because a large proportion of the utterances in the human-human interview dialog is backchannels [7]. The backchannel words were chosen from the transcriptions of the human-human dialogs. We collected 637 utterances and used the transcription of the utterances for the experiment.

One person (the first author of the paper) annotated the question type and focus tags to the system's utterance. Multiple labels were annotated to the system's next utterance when several question types are acceptable. Table 2 shows an example of the annotation of the question type.

Table 2. Example of question type annotation

S1	U1	Question type of S2
Have you seen any movies lately?	I saw *Godzilla.*	DQ Pos, DQ Neu

4 Focus Detection by Conditional Random Field

4.1 Conditions of Focus Detection

We employed the Conditional Random Field (CRF) for the focus detection test. The features shown in Table 3 were extracted to detect the user focus. Focus detection is also examined in the conventional study [6]. Yoshino et al. proposed the method to detect the user focus at a phrase level using the 12 features, such as order of the phrase in the sentence and that in the predicate-argument (P-A) structure, POS tags in the phrase, POS tag sequence, pair of POS tags, semantic role of the phrase, semantic roles used in the sentence, pair of semantic roles, P-A template score, words in the phrase, word pairs in the phrase and order of the word pairs in the phrase. We compared the performance of the focus detection between the proposed feature set and the conventional method. The experiments were conducted by 10-fold cross-validation.

Table 3. Proposed feature set for focus detection

Feature type	Feature
WORD	Uni-gram and bi-gram of the previous, current, and next word
POS	Uni-gram and bi-gram of POS tags of the previous, current, and next word
Sub-POS	Uni-gram and bi-gram of sub-categories of the POS tags of the previous, current, next word

4.2 Results of Focus Detection

Figure 3 shows the result of the focus detection. The figure shows the average precision and recall of each label. PROPOSED and CONVENTIONAL in the figure represent the result when using the proposed feature set and conventional method. As shown in the figure, the proposed method obtained the better performance compared with the

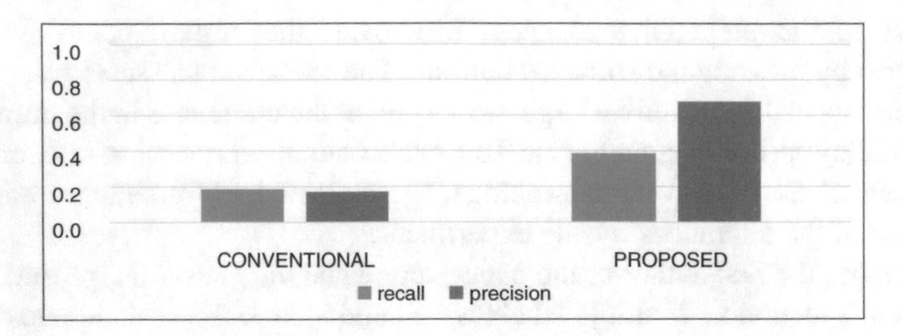

Fig. 3. Result of focus detection

conventional method. One of the possible reasons for this result is that it is difficult to extract the features of the conventional method from the conversational style dialog like the experimental data. On the other hand, the method using the proposed feature set detects 40% of the user focus. In future work, we are going to collect a large scale dialog data to improve the focus detection.

5 Classification of Question Types by Support Vector Machine

5.1 Feature Extraction from Utterance

We used the word2vec [8] to extract the semantic information of the utterance. Documents of the Japanese Wikipedia and Corpus of Spontaneous Japanese (CSJ) were used to train word2vec. All texts were tokenized by MeCab[1] with the NEologd dictionary [10]. The word2vec was trained by skip-gram while the window size was 5. The average vector of the word embedding vector was used as the feature of the utterance.

5.2 Conditions of Question Type Classification

We employed the Support Vector Machine (SVM) with the RBF-kernel as the classifier of the question types. Features of the question type classification are embedding vectors of the system's utterance and user's utterance, existence of the user focus, and the question type of the system's previous utterance. The existence of the user focus was represented by the binary value, and the question type was represented by the 4-dimensional vector. The classification was the subject-open cross-validation. The hyper-parameters were decided by the grid-searching. The samples that have multiple labels were treated as the different samples having different label. The dimension of each embedding vector was decided to 200 by the preliminary experiments. The reference labels were used for the existing of the focus value to train the classifier.

[1] http://taku910.github.io/mecab/.

5.3 Classification Results of Question Labels

The classification was conducted while changing the combination of the features. The results are summarized in Table 4. The performance was evaluated by the precision. When the predicted label was matched with any of the reference labels, the result was regarded as correct. As shown in the figure, we obtained the best classification performance when using the all of the features. In particular, precision of the DQ Neu and DQ Neg was improved by using the embedding vector of the utterances. This result seems to indicate the features extracted by word2vec is efficient to represent the information related to the semantic orientation.

Table 4. Classification results of question types by combination of features (S: system's utterance, U: user's utterance, F: existing of the focus, Q: question type of system's previous utterance)

	OQ	DQ Pos	DQ Neu	DQ Neg	Ave.
F + Q	0.980	0.958	0.000	0.211	0.537
U + F + Q	0.980	0.934	0.760	0.592	0.817
S + F + Q	0.970	0.934	0.793	0.570	0.817
S + U + F + Q	0.973	0.922	0.736	0.738	0.842

Finally, we conducted the classification experiments considering the focus detection result. We used the classifier trained by the reference labels of the user focus. Figure 4 shows the classification results. ORACLE in the figure shows the result when using the reference of the focus detection for test data, and FD in the figure shows the results when using the focus detection results for test data. These results suggest that the performance of the focus detection is also important to the result of the question type classification.

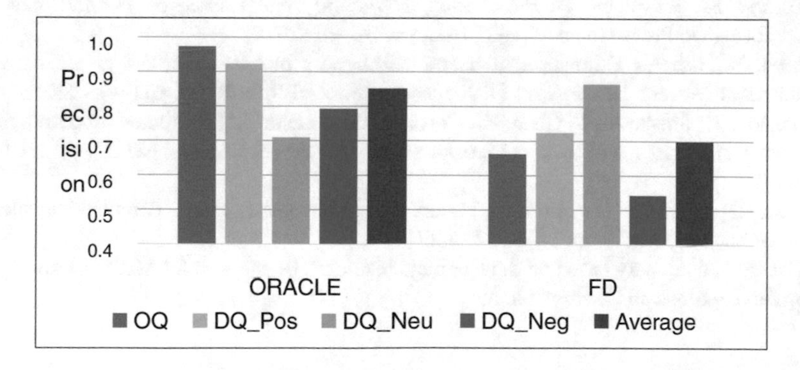

Fig. 4. Classification results of question type considering focus detection

6 Conclusion

In this paper, we examined the method to select the system's utterance based on the user focus and user's emotion. The system detects the user focus by using the word-level features. The four question types corresponding to the semantic orientation are estimated by SVM using embedding vectors of user's utterance and system's utterance, existing of the user focus, and the question type of the system's previous utterance. In the classification experiment, we obtained the best result by using all of the features.

One of the remaining problems is the performance of the focus detection. In future work, we are going to prepare a large-scale dataset for the focus detection to improve the detection performance. In addition, we will also examine the naturalness of the utterance selected by the proposed method.

References

1. Bickmore, T.W., Picard, R.W.: Establishing and maintaining long-term human-computer relationships. ACM Trans. Comput.-Human Interact. **12**(2), 293–327 (2005)
2. Yu, Z., Nicolich-Henkin, L., Black, A., Rudnicky, A.I.: A wizard-of-Oz study on a non-task-oriented dialog systems that reacts to user engagement. In: Proceedings of SIGDIAL, pp. 55–63 (2016)
3. Pargellis, A.N., Kuo, H.J., Lee, C.: An automatic dialogue generation platform for personalized dialogue applications. Speech Commun. **42**(3–4), 329–351 (2004)
4. Stent, A., Stenchikova, S., Marge, M.: Dialog systems for surveys: the rate-a-course system. In: Proceedings of IEEE/ACL Spoken Language Technology Workshop, pp. 210–213 (2006)
5. Johnston, M., Ehlen, P., Conrad, F.G., Schober, M.F., Antoun, C., Fail, S., Hupp, A., Vickers, L., Yan, H., Zhang, C.: Spoken dialog systems for automated survey interviewing. In: Proceedings of SIGDIAL, pp. 329–333 (2013)
6. Yoshino, K., Kawahara, T.: Information navigation system based on POMDP that tracks user focus. In: Proceedings of SIGDIAL, pp. 32–40 (2014)
7. Chiba, Y., Ito, A.: Estimation of user's willingness to talk about the topic: analysis of interviews between humans. In: Dialogues with Social Robots, pp. 411–419 (2016)
8. Mikolov, T., Sutskever, I., Chen, K., Corrado, G.S., Dean, J.: Distributed representations of words and phrases and their compositionality. In: Proceedings of NIPS, pp. 3111–3119 (2013)
9. Maekawa, K., Koiso, H., Furui, S., Isahara, H.: Spontaneous speech corpus of Japanese. In: Proceedings of LREC, pp. 947–952 (2000)
10. Neologism dictionary based on the language resources on the web for MeCab, https://github.com/neologd/mecab-ipadic-neologd

Development and Evaluation of Julius-Compatible Interface for Kaldi ASR

Yusuke Yamada[1(✉)], Takashi Nose[1], Yuya Chiba[1], Akinori Ito[1], and Takahiro Shinozaki[2]

[1] Graduate School of Engineering, Tohoku University, Aramaki Aza-Aoba 6–6–05, Aoba-ku, Sendai-shi, Miyagi 980–8579, Japan
yusuke.yamada.t1@dc.tohoku.ac.jp, tnose@m.tohoku.ac.jp, {yuya,aito}@spcom.ecei.tohoku.ac.jp

[2] Interdisciplinary Graduate School of Science and Engineering, Tokyo Institute of Technology, 4259 Nagatsuta-cho, Midori-ku, Yokohama 226-8502, Japan

Abstract. In recent years, the use of Kaldi has rapidly grown because it has adopted various technologies of DNN-based speech recognition in succession and has shown high recognition performance. On the other hand, the speech recognition engine, Julius, has been widely used especially in Japan. Julius is also attracting attention since DNN-HMM is implemented in it. In this paper, we describe the design plan of interfaces that make Kaldi speech recognition engine be compatible with Julius, a system overview, and the details of the speech input unit and the recognition result output unit. We also refer to the functions that we are planning to implement.

Keywords: DNN-based speech recognition · Kaldi · Julius

1 Introduction

Nowadays, speech interface is rapidly growing on many devices, such as PC, mobile phone, car navigation system and so on. Speech recognition is one of the most important technologies for such interface, and there have been studied for spontaneous and emotional speech [3,4].

In the background, research of speech recognition area is advancing using deep learning same as other research areas categorized into pattern recognition. Traditionally in this research area, HTK [1] has been widely used as a de facto standard in the world. Because source program of HTK is published freely by Microsoft and HTK supports and maintains many usable speech technology.

In Japan, an LVCSR system, Julius [5] is widely used. Julius is provided on Linux, Windows, macOS. Julius is now developed by Nagoya Institute of Technology. Julius supports HTK format acoustic model and ARPA format language model. And development team of Julius provides Japanese acoustic models

J.-S. Pan et al. (eds.), *Advances in Intelligent Information Hiding and Multimedia Signal Processing*, Smart Innovation, Systems and Technologies 82,
DOI 10.1007/978-3-319-63859-1_12

and Japanese language models. In addition, Julius has a function called module mode. In module mode, Julius behaves as speech recognition server. Client can control Julius via network using TCP/IP. Thus, in Japan, many developers make applications with speech interface using Julius as speech recognition server.

Acoustic model on ASR was modeled by GMM-HMM until few years ago. In GMM-HMM, Gaussian mixture represents fluctuation of speech and HMM represents time transition of speech, well. While now, DNN-HMM is used as acoustic model, and is achieved higher performance than GMM-HMM in ASR. DNN-HMM is a model that uses deep neural network (DNN) instead of GMM in GMM-HMM.

One of speech recognition toolkits that implement DNN-HMM is Kaldi [6]. Kaldi project provides training script corresponding to many corpora. This training script is called "recipe". When corpus is owned and recipe is prepared, recipe script trains acoustic model and language model automatically. Furthermore, Kaldi ASR system is ready to use.

Now, Julius is also support DNN-HMM acoustic model [2]. Using Japanese acoustic model trained by Japanese corpus, DNN-HMM model acquired higher performance than DNN-HMM model in benchmark while Kaldi recipe for Corpus of Spontaneous Japanese is published by Shinozaki et al., Julius and Kaldi can be used as a Japanese ASR system. But to use Kaldi is more difficult than Julius because of complexity. In our research, we developed wrapper interface on Kaldi that is compatible with Julius module mode. Any systems using Julius by module mode can be used Kaldi ASR without any changes.

2 Implement of Julius Compatible Wrapper on Kaldi

In this section, we describe how to implement Julius compatible wrapper on Kaldi. First, we explain module mode on Julius.

2.1 Module Mode on Julius

In this section, we explain module mode on Julius. Figure 1 shows overview of module mode on Julius. In module mode, client connects Julius via network with TCP/IP. Julius has two interfaces, named control interface and speech interface. The control interface is used to control Julius by client. The client can start, stop, and pause the recognizer. And client can get recognition results and recognizer status. Speech interface is used to send speech to Julius. The client program using speech interface is named adintool. Adintool can get input speech from audio device or file or netaudio device.

We build control interface and speech interface to Kaldi as wrapper. In the next section, we describe the wrappers.

2.2 Wrapper of Speech Input Interface

In Julius package, a program named adintool is included. Adintool gets input speech from device or file and send to Julius using adinnet protocol.

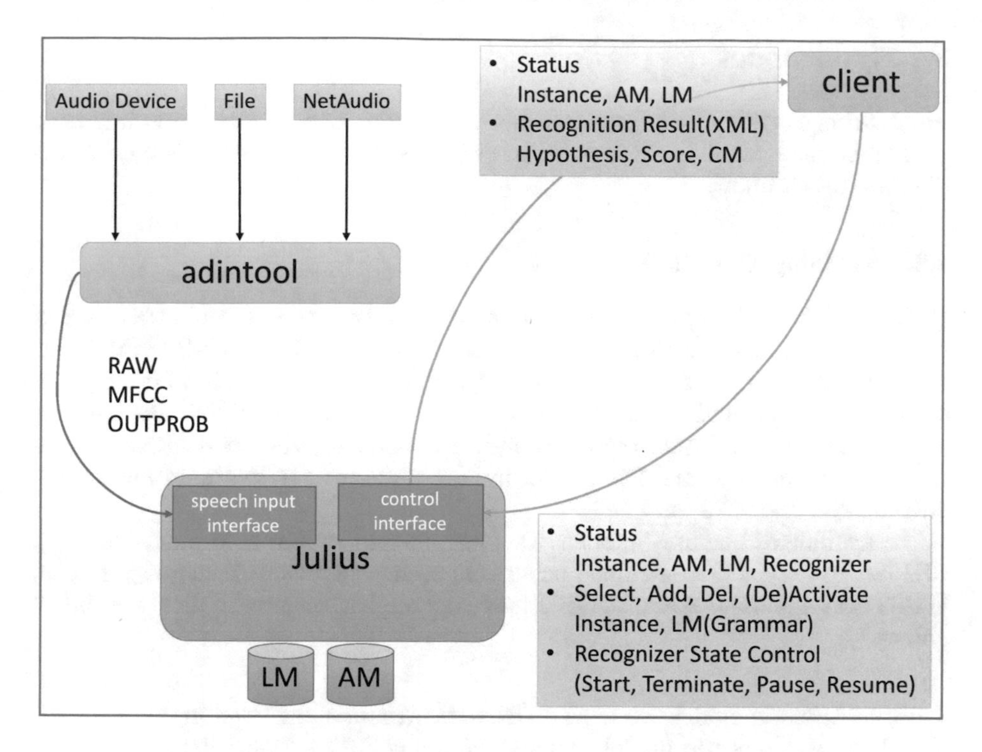

Fig. 1. Overview of module mode on Julius

We implemented wrapper as server that get input speech from adintool using adinnet protocol and send to Kaldi. Now, we only implemented speech input from adintool.

2.3 Wrapper of Control Interface

Control interface gets control command from client. The wrapper needs to interpret control command and control Kaldi recognizer. And the wrapper returns status of recognizer or recognition results to client using Julius format.

2.4 Implementation

The wrappers were designed to run on multi platforms. We implemented the wrappers using Qt framework and C++. Qt was used to implement network process and thread. Function of speech input and output of recognition result is implemented as module. We build Kaldi as library and build an execution file that includes library and wrappers. Now, we developed the system only on macOS and Ubuntu Linux. We are planning to develop the system for Windows in a future.

3 Experiment

In experiment, WER was compared between Julius and Kaldi. Julius was used in module mode. Kaldi was used with wrappers. In both systems, adintool was used as speech input.

3.1 Training Conditions

Kaldi. In the training of the acoustic model, conference talk and other talk in CSJ were used. Total length of training data was 294 h. First, GMM-HMM was trained from data. Next, DNN was trained. Feature of a frame is 40 dimensions of spliced MFCC and 100 dimensions of i-vector. Input of DNN was features of 7 predecessor and 4 successor frames. Structure of DNN is 4 hidden layers with 2000 input units and 250 output units with p-norm activation function [7], output layer has 9270 units.

In training of language models, CSJ corpus and BCCWJ corpus were used. CSJ language model was trained using CSJ recipe. In BCCWJ language model, WFST was expanded from BCCWJ language model included in dictation-kit of Julius.

Julius. Acoustic model was used a DNN-HMM model included in dictation-kit of Julius. The acoustic model was trained using ASJ and simulated talk from CSJ. The length of training data was 378 h. Feature of a frame was 40 dimensions of FBANK with delta and acceleration term. Input of DNN was features of 5 predecessor and 5 successor frames. Structure of DNN was 7 hidden layers with 2048 input/output units, output layer has 4874 units.

In training of language model, CSJ language model was trained with vocabulary and dictionary from CSJ recipe using SRILM. CSJ language model was smoothed using Knesear-Ney method.

3.2 Evaluation Conditions

Evaluation was performed using CSJ excludes training data of acoustic model of Kaldi. Total length of evaluation data was 5 h and 16 min.

3.3 Result

Experimental result was shown in Fig. 2. In comparison of WER, Kaldi is lower than Julius with any language models. Because, in Kaldi, decoder parameter was optimized with CSJ recipe. Decoder parameter of Julius needs optimization. When BCCWJ language model was used, WER is higher than CSJ language model. Because BCCWJ was a corpus from written words. And unknown words were involved.

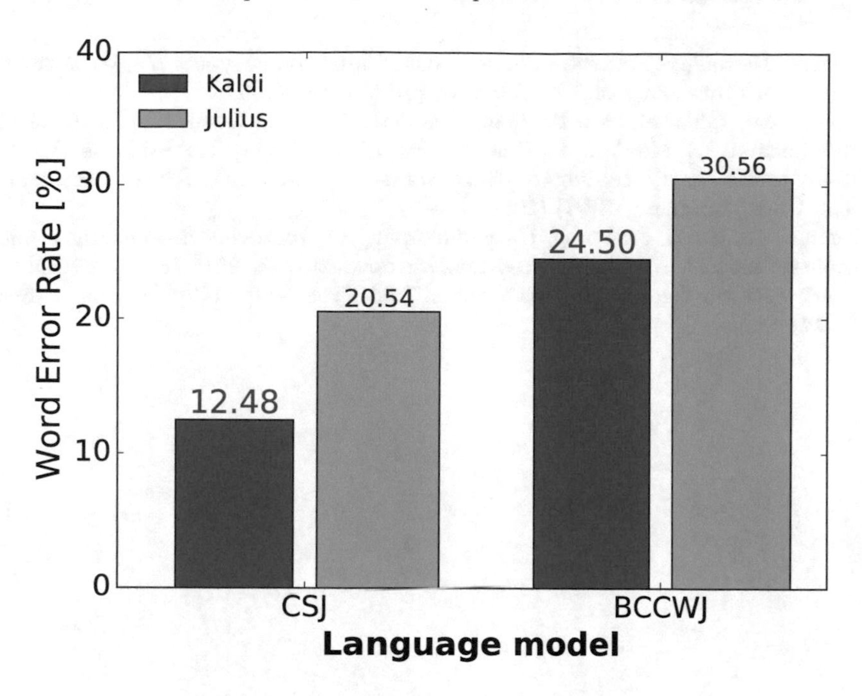

Fig. 2. Recognition results (WER [%])

4 Conclusion

In this paper, we describe Julius compatible wrapper for Kaldi. From experimental result, in WER, Kaldi was lower than Julius. In future work, we will implement to achieve more compatibility.

Acknowledgment. Part of this work was supported by JSPS KAKENHI Grant Number JP26280055 and JP15H02720.

References

1. The Hidden Markov Model Toolkit (HTK), http://htk.eng.cam.ac.uk/
2. Glas, D.F., Minato, T., Ishi, C.T., Kawahara, T., Ishiguro, H.: Erica: the erato intelligent conversational android. In: Proceedings of the 25th IEEE International Symposium on Robot and Human Interactive Communication (RO-MAN), pp. 22–29 (2016)
3. Ijima, Y., Nose, T., Tachibana, M., Kobayashi, T.: A rapid model adaptation technique for emotional speech recognition with style estimation based on multiple regression HMM. IEICE Trans. Inf. Syst. **93**(1), 107–115 (2010)
4. Kawahara, T., Nanjo, H., Shinozaki, T., Furui, S.: Benchmark test for speech recognition using the corpus of spontaneous japanese. In: ISCA & IEEE Workshop on Spontaneous Speech Processing and Recognition, pp. 1–4 (2003)

5. Lee, A., Kawahara, T.: Recent development of open-source speech recognition engine julius. In: Proceedings of APSIPA ASC, pp. 131–137 (2009)
6. Povey, D., Ghoshal, A., Boulianne, G., Burget, L., Glembek, O., Goel, N., Hannemann, M., Motlicek, P., Qian, Y., Schwarz, P., et al.: The kaldi speech recognition toolkit. In: Proceedings of IEEE Workshop on Automatic Speech Recognition And Understanding (ASRU) (2011)
7. Zhang, X., Trmal, J., Povey, D., Khudanpur, S.: Improving deep neural network acoustic models using generalized maxout networks. In: 2014 IEEE International Conference on Acoustics, Speech and Signal Processing (ICASSP), pp. 215–219 (2014)

Voice Conversion from Arbitrary Speakers Based on Deep Neural Networks with Adversarial Learning

Sou Miyamoto[1(✉)], Takashi Nose[1], Suzunosuke Ito[1,2], Harunori Koike[1,2], Yuya Chiba[1], Akinori Ito[1], and Takahiro Shinozaki[2]

[1] Graduate School of Engineering, Tohoku University, Aramaki Aza Aoba 6–6–05, Aoba-ku, Sendai-shi, Miyagi 980–8579, Japan
sou.miyamoto.s8@dc.tohoku.ac.jp, tnose@m.tohoku.ac.jp, {yuya,aito}@spcom.ecei.tohoku.ac.jp

[2] Department of Information and Communication Engineering, School of Engineering, Tokyo Institute of Technology, Nagatsuta-cho 4259, Midori-ku, Yokohama-shi, Kanagawa 226-8502, Japan

Abstract. In this study, we propose a voice conversion technique from arbitrary speakers based on deep neural networks using adversarial learning, which is realized by introducing adversarial learning to the conventional voice conversion. Adversarial learning is expected to enable us more natural voice conversion by using a discriminative model which classifies input speech to natural speech or converted speech in addition to a generative model. Experiments showed that proposed method was effective to enhance global variance (GV) of mel-cepstrum but naturalness of converted speech was a little lower than speech using the conventional variance compensation technique.

Keywords: DNN-based voice conversion · Adversarial learning · Spectral differential filter · Model training

1 Introduction

Voice conversion is a technique for change the global property of speech, e.g., speaker individuality, to the target speech. There have been many studies for voice conversion [14]. A variety of approaches were proposed using Gaussian mixture models [6], hidden Markov models [10], neural networks [1], and Gaussian process regression [12]. In the statistical parametric speech synthesis including voice conversion, over-smoothing effect of acoustic features causes quality degradation of converted speech. In order to improve over-smoothing effect, various methods were proposed such as directly emphasizing spectral peak information [8], compensating global variance (GV) of mel-cepstrum [15] and so on. Each of the methods shows the effect of improving deterioration in the quality of converted speech. The voice conversion method using adversarial learning (AL) [3]

J.-S. Pan et al. (eds.), *Advances in Intelligent Information Hiding and Multimedia Signal Processing*, Smart Innovation, Systems and Technologies 82,
DOI 10.1007/978-3-319-63859-1_13

for use in learning the conversion model was also proposed and shows the effect of improving the quality of converted speech [13]. In this study, we introduce a learning algorithm of the conversion model using adversarial learning to speaker-independent voice conversion from arbitrary speakers (SI-AL). Variance compensation based on affine transformation [11] we used in speaker-independent voice conversion based on neural networks (SI-GV) [7] is not suitable for real-time conversion on account of variance compensation as a postprocessing after whole speech converted. But using adversarial learning instead of variance compensation, has a possibility to enable us to improve the quality of converted speech without postprocessing. And we consider the effects of the proposed method through objective and subjective evaluations.

2 Speaker-Independent Voice Conversion Using Spectral Differential Filter Based on Neural Networks

In speaker-independent voice conversion using spectral differential filter based on neural networks, we train NNs to represent the relationship between multiple source speaker's acoustic features and a target speaker's one so that we can convert speech of an arbitrary input speaker into a specified speaker's one. As a first step of conversion, we extract mel-cepstrum from input speech. After extraction, we convert mel-cepstrum by trained NNs so that it becomes similar to the target speaker's mel-cepstrum, and compensate the GV by variance compensation based on affine transformation. Next, we design a filter based on the difference of the mel-cepstra of the input and the target speech so that we can directly convert waveform. Finally, we apply the filter to input waveform. Therefore, we can convert input waveform to target speaker's one. An advantage of the this method is that the direct waveform conversion alleviates the quality degradation caused by the F0 extraction error.

3 Introducing Adversarial Learning to Conversion of Acoustic Features Based on DNNs

We propose speaker-independent voice conversion based on deep neural networks using adversarial learning. In this study, we introduce discriminative model based on adversarial learning [13] to method explained in Sect. 2 instead of using variance compensation. We use mel-cepstrum as input of DNNs. In the voice conversion using adversarial learning, we train DNNs using two models. One is a generative model, which converts input speaker's mel-cepstrum to target speaker's one. Another is discriminative model, it classifies input mel-cepstrum into one extracted from natural speech and converted one by the generative model. First, we train the generative model so as to minimize mean squared error between converted features and natural features as initialization. The discriminative model outputs the probability that input feature is extracted from natural speech.

In the training of the discriminative model, the generated mel-cepstrums and those of the natural speech samples are used as the input. The model is trained so as to minimize the cross-entropy error between the output probabilities and training labels. After initial training of both models, we update the generative model pitted against discriminative model to minimize the following equation.

$$L(\mathbf{c}, \hat{\mathbf{c}}) = L_G(\mathbf{c}, \hat{\mathbf{c}}) + \omega \frac{E_{L_D}}{E_{L_G}} L_D(\hat{\mathbf{c}}) \tag{1}$$

In Eq. 1, $\mathbf{c}$ is sequential features of natural speech, and $\hat{\mathbf{c}}$ is ones of converted speech. The first term on the right side in Eq. 1 represents mean squared error between features of target speaker's natural speech and ones of converted speech. The second term represents cross-entropy error between the output of the discriminator when converted features and training labels of natural features are inputted. The ω controls the weight of discriminator, E_{L_G}, E_{L_D} represents expected values of $L_G(\mathbf{c}, \hat{\mathbf{c}}), L_D(\hat{\mathbf{c}})$. The ratio $\frac{E_{L_G}}{E_{L_D}}$ is the scale normalization term between $L_G(\mathbf{c}, \hat{\mathbf{c}})$ and $L_D(\hat{\mathbf{c}})$. After that, we iterate updating the discriminator and the generative model. This training method enables us to convert inputed features into ones more similar to ones extracted from target's natural speech.

4 Experiments

4.1 Experimental Conditions

In the experiment, we used one male and one female professional narrator of ATR Japanese speech database set B as the voice of the target speakers. Also, we conducted conversion between the same gender using two general speakers, male and female, of set C as the input speaker. We used a total of 32 speakers, 16 males and 16 females, who were not included in the same set of input speakers as source speakers to train DNNs. From the 150 sentences of the ATR set C, we used 100 sentences for training data and 50 sentences for evaluation data. Speech signals were sampled at a rate of 16 kHz. The acoustic features were extracted by WORLD [9] with a 5 ms frame shift, and consisted of 25 mel-cepstral coefficients. From the result of preliminary experiments, we decided that the structure of the generative model consists of three hidden layers and the number of units was 512. The structure of the discriminative model had one hidden layer, and the number of units was 128. In the training of the model, we trained the generative model with 25 epochs as initial learning, the discrimination model with 5 epochs, and then adversarial learning was performed with 25 epochs. There was a problem that adversarial learning could not be performed normally if $\frac{E_{L_G}}{E_{L_D}}$ in Eq. 1 increased as training of the discriminative model proceeded, when $\frac{E_{L_G}}{E_{L_D}}$ is 100 or more, we used 100.

4.2 Objective Evaluations

In this paper, we compared proposed method SI-AL and SI-Base that learns to minimize the mean squared error, and SI-GV that made variance compensation

Table 1. The results of objective evaluation.

Method	MCD (dB)	GV-RMSE	Spoofing-rates
Natural	-	-	1.0
SI-Base	5.07	0.223	0.0
SI-GV	5.47	0.081	0.995
SI-AL	5.69	0.127	1.0

for SI-Base. As the measures for the evaluation, we used the mel-cepstral distance (MCD) between the converted mel-cepstrum and mel-cepstrum of the natural speech of the target speaker, the GV-RMSE representing RMSE of GV between converted mel-cepstrum and of natural speech, spoofing-rates representing the ratio of being discriminated as natural speech by inputting the speech converted by each method. Table 1 shows the results of objective evaluation, each value is the average of four source speakers. Natural speech (Natural) is the average of two target speakers' natural speech.

In Table 1, it was found that SI-GV is the smallest in GV-RMSE. In variance compensation based on affine transformation, it is considered that the value is the smallest among the three methods because compensation is performed so that the GV of training data is equal to the natural speech. Since GV-RMSE of SI-AL is smaller than GV-RMSE of SI-Base, we confirmed that the proposed method has the effect of compensating for GV. The spoofing-rate is 1.0 for natural speech and 0.0 for SI-Base, confirming that the discriminative model correctly classifies into natural speech and converted speech. It is 1.0 in the proposed method and it can be said that we were able to convert input speech to speech recognized as natural speech on the discriminator. In addition, the SI-GV is also about 1, and it is possible to spoof the discriminator. As a reason for this, since SI-GV performs only manipulation of emphasizing spectrum envelope for SI-Base, it is conceivable that GV is greatly involved as a factor for determining the difference between natural speech and converted speech. In the value of MCD, SI-Base is smallest in 3 methods. SI-GV and SI-AL were worse than SI-Base. This shows that the distance between mel-cepstra has increased due to emphasis of spectral envelope, but the value of SI-AL is worse than SI-GV with a large improvement range of GV.

Next, Fig. 1 shows the drawing of the spectral envelopes of the speech converted by each method. It can be confirmed that the peaks of the spectrum of SI-Base is over-smoothed. SI-GV has the effect of emphasizing the peak of SI-Base while keeping its original shape. In SI-AL, some parts of spectral peaks around 4 kHz are not emphasized, but it was confirmed that the effect of emphasizing the peaks of the spectral envelope was obtained as in the case where variance compensation was performed.

The MCD and GV-RMSE and spoofing-rates are shown in Fig. 2 when ω in the Eq. 1 is changed from 0.1 to 1.5 in increments of 0.1. Each value was the average of two male speakers on newly trained model. Even when changing

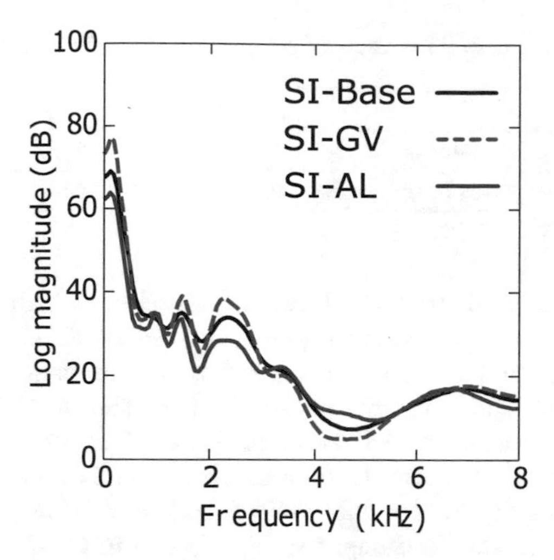

Fig. 1. Spectral envelopes of converted speech.

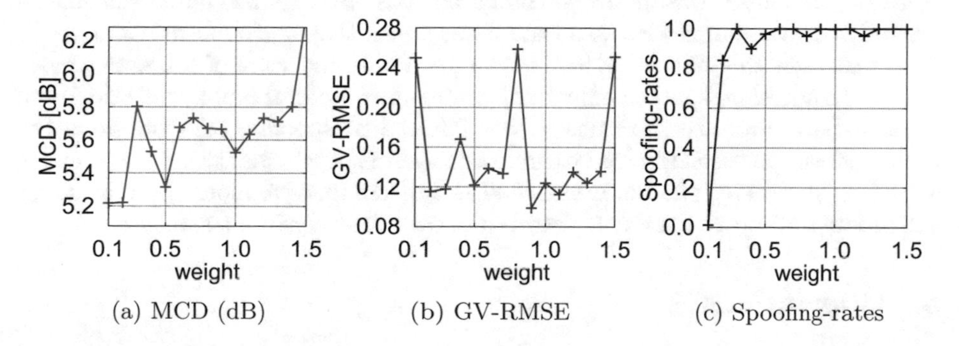

Fig. 2. The effect of changing the ω on MCD, GV-RMSE, and spoofing-rates.

the weight of the discriminative model, GV-RMSE did not change continuously, and also did not change in auditory quality. The spoofing-rates increased by increasing the weight ω applied to the discriminative model, and it was confirmed that it follows the change of GV-RMSE to some extent. Since the value of MCD tends to get worse by increasing the ω, it is necessary to determine an appropriate weight value in consideration with other parameters.

4.3 Subjective Evaluation

In order to evaluate the performance of the proposed method, paired comparison experiment was conducted by subjective evaluation of seven subjects. A paired comparison experiment was carried out using speech samples converted by three methods of SI-AL, SI-Base, and SI-GV. In the evaluation of naturalness, each

Table 2. The result of subjective evaluation (%).

	SI-Base	SI-GV	SI-AL
Naturalness	21.6	56.5	21.8
Speech individuality	26.0	49.2	24.8

voice was played and evaluated by AB test which selects which one was more natural as speech. Speaker individuality was evaluated by XAB test, in which each speech was played after playing natural speech of the target speaker, and the sample having closer individuality was chosen. Table 2 shows the percentage of the overall score when evaluating in all combinations of 3 methods. Since SI-GV had the highest score both in naturalness and speaker individuality score, SI-GV is superior both in quality improvement effect and conversion performance. Compensation effect of feature is thought to be large from GV-RMSE of Table 1, in order of SI-GV, SI-AL, SI-Base, but the score of subjective evaluation of SI-AL and SI-Base did not show noticeable difference. Although the clarity of the speech converted by SI-AL was improved compared to SI-Base, it was found that background noise was emphasized, and there was a part that was unnatural as speech. It was thought that this leads to the degradation in the score of subjective evaluation. In addition, speaker individuality conversion performance was also lower than SI-Base. From these results, it was found that spoofing the discriminative model does not necessarily lead to improvement of speech quality. As this may be solved by improving the performance of the discriminative model by increasing the number of source speakers for training, we will study it in the future.

5 Conclusions

In this paper, we proposed voice conversion from arbitrary speakers using adversarial learning (SI-AL) as a method to improve the quality of converted speech based on DNNs, and compared the proposed method with variance compensation (SI-GV) and SI-Base techniques by evaluation experiments. It was found that SI-GV is the best among the three methods for the conversion performance and the quality improvement effect. The score of subjective evaluation was about the same as SI-Base, become a problem that it is not always possible to obtain the improvement effect of subjective quality by spoofing the discriminative model.

This is a future task to study for quality improvement. Specifically, in the experiment conducted in this study, since learning and conversion were performed frame by frame, time fluctuation of mel-cepstrum was not taken into consideration. Therefore, it is thought that it is necessary to examine the case using the dynamic feature [2] representing the amount of change in the time direction at each time point of the feature amount and using Long short-term memory (LSTM) [4] for the DNNs. It is also designed to use Batch Normalization [5] which normalizes feature to be input for each mini-batch as a method for improving stability at adversarial learning.

Acknowledgment. Part of this work was supported by JSPS KAKENHI Grant Number JP26280055 and JP15H02720.

References

1. Desai, S., Raghavendra, E.V., Yegnanarayana, B., Black, A.W., Prahallad, K.: Voice conversion using artificial neural networks. In: Proceedings of the ICASSP, pp. 3893–3896 (2009)
2. Furui, S.: Speaker-independent isolated word recognition using dynamic features of speech spectrum. IEEE Trans. Acoust. Speech Sig. Process. **34**(1), 52–59 (1986)
3. Goodfellow, I., Pouget-Abadie, J., Mirza, M., Xu, B., Warde-Farley, D., Ozair, S., Courville, A., Bengio, Y.: Generative adversarial nets. In: Advances in Neural Information Processing Systems, pp. 2672–2680 (2014)
4. Hochreiter, S., Schmidhuber, J.: Long short-term memory. Neural comput. **9**(8), 1735–1780 (1997)
5. Ioffe, S., Szegedy, C.: Batch normalization: accelerating deep network training by reducing internal covariate shift. arXiv preprint (2015). arXiv:1502.03167
6. Kain, A., Macon, M.: Spectral voice conversion for text-to-speech synthesis. In: Proceedings of the ICASSP, pp. 285–288 (1998)
7. Koike, H., Nose, T., Shinozaki, T., Ito, A.: Improvement of quality of voice conversion based on spectral differential filter using straight-based mel-cepstral coefficients. J. Acoust. Soc. Am. **140**(4), 2963–2963 (2016)
8. Ling, Z.H., Wu, Y.J., Wang, Y.P., Qin, L., Wang, R.H.: USTC system for blizzard challenge 2006 an improved HMM-based speech synthesis method. In: Blizzard Challenge Workshop (2006)
9. Morise, M., Yokomori, F., Ozawa, K.: World: a vocoder-based high-quality speech synthesis system for real-time applications. IEICE Trans. Inf. Syst. **99**(7), 1877–1884 (2016)
10. Nose, T., Ota, Y., Kobayashi, T.: HMM-based voice conversion using quantized F0 context. IEICE Trans. Inf. Syst. **E93–D**(9), 2483–2490 (2010)
11. Nose, T.: Efficient implementation of global variance compensation for parametric speech synthesis. IEEE/ACM Trans. Audio Speech Lang. Process. **24**(10), 1694–1704 (2016)
12. Pilkington, N.C., Zen, H., Gales, M.J., et al.: Gaussian process experts for voice conversion. In: Proceedings of the INTERSPEECH, pp. 2772–2775 (2011)
13. Saito, Y., Takamichi, S., Saruwatari, H.: Training algorithm to deceive anti-spoofing verification for DNN-based speech synthesis. In: Proceedings of the ICASSP
14. Stylianou, Y.: Voice transformation: a survey. In: Proceedings of the ICASSP, pp. 3585–3588 (2009)
15. Tomoki, T., Tokuda, K.: A speech parameter generation algorithm considering global variance for HMM-based speech synthesis. IEICE Trans. Inf. Syst. **90**(5), 816–824 (2007)

Evaluation of Nonlinear Tempo Modification Methods Based on Sinusoidal Modeling

Kosuke Nakamura[1,2(✉)], Yuya Chiba[1,2], Takashi Nose[1,2], and Akinori Ito[1,2]

[1] Faculty of Engineering, Tohoku University, Sendai, Japan
kosuke.nakamura.p6@dc.tohoku.ac.jp,
{yuya,aito}@spcom.ecei.tohoku.ac.jp,
tnose@m.tohoku.ac.jp
[2] Graduate School of Engineering, Tohoku University, Sendai, Japan

Abstract. Modifying tempo of musical signal is one of the basic signal processing for music signal, and many methods have been proposed so far. Nishino et al. proposed a tempo modification method of nonlinear modification based on sinusoidal model, but the evaluation of the methods was insufficient. In this paper, we evaluated the tempo modification methods with sinusoidal model and nonlinear signal stretch and compression. Namely, we compared effectiveness of use of residue signal and methods of determination of stretchable parts. From the experimental result, we could confirm the efficiency of the nonlinear tempo modification. We also compared several methods of determining the stretchable parts as well as the use of residue signal. As a result, the effect of the methods depended on the input signal.

Keywords: Music signal processing · Tempo modification · Sinusoidal model

1 Introduction

Changing tempo of music signal without changing its pitch has been one of basic signal processing methods for manipulating music signal. The tempo modification technology was used for variety of music manipulation such as remix [1], virtual conducting [2], or modification of musical expression [3]. It is also a basic technology for developing new digital instruments [4].

Various techniques have been used to realize the tempo modification, such as the phase vocoder [5], TD-PSOLA [6], WSOLA [7], TDHS [8], etc. Most of those techniques stretch or compress the music signal linearly, which sometimes makes degradation because the temporal structure of the instrumental sound such as attack and decay is not kept of the tempo is modified linearly.

We have been developing a method to modify tempo of music signals based on sinusoidal model, which nonlinearly stretches or shrinks the music signal to keep the structure of instrumental sounds [9–11]. However, systematic evaluation of the method and comparison between various conditions have not been conducted yet. Thus, in this paper, we conducted systematic evaluations of the tempo modification methods based on the sinusoidal modeling.

J.-S. Pan et al. (eds.), *Advances in Intelligent Information Hiding and Multimedia Signal Processing*, Smart Innovation, Systems and Technologies 82,
DOI 10.1007/978-3-319-63859-1_14

2 Tempo Modification Based on Sinusoidal Model

2.1 Sinusoidal Representation of a Signal

First, we briefly describe the tempo modification method based on the sinusoidal model [11]. The sinusoidal model is to express a signal using sum of sinusoids with time-varying amplitude and phase [12], as follows.

$$x(t) = \sum_{k=1}^{K} \exp(a_k(t)) \cos(p_k(t)) \tag{1}$$

Here, $a_k(t)$ and $p_k(t)$ are instantaneous amplitude and phase, respectively. We exploited the local vector transform (LVT) method [13] for estimating the instantaneous amplitude and phase.

In addition to the sinusoids, the original signal can be expressed more precisely by adding the residue signal [14].

$$x(t) = \sum_{k=1}^{K} \exp(a_k(t)) \cos(p_k(t)) + \varepsilon(t) \tag{2}$$

2.2 Overview of the Tempo Modification Method

Once the signal is expressed as parametric representation, we can manipulate the signal so that the tempo is changed without changing its pitch. Ito et al. proposed a method to change the tempo of the signal using sinusoidal representation [10, 16]. Figure 1 shows the block diagram of the method. The input signal is analyzed using the sinusoidal model, and divided into the harmonic part (represented as sum of sinusoids) and the residue. Then the sinusoidal part and the residue are stretched or compressed individually, and then those parts are combined again to synthesize the output signal.

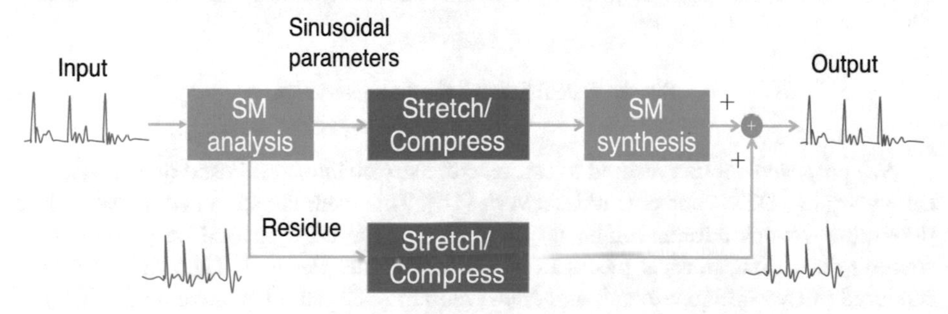

Fig. 1. Tempo modification based on the sinusoidal model

To realize the tempo modification that keeps the structure of instrumental sounds, the "stretchable parts" are estimated from the signal. Only stretchable parts of the signal are stretched or compressed. The stretchable parts are regarded as steady, and the values of the residue signal in the parts are replaced with zeroes. Then nonlinear tempo modification of the residue is realized by increasing or decreasing number of zeroes in the parts.

2.3 Estimation of Stretchable Parts

Nishino et al. used the *transient index* for determining the stretchable parts [11], which was proposed for beat tracking [16]. Figure 2 shows an example of calculation of the transient index. First, the input signal (Fig. 2(a)) is analyzed by the short time Fourier transform (STFT) for generating a spectrogram (Fig. 2(b)). Then the temporal derivatives of the STFT are calculated for each of the frequency bin. Finally, the absolute value of the temporal derivative values of each frame are integrated to calculate the transient index (Fig. 2(c)). The transient index gets larger when any of the frequency components have rapid change. We apply a threshold on the transient index and regard the frames that are lower than the threshold as the stretchable parts.

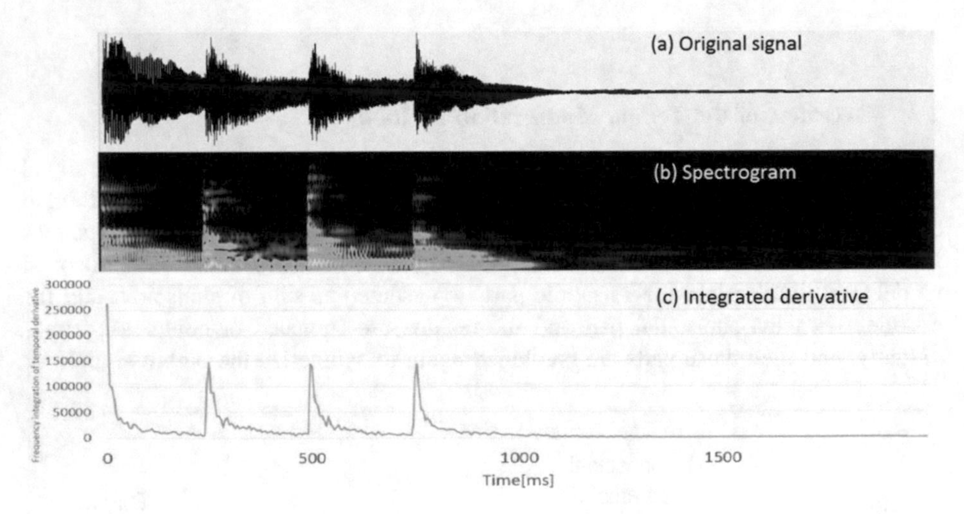

Fig. 2. Calculation of the transient index

We proposed another method to estimate the stretchable parts based on the dynamic time warping (DTW) and a neural network [16]. This method uses a neural network to determine whether a frame can be stretched or not. The DTW is used for preparing the training data. First, musical pieces in MIDI format are prepared. Then the pieces are rendered in two different tempi, and Mel-Frequency Cepstral Coefficients (MFCC) of them are calculated. After aligning the two series of MFCC using DTW, we can know whether a frame in the shorter signal is stretched or not when aligning to the longer signal.

Finally, we label each frames of the shorter signal as either "stretchable" or "not stretchable", which are used for training the neural network.

3 Comparison of Tempo Modification Methods

3.1 Overview

In this paper, we compared the tempo modification methods in the following aspects.

1. Modification methods: (a) WSOLA, (b) Sinusoidal model
2. Linearity: (a) Linear modification, (b) Nonlinear modification
3. Use of residue of sinusoidal modeling: (a) Use residue, (b) Do not use residue
4. Estimation of the stretchable parts: (a) transition index, (b) DTW

Here, 4(b) (DTW) means the method of determining the stretchable parts using the DTW. The method proposed in [16] estimates the stretchable parts using a neural network. To investigate the efficacy of this method, we used the training data of the neural network as the label of the stretchable parts, which gives the upper limit of the neural-network-based method.

3.2 Experimental Conditions

We used three kinds of data. The first one (set A) was total of 136 monophonic music signals from four instruments: piano, guitar, clarinet and violin. Each instrument had 34 different music signals. The second one (set B) were 20 vocal signals. The third one (set C) were 48 polyphonic music signals extracted from pops and classic music. Length of the all signals was 3 s. The pieces in set B and the pops of set C were the same.

All of the music signals were generated from MIDI representation of the music. MuseScore2 was used for generating the instrumental sounds, and Vocaloid3 was used for synthesizing vocal signals. Sampling frequency of all signals was 44.1 kHz, and analysis frame shift was 5 ms. All signals were stretched twice as slower as the original signal. This stretch rate is considered to be large enough to evaluate the method so that difference between the qualities of tempo-modified signals by different methods becomes clear. We used the sound manipulation tool SoX for modification using WSOLA.

The objective difference grade (ODG) based on perceptual evaluation of audio quality (PEAQ) [17] was used for evaluating the modified signal. Peaqb [18] was used to calculate the ODG.

3.3 Experimental Results

Figure 3 shows the experimental results with different methods of determining the stretchable parts. In this experiment, the sinusoidal model and nonlinear modification were employed. In this figure, "TI" and "DTW" mean determination of the stretchable

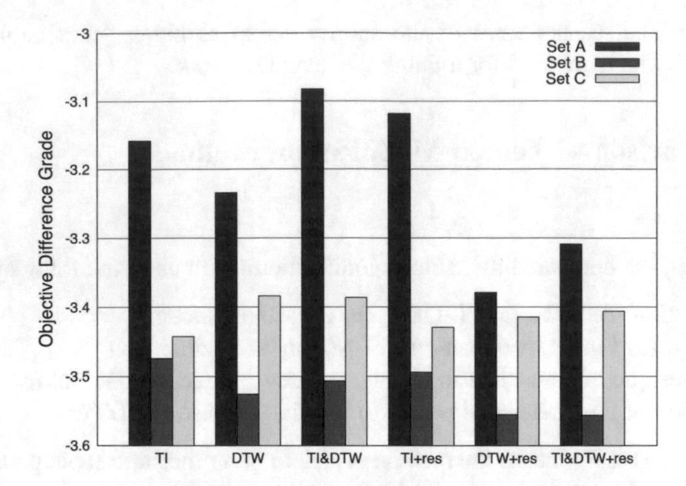

Fig. 3. Experimental results for different methods of stretchable part determination

parts using the transient index and DTW, respectively. The label "TI&DTW" means the result where the intersection frames of the stretchable parts determined by TI and DTW were regarded as the stretchable parts. The labels with "+res" means the result with residue signal.

The absolute difference of all methods was not large, and the best result depended on the input signal. As for the polyphonic signal (Set C), "TI&DTW" gave the best result. TI gave better scores than DTW for Set A and B, while DTW gave better scores as for Set C. This seems to be caused by difficulty of determining stretchable parts of the polyphonic signal by TI. In the polyphonic signal, several instrumental sounds start and stop at different times, which makes determination of transient sound by TI more difficult.

Use of the residue affected the performance of the methods negatively in all conditions except TI. This seems to be caused by estimation errors of the stretchable parts. Figure 4 shows how the residue signal causes degradation of the stretched signal. First, the original signal is split into the sinusoidal part and the residue (Fig. 4(a)). If a stable part is determined as a single stretchable part, that part is stretched and the residue signal is not added to the stretched part (Fig. 4(b)). Here, if the stable part is determined as two stretchable parts and a non-stretchable part, the two stretchable parts are stretched individually and the non-stretchable parts remain unchanged. Since the residue signal is added to the sinusoidal signal of the non-stretchable parts, the residue sounds like impulsive noise (Fig. 4(c)).

Next, different methods of tempo modification were compared using the three evaluation sets. Different combination of the modification methods, linearity, and use of residue were investigated. The stretchable parts of the nonlinear modification were determined by the best methods obtained in the previous experiment. Figure 5 shows the results. From these results, we can see that the sinusoidal model was slightly better than WSOLA when the signal was modified linearly. Next, the nonlinear modification was effective compared with the linear modification, especially for monophonic signal (set A).

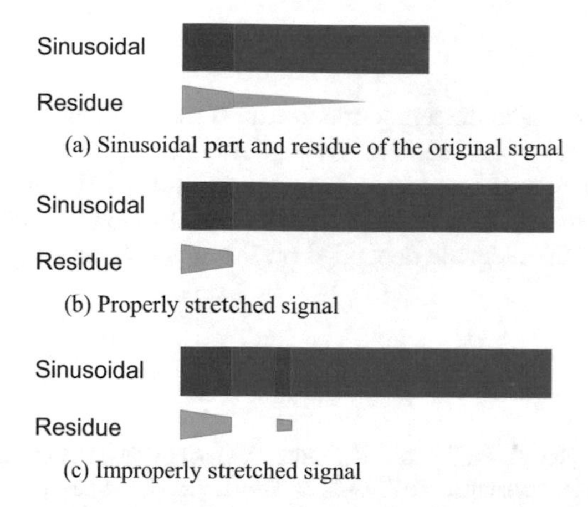

Fig. 4. How the residue signal harms the quality of the stretched signal

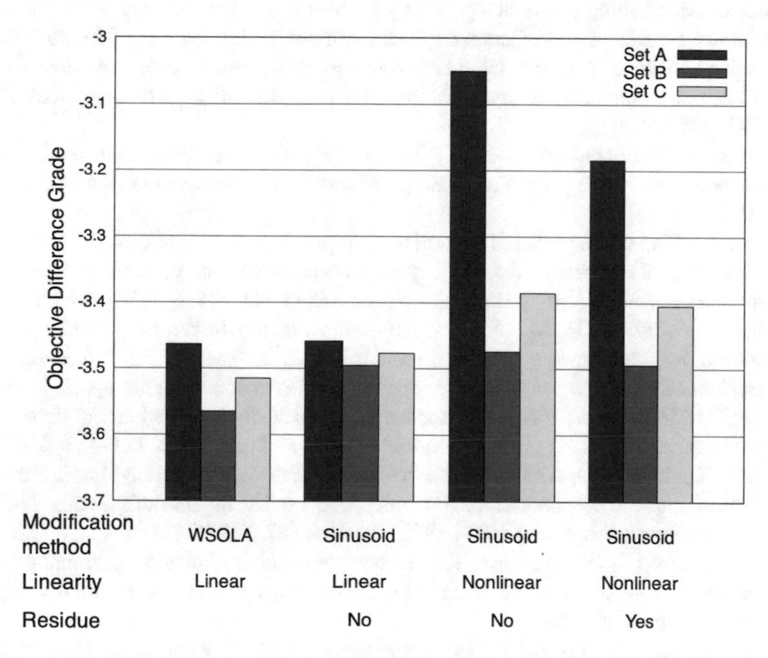

Fig. 5. Comparison results of modification methods

The cause of this might be that monophonic instrumental sounds were simpler than the other signals such as vocal signals (set B) and polyphonic signals (set C). Third, use of the residue signal did not improve the signal quality.

4 Conclusion

In this paper, we evaluated the tempo modification methods with sinusoidal model and nonlinear signal stretch and compression. From the experimental result, we could confirm the efficiency of the nonlinear tempo modification. We also compared several methods of determining the stretchable parts as well as the use of residue signal. As a result, the effect of the methods depended on the input signal.

References

1. Davies, M.E.P., Hamel, P., Yoshii, K., Goto, M.: AutoMashUpper: automatic creation of multi-song music mashups. IEEE/ACM Trans. Audio Speech Lang. Proc. **22**(12), 1726–1737 (2014)
2. Bruegge, B., Teschner, C., Lachenmaier, P., Fenzl, E., Schmidt, D., Bierbaum, S.: Pinocchio: conducting a virtual symphony orchestra. In: Proceedings of the International Conference on Advances in Computer Entertainment Technology, pp. 294–295 (2007)
3. Fabiani, M., Friberg, A.: Rule-based expressive modifications of tempo in polyphonic audio recordings. In: International Symposium on Computer Music Modeling and Retrieval, pp. 288–302 (2007)
4. Berthaut, F., Desainte-Catherine, M., Hachet, M.: Drile: an immersive environment for hierarchical live-looping. In: Proceedings of the New Interface for Musical Expression, p. 192 (2010)
5. Dolson, M.: The phase vocoder: a tutorial. Comput. Music J. **10**(4), 14–27 (1986)
6. Moulines, E., Charpentier, F.: Pitch-synchronous waveform processing techniques for text-to-speech synthesis using diphones. Speech Commun. **9**(5–6), 453–467 (1990)
7. Verhelst, W., Roelands, M.: An overlap-add technique based on waveform similarity (WSOLA) for high quality time-scale modification of speech. In: Proceedings of the International Conference on Acoustics, Speech, and Signal Processing, pp. 554–557 (1993)
8. Malah, D.: Time-domain algorithms for harmonic bandwidth reduction and time scaling of speech signals. IEEE Trans. Acoust. Speech Signal Process. **27**(2), 121–133 (1979)
9. Igarashi, Y., Ito, M., Ito, A.: Evaluation of sinusoidal modeling for polyphonic music signal. In: Proceedings of the International Conference on Intelligent Information Hiding and Multimedia Signal Processing (IIH-MSP), pp. 464–467 (2013)
10. Ito, A., Igarashi, Y., Ito, M., Nose, T.: Tempo modification of music signal using sinusoidal model and LPC-Based residue model. In: Proceedings of the International Congress on Sound and Vibration (2014)
11. Nishino, T., Nose, T., Ito, A.: Tempo modification of mixed music signal by nonlinear time scaling and sinusoidal modeling. In: Proceedings of the International Conference on Intelligent Information Hiding and Multimedia Signal Processing (IIH-MSP), pp. 146–149 (2015)
12. McAulay, R.J., Quatieri, T.F.: Speech analysis/synthesis based on a sinusoidal representation. IEEE Trans. Acoust. Speech Sig. Process. **34**(4), 744–754 (1986)
13. Ito, M., Yano, M.: Sinusoidal modeling for nonstationary voiced speech based on a local vector transform. J. Acoust. Soc. Am. **121**, 1717 (2007)
14. Ding, Y., Qian, X.: Processing of musical tones using a combined quadratic polynomial phase sinusoid and residual (QUASER) signal model. J. Audio Eng. Soc. **45**(7/8), 571–584 (1997)

15. Alonso, M., David, B., Richard, G.: Tempo and beat estimation of music signals. In: Proceedings of the ISMIR, pp. 158–164 (2004)
16. Nishino, T., Nose, T., Ito, A.: Deciding expandable sections for nonlinear changing play back speed of music signals. In: Proceedings of the ASJ Spring Meeting, 2-10-13 (2016). (in Japanese)
17. International Telecommunication Union, "Method for objective measurements of perceived audio quality," ITU-R BS. 1387-1 (2001)
18. Peaqb-fast. http://github.com/akinori-ito/peaqb-fast/. Accessed 1 Mar 2017

A Study on 2D Photo-Realistic Facial Animation Generation Using 3D Facial Feature Points and Deep Neural Networks

Kazuki Sato[1](✉), Takashi Nose[1], Akira Ito[1], Yuya Chiba[1], Akinori Ito[1], and Takahiro Shinozaki[2]

[1] Graduate School of Engineering, Tohoku University, 6-6-05 Aramaki Aza Aoba, Aoba-ku, Sendai, Miyagi 980-8579, Japan
{kazuki.satou.p3,akira.ito.s1}@dc.tohoku.ac.jp, tnose@m.tohoku.ac.jp, {yuya,aito}@spcom.ecei.tohoku.ac.jp

[2] Interdisciplinary Graduate School of Science and Engineering, Tokyo Institute of Technology, 4259 Nagatsuta-cho, Midori-ku, Yokohama 226-8502, Japan

Abstract. This paper proposes a technique for generating a 2D photo-realistic facial animation from an input text. The technique is based on the mapping from 3D facial feature points with deep neural networks (DNNs). Our previous approach was based only on a 2D space using hidden Markov models (HMMs) and DNNs. However, this approach has a disadvantage that generated 2D facial pixels are sensitive to the rotation of the face in the training data. In this study, we alleviate the problem using 3D facial feature points obtained by Kinect. The information of the face shape and color is parameterized by the 3D facial feature points. The relation between the labels from texts and face-model parameters are modeled by DNNs in the model training. As a preliminary experiment, we show that the proposed technique can generate the 2D facial animation from arbitrary input texts.

Keywords: Photo-realistic facial animation · Face image synthesis · Deep neural network · Kinect

1 Introduction

In recent years, there have been many studies related to visual-speech synthesis for the realization of a real human-like computer agent, which is expected to lead to richer and more advanced human computer interaction. Anderson et al. [1] expanded the active appearance models (AAMs) [3], that is known as one of the general 2D face models, and proposed a technique for generating high-quality facial animations with multiple emotional expressions using cluster adaptive training [4] and hidden Markov models (HMMs), which was inspired by a style control technique in speech synthesis [8]. However, this technique has a

J.-S. Pan et al. (eds.), *Advances in Intelligent Information Hiding and Multimedia Signal Processing*, Smart Innovation, Systems and Technologies 82,
DOI 10.1007/978-3-319-63859-1_15

problem that the data collection cost is relatively high because manual labeling of training data with feature points is partly required for constructing AAMs. Wang et al. proposed another technique that combines 3D model generation from a 2D face image and sample-based image generation. Although this technique achieved 3D facial animations with good naturalness by rendering the regions difficult to reproduce, e.g., a mouth region, using sample-based approach, the technique is sometimes difficult to be implemented in mobile devices because the computational cost for 3D rendering is high.

Our previous approach [11] was based on the HMM-based parameter generation and DNN-based parameter mapping. In the technique, animation units (AUs) were used as intermediate parameters between labels (from texts) and output pixel images. The technique has two stages: the first stage is the parameter generation of an AU sequence from given labels, and the second stage is the conversion from AUs to pixels using DNN-based mapping frame by frame. The result of objective and subjective evaluations showed that our technique outperformed the conventional PCA-based visual-speech synthesis [10]. However, our conventional technique suffers from the critical degradation of the synthesized quality when the degree of the head movements are relatively large.

In this paper, we propose a novel technique for generating 2D facial animations for visual-speech synthesis from texts. The technique uses the 3D facial feature points obtained by Microsoft Kinect [13] to explicitly model the face shapes and colors. Specifically, 3D face parameters are obtained from 2D face images and 3D facial feature points, and the mapping from linguistic labels to 3D face parameters are trained using DNNs where a technique similar to DNN-based speech synthesis [12]. We show a preliminary result of facial animation generation using the proposed technique.

2 3D Facial Parameterization Using 3D Feature Points and 2D Image

The main weakness of our conventional visual-speech synthesis [11] is that the face images with different head angles cannot be appropriately modeled because there is no rotation information in the modeling process. This section describes how we attempts to solve the problem by explicitly modeling a 3D face with face parameters, i.e., polygons with color information.

In the proposed technique, 2D face images are parametrized using 3D facial feature points obtained by Kinect v2. 1,347 feature points can be extracted by using High definition face tracking API [5] of Kinect v2. For each frame of the sequence of the extracted 3D feature points, the resolution of the entire data is fixed by determining the number of pixels in each triangular polygon constructed by the referential feature points which are chosen from mouth-opening frames of the datasets randomly. The number of pixels in each polygon is determined as follows. Let

$$\boldsymbol{p}_n = (\boldsymbol{p}_{n1}, \boldsymbol{p}_{n2}, \boldsymbol{p}_{n3}) \tag{1}$$

be a set of apexes of the nth polygon, and the point $\boldsymbol{p}$ in the polygon is given by

$$\boldsymbol{p} = s(\boldsymbol{p}_{n2} - \boldsymbol{p}_{n1}) + t(\boldsymbol{p}_{n3} - \boldsymbol{p}_{n1}) + \boldsymbol{p}_{n1}, \tag{2}$$

where

$$0 \leq s \leq 1 \tag{3}$$
$$0 \leq t \leq 1 \tag{4}$$
$$0 \leq s + t \leq 1. \tag{5}$$

Since s and t can be regarded as texture coordinates, we introduce a shift width parameter, h_n, to split the texture into small blocks. s and t are redefined as follows:

$$s = lh_n \tag{6}$$
$$t = mh_n, \tag{7}$$

where l and m have integer values and their ranges are uniquely determined from the ranges of s and t. The resolution of the nth polygon is determined by the shift width h_n. Let the number of pixels of the nth polygon N_n, and Eq. (8) is established between N_n and h_n.

$$N_n = \frac{1}{2}\lfloor\frac{1}{h_n}\rfloor(\lfloor\frac{1}{h_n}\rfloor + 1) \tag{8}$$

Suppose the area ratio is equal to the ratio of the number of pixels between polygons. Let the area of the nth polygon S_n, the area of the minimum polygon S_{min} and the number of pixels of the minimum polygon N_{min}, Eq. (9) is established.

$$N_n = \frac{S_n N_{min}}{S_{min}} \tag{9}$$

Therefore if N_{min} is determined, N_n and h_n can be calculated. In this study, the number of pixels of the smallest polygon was set to unity. Hereafter, the parameter set of N_n and $\boldsymbol{p}_n$ is called *polygon property*. N_n is fixed for all frames of the training data in order to obtain the same size of the textures from the data. The pixel values in each point of the texture coordinates are obtained using bicubic interpolation with the 2D pixel values mapped from 3D coordinates to 2D plane. In this study, these 3D feature points and textures are used as the 3D model parameters.

3 Proposed Photo-Realistic Facial Animation Generation Using DNNs

3.1 Facial Features for DNNs

Since texture data has high dimensionality and is difficult to be processed, we apply principal component analysis (PCA) to the texture data for dimension reduction. The 3D feature points obtained by Kinect have the information of the facial direction. Therefore, we align the face direction and remove the rotation effect by fitting the set of 3D facial feature points to referential feature points used to determine the number of pixels in each polygon. Hereby, we overcome the weakness of our previous approach. The rigid transformation matrix for the fitting is estimated using Iterative Closest Point (ICP) algorithm [2]. At the estimation, Kalman filter is applied to suppress the vibration by the noise occurring at fitting. Finally, PCA coefficients are calculated for the textures, and 3D facial feature points and the PCA coefficients are normalized to $[0, 1]$ using the minimum and maximum values of each coefficient. After the normalization, Δ and Δ^2 are calculated from the static coefficient sequences. These static and dynamic features are combined and used as facial features.

3.2 Mapping from Linguistic Labels to Facial Features Using DNNs

In the training step, DNNs are used for the mapping from context-dependent labels created from the texts to the facial features described in Sect. 3.1. In the synthesis step, first phone duration is predicted for the labels of the input text using duration models trained in advance. The predicted durations are used for creating the context-dependent labels including frame position information. The labels are inputted to the DNNs, and the frame sequence of the facial features are predicted. A 2D face image is obtained from the facial features. Finally, a complete movie is created by pasting the face image to a original movie including a background image in each frame.

4 Experiments

4.1 Experimental Conditions

For the experiment, we recorded color video samples of a male speaker who uttered 11 sentences using Kinect v2. The sentences were selected from the subsets A of 503 phonetically balanced sentences of the ATR Japanese speech database set B [6]. 10 sentences were chosen for the candidates of the model training for DNN, and 1 sentences were chosen for the evaluation test. The size of the images was 400×400 pixels. The sampling frequency of speech was 16 kHz. We selected 128 frames for PCA from the training data randomly, and used all principal components. To reduce the computational cost, we used Incremental PCA. As the feature vectors, we used 12,507 dimensional vectors including normalized coordinates of 1,347 3D feature points, 128 dimensional normalized PCA coefficients of textures, and their Δ and Δ^2 parameters. We used the fully connected networks. The conditions for the training of DNNs are listed in Table 1. The context consists of relative frame position (number from 0 to 1), and answers (Yes. 1, No: 0) for 412 questions. The questions were about the type of phonemes, accent positions, breath group and the length of sentence. They are used in the decision tree based context clustering in HMM-based speech synthesis. In this study, we used the phoneme duration of natural speech for synthesis.

Table 1. Structure of DNNs

Number of units for input layer	413
Number of units for output layer	12,507
Number of hidden layers	3
Number of units for hidden layers	1,024
Optimizer	Adam
Activation function	tanh
Batch size	100
Number of epochs	1,000
Dropout rate	0.5

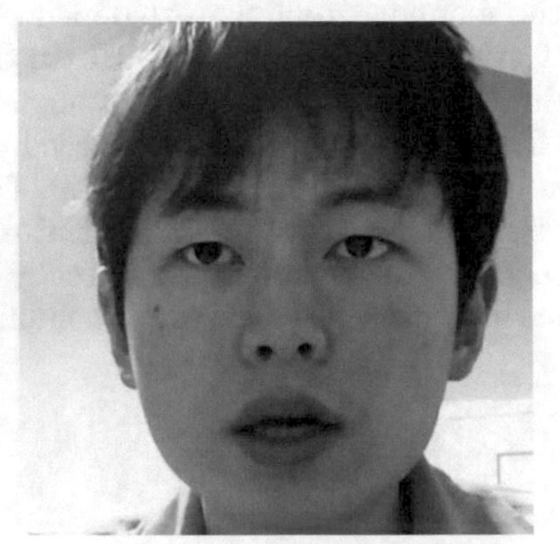
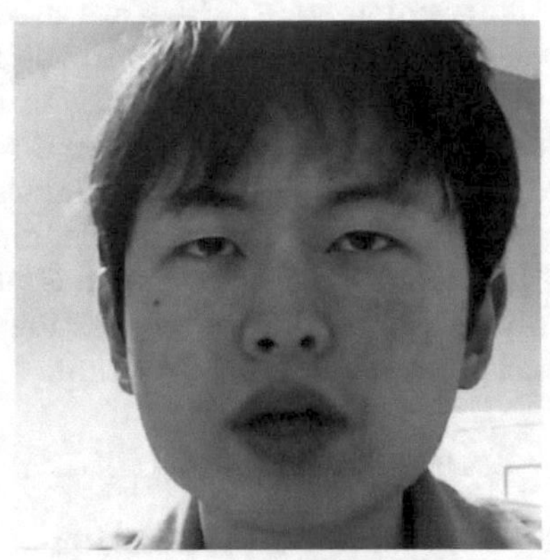

(a) Mouth-open frames

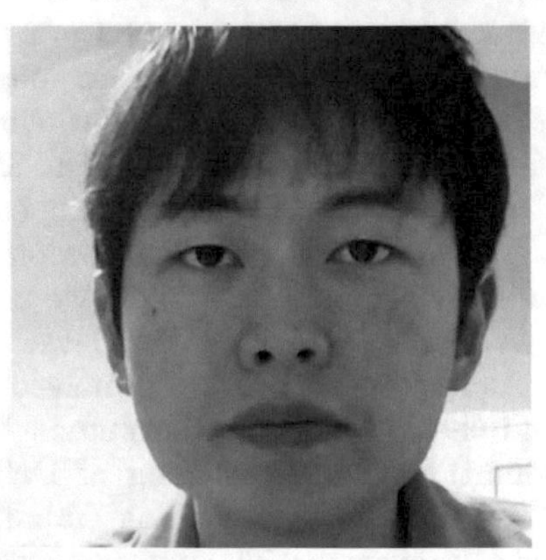
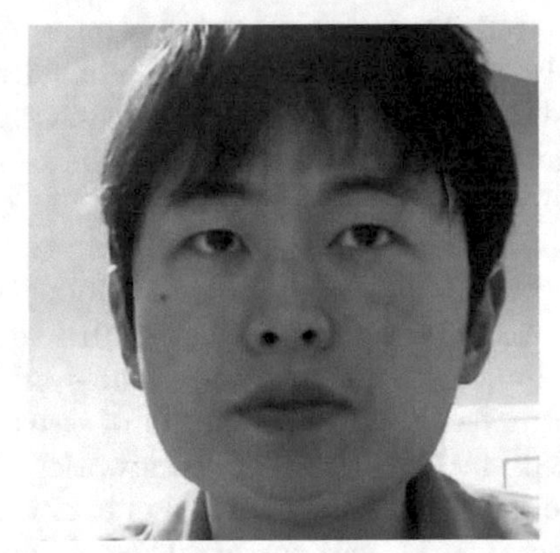

(b) Mouth-closed frames

Fig. 1. Comparison of the captured and synthesized frames. From left: captured, proposal method.

4.2 Facial Animation Generation with Proposal Technique

Figure 1 shows the results. It shows examples of the mouth open and closed face image extracted from the facial animations of synthetic and original sample. We see no difference in the naturalness of the face between each frames. On the other hand, the accuracy of lip sync is not high and there is still room for improvement. Also it is not natural because we can see the boundary between original and synthetic facial region. It is necessary to improve the way of attaching synthetic facial region.

5 Conclusions

In this paper, we proposed a technique for synthesizing a 2D photo-realistic talking face animation using 3D face features with DNNs. We show that the proposed technique can generate the 2D facial animation from arbitrary input texts. In future work, we plan to expand the technique to expressive faical animation generation to suppress the cost for training in the previous study [1]. The model adaptation technique in style control [7] can be introduced to decrease the training data. Improving the perceived quality is also important using contrast enhancement, e.g., variance compensation for parametric speech synthesis [9]. In addition, we need to evaluate the naturalness of generated movies and compare with conventional techniques.

Acknowledgment. Part of this work was supported by JSPS KAKENHI Grant Number JP15H02720 and JP26280055.

References

1. Anderson, R., Stenger, B., Wan, V., Cipolla, R.: Expressive visual text-to-speech using active appearance models. In: Proceedings of the Computer Vision and Pattern Recognition (CVPR), pp. 3382–3389 (2013)
2. Besl, P.J., McKay, N.D.: Method for registration of 3-D shapes. In: Robotics-DL tentative, pp. 586–606. International Society for Optics and Photonics (1992)
3. Cootes, T.F., Edwards, G.J., Taylor, C.J.: Active appearance models. IEEE Trans. Pattern Anal. Mach. Intell. **23**(6), 681–685 (2001)
4. Gales, M.J.: Cluster adaptive training of hidden Markov models. IEEE Trans. Speech Audio Process. **8**(4), 417–428 (2000)
5. Kinect for Windows SDK 2.0 Programming Guide: High definition face tracking. https://msdn.microsoft.com/en-us/library/dn785525.aspx
6. Kurematsu, A., Takeda, K., Sagisaka, Y., Katagiri, S., Kuwabara, H., Shikano, K.: ATR Japanese speech database as a tool of speech recognition and synthesis. Speech Commun. **9**(4), 357–363 (1990)
7. Nose, T., Tachibana, M., Kobayashi, T.: HMM-based style control for expressive speech synthesis with arbitrary speaker's voice using model adaptation. IEICE Trans. Inf. Syst. **E92–D**(3), 489–497 (2009)
8. Nose, T., Yamagishi, J., Masuko, T., Kobayashi, T.: A style control technique for HMM-based expressive speech synthesis. IEICE Trans. Inf. Syst. **E90–D**(9), 1406–1413 (2007)

9. Nose, T.: Efficient implementation of global variance compensation for parametric speech synthesis. IEEE/ACM Trans. Audio Speech Lang. Process. **24**(10), 1694–1704 (2016)
10. Sako, S., Tokuda, K., Masuko, T., Kobayashi, T., Kitamura, T.: HMM-based text-to-audio-visual speech synthesis. In: Proceedings of the INTERSPEECH, pp. 25–28 (2000)
11. Sato, K., Nose, T., Ito, A.: Synthesis of photo-realistic facial animation from text based on HMM and DNN with animation unit. In: Proceeding of the Twelfth International Conference on Intelligent Information Hiding and Multimedia Signal Processing (IIH-MSP), pp. 29–36 (2017)
12. Zen, H., Senior, A., Schuster, M.: Statistical parametric speech synthesis using deep neural networks. In: Proceedings of the ICASSP, pp. 7962–7966 (2013)
13. Zhang, Z.: Microsoft kinect sensor and its effect. IEEE Multimedia **19**(2), 4–10 (2012)

Recent Advances in Information Hiding and Signal Processing for Audio and Speech Signals

Towards an Interrogation Speech Manipulation Detection Method Using Speech Fingerprinting

Shinnya Takahashi and Kazuhiro Kondo(✉)

Graduate School of Science and Engineering, Yamagata University,
4-3-16 Jonan, Yonezawa, Yamagata 9928510, Japan
nyamo.klab@gmail.com, kkondo@yz.yamagata-u.ac.jp

Abstract. We proposed a manipulation detection method for interrogation speech. We used a robust fingerprinting method optimized for speech since our intended target is interrogation speech recorded during a police investigation. The fingerprint uses line spectral pairs (LSP) to measure the spectral envelope of the speech, and is coarsely quantized so that the fingerprint will not be altered by small degradation in the signal, but will be altered enough by malicious modifications to the speech content. This fingerprint is embedded in the speech signal using conventional spread-spectrum watermarks. To detect manipulation, the watermarked fingerprint is detected, and compared to the fingerprint extracted from the speech itself. If the fingerprints match within the predetermined tolerance, it can be authenticated to be unaltered. Otherwise, manipulation should be suspected. We conducted initial experiments to verify the feasibility of the proposed method, and confirmed that at the utterance level, we can identify all substitution manipulated speech utterances successfully.

Keywords: Interrogation speech · Manipulation detection · Audio watermark · Speech fingerprinting · Line spectral pairs

1 Introduction

Recently, the Japanese Judicial System decided to record, both video and audio, all police interrogations that are subject to lay judge trials [5]. It is expected that eventually, all criminal interrogations will be fully recorded, although there are still some strong objections from the police authorities. This trend seems to be apparent in many of the developed Western countries, including the U.S. [4].

These recordings will be done mostly in the modern digital form. Obviously, it is very easy to make perfect modifications to these recordings in digital format. One can easily imagine the police authorities altering and falsifying the recordings to match their theory as to how the crime occurred, ensuring fast and incriminating trials. In order to avoid such modifications, we need a robust mechanism to guarantee the authenticity of the recordings.

One of the most famous alterations of audio recordings submitted for evidence is the Watergate Tapes. An 18 and a half minute silence was found in what seems

J.-S. Pan et al. (eds.), *Advances in Intelligent Information Hiding and Multimedia Signal Processing*, Smart Innovation, Systems and Technologies 82,
DOI 10.1007/978-3-319-63859-1_16

to be a critical portion of the conversation between the former President Nixon and his Chief of Staff Harry Hadelman [8]. From the context of the conversation, many believed that the silence was actually the result of someone erasing this portion to cover up the involvement of the President in the Watergate scandal. A scientific investigation was conducted by a panel of experts to determine the cause of the 18 and a half minute gap, and it was concluded by a laborious analysis that the gap was the result of erasing this portion by a tape recording device different from the one that made the original recordings [1].

This example shows that some system is necessary to ensure the authenticity of the recordings submitted for evidence to the court, including the police interrogations, which does not require sophisticated or laborious analysis. Accordingly, we propose an authentication method for speech signals which fingerprints the speech content using a sequence of quantized line spectral pairs (LSP), and embeds this fingerprint using audio watermarks. The speech can be checked for its authenticity by extracting the embedded watermark, and comparing this to the extracted fingerprint from the received speech signal. With this method, any form of manipulation, including deletion, insertion of other speech segments, or substitution from other portions of the speech, will result in a mismatch in the fingerprint and the watermarked fingerprint, which can be used to warn the existence of manipulation. This paper proposes an initial configuration of such speech manipulation detection system, and conducts an initial experimentation to investigate the feasibility of this proposal.

This paper is organized as follows. In the next section, we will describe the interrogation speech manipulation detection method. Then the optimization of the manipulation detection parameters will be described. This will be followed by the experimental results for the initial evaluation of the proposed method and discussions. Finally, conclusions and issues will be given.

2 An Interrogation Speech Manipulation Detection Method Using Speech Fingerprinting

Figure 1 shows the overall configuration of the proposed interrogation speech manipulation detection method. At the police station, where interrogation of a suspect is being conducted, the interrogation speech is recorded using a sealed digital recorder. Inside this recorder, a fingerprint which roughly measures the time sequence of the characteristics of the speech input is calculated. This fingerprint is then encrypted so that a malicious third party will not be able to tamper with this fingerprint, and embedded in the speech using a digital watermark. The recorded speech can be stored onto a medium, or transmitted using a network.

When this interrogation recording is being examined as evidence, for instance at a court house before a jury, the watermarked fingerprint can be extracted, and decrypted using the secret key. A new set of fingerprint is extracted from the speech being played out. These two sets of fingerprints can be matched to detect its authenticity. If the two sets of fingerprints match within a predetermined

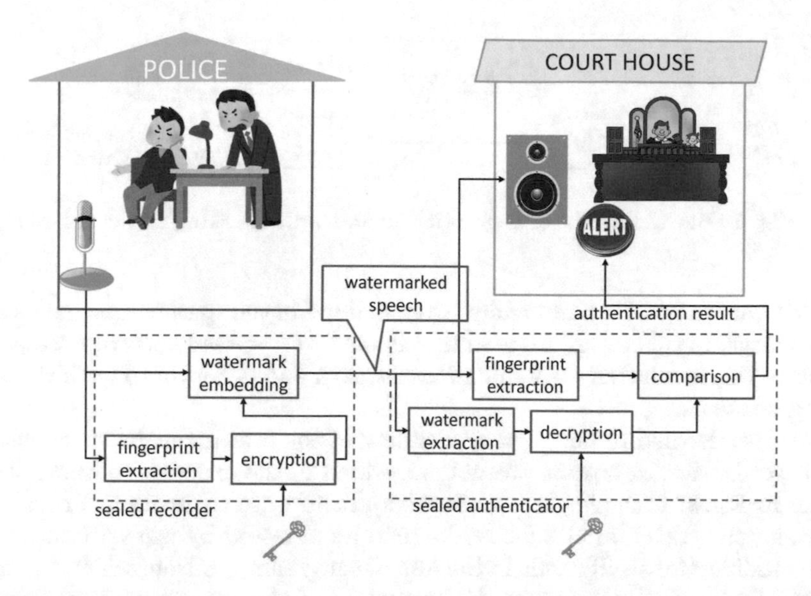

Fig. 1. Block diagram of the interrogation speech manipulation detection method

tolerance, for instance by measuring the vector distance between the two sets and comparing this distance to the preset threshold, the speech can be authenticated to have no modifications. However, if the two fingerprints do not match, i.e., the vector distance is over the threshold, they are marked to include modifications and alerts should be displayed as so. All the above processing should be done using a sealed authenticator unit.

Figure 2 shows a more detailed configuration of the fingerprinting and watermarking process at the recorder. The figure depicts the model used to evaluate the feasibility of the proposed, and focuses on the fingerprinting and its watermarking process. The encryption process has been left out intentionally here for brevity.

The speech signal is first divided into frames, and its fingerprint is extracted from each frame. As shown in the figure, the fingerprint we used is based on the Line Spectral Pairs (LSP) [3,7] of the signal. LSPs are one of the forms commonly used to represent the spectral envelope of the signal. It is known to be one of the most efficient format to express the spectral envelope, and is known to be robust to quantization. In our proposal, this LSP is coarsely quantized to give a very crude description of the speech spectrum envelope. The quantization should be coarse enough to be robust to non-malicious signal degradation, but fine enough that malicious alteration of the signal can be detected as changes in its values. Through preliminary experiments, we decided that 10 LSP coefficients are optimum to measure the speech spectrum for our purpose, and each of the coefficients should be linearly quantized using 4 bits, or 16 levels, for a total of 40 bits per frame.

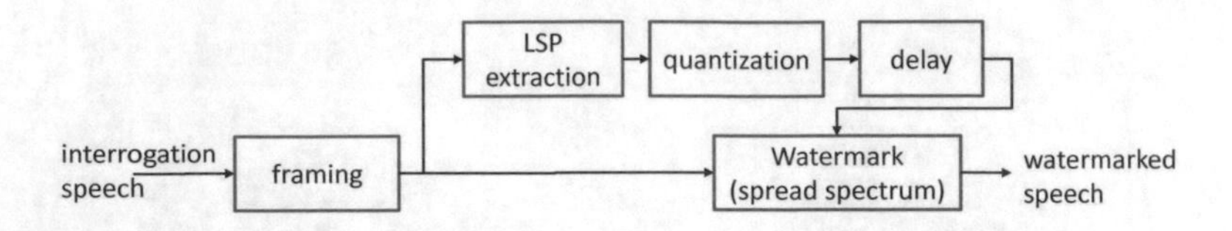

Fig. 2. Block diagram of fingerprinting and watermarking at the recorder

The quantized LSP parameters are embedded in the speech signal using audio watermarks. In this paper, we used the classic direct spread spectrum watermark [2], since this is relatively a standard watermark and is known to be fairly robust to degradations.

We also decided to delay the embedding of the fingerprint by some offset. In other words, the fingerprint was not embedded in the exact same frame. This is shown in Fig. 3. If the fingerprint frame and the watermark frame are offset by some samples, deletion or substitution can be detected by the watermark since these modifications will result in the loss of some samples from which the watermarked fingerprint was extracted. Conversely, if the watermark frame and the fingerprint frame were in the exact same timing, we would not be able to detect substitution since the substituted samples will result in the same fingerprint as the watermarked fingerprint. We arbitrarily used a half frame as the offset in this paper.

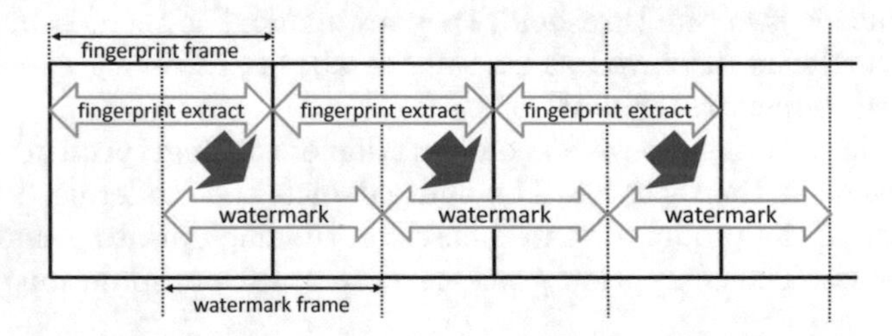

Fig. 3. Fingerprint frame and watermark frame

Figure 4 shows the configuration of the manipulation detection. The speech signal with embedded fingerprint is first synchronized to the frame position, then divided into frames, and the same fingerprinting process is applied. The set of fingerprints in the watermark is also extracted, and compared to the set of fingerprints extracted from the speech signal. The distance between these two sets are then calculated, frame by frame. The distance we used here has the simple Euclidean distance shown below.

$$L = \sqrt{\sum_{i=1}^{10} (l_i - l'_i)^2} \tag{1}$$

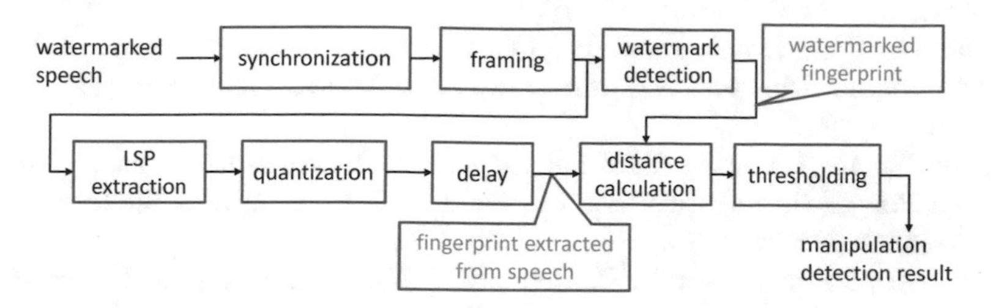

Fig. 4. Block diagram of watermark detection and fingerprint comparison at the detector

Here, l_i is the fingerprint extracted from the speech, and l_i' is the fingerprint extracted from the watermark. Note that both l_i and l_i' were scaled as 16 equi-spaced 16 bit unsigned integers with pre-optimized maximum levels for each order in the distance calculation. The distance is compared to a threshold value. The input speech is judged to be without manipulation if the distance is below this threshold, but decided to be manipulated if above. In the next section, we conducted some experiments to determine this threshold value.

3 Optimization of the Detection Threshold

In this section, we will describe the preliminary experiments we conducted to determine the optimum threshold value. In this experiment, we substituted a segment of the speech utterance with another speaker with the same length. We evaluated the False Negative Ratio (FNR), the False Positive Ratio (FPR), and the Detection Rate (DR) defined below for these speech utterances with various threshold values. We attempted to minimize the FNR and the FPR while maximizing the DR.

$$\mathrm{FNR} = \frac{\text{number of speech utterances decided not to be manipulated}}{\text{number of manipulated utterances}} \tag{2}$$

$$\mathrm{FPR} = \frac{\text{number of utterances decided to be manipulated}}{\text{number of unmanipulated utterances}} \tag{3}$$

$$\mathrm{DR} = \frac{\text{number of utterances decided to be manipulated}}{\text{number of manipulated utterances}} \tag{4}$$

As stated in the previous section, we calculated the distance between the extracted fingerprint from the signal and the extracted fingerprint from the watermark, frame by frame, compare this to the threshold, and decided the utterance as manipulated if any of the frames show distances above the threshold.

We used 300 speech utterances from the ASJ speech database for research [6]. The utterances were read Japanese phonetically-balanced sentences. We used one male and one female speaker, three sets of 50 sentences for a total of 300 read sentences in this test. All were sampled at 16 kHz, 16 bits per sample, and monaural.

Three 50 ms segments were substituted with segments of other speakers. The locations of the substitutions were at 1/4, 1/2 and 2/3 of the length of the utterances, respectively.

Figures 5, 6 and 7 show FPR, FNR, and DR vs. the threshold, respectively. As can be seen in these figures, the FPR can be kept at 0% with a threshold above 200, the FNR can be kept at 0% a threshold below 1000, and the DR can be kept at 100% with a threshold below 1000. Based on these observations, we will use a threshold value of 500 in all further experiments.

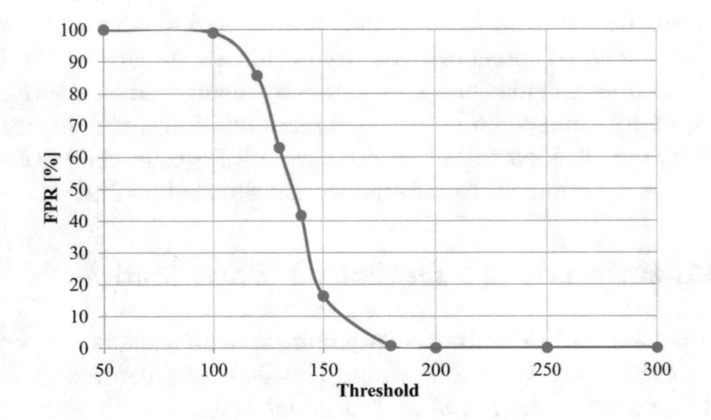

Fig. 5. FPR vs. Threshold

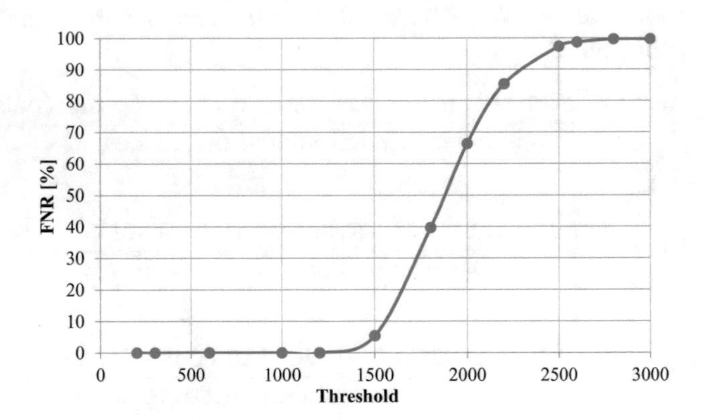

Fig. 6. FNR vs. Threshold

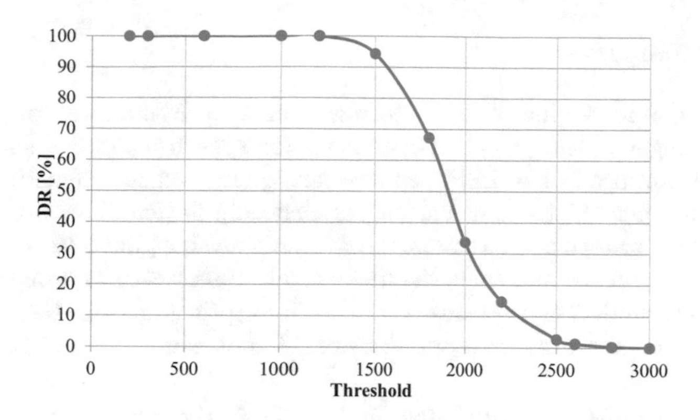

Fig. 7. DR vs. Threshold

4 Manipulation Detection Accuracy Evaluations

We evaluated the detection rate with the proposed detection method and threshold values stated in the previous sections. In this experiment, we used 700 utterances from the ASJ database [6] spoken by 6 speakers, 3 male and 3 female. Two speakers, one per gender, spoke 3 sentence sets, while the others spoke 2 sets, 50 sentences per set. All were phonetically balanced sentences, 16 kHz sampling, 16 bits per sample, monaural.

The watermark used was a standard direct-spread spectrum watermark [2]. We used an M-sequence for the pseudo-random code, and a frame length of 512 samples was used. The chip rate used was 10.

We tested two types of substitution manipulations: substitution of segments of the same speaker, and substitutions of segments from a different speaker. Substitutions were done in the same manner as stated in the previous section, i.e., three substitutions at three locations, at 1/4, 1/2 and 2/3 of the length of the utterances, each in 50 ms segments.

During the manipulation detection, if the distance between the fingerprints exceeds the threshold in at least one frame, then the speech utterance was marked as being manipulated. If all frames are below threshold, then the utterance is marked as manipulation-free.

The following were found from the evaluation tests.

- All utterances with substitutions were correctly marked as manipulated. This was true for substitutions with the same speaker and a different speaker.
- All utterances without substitutions were correctly marked as manipulation-free.

Thus, we can conclude that the proposed method can correctly identify manipulated utterances and manipulation-free utterances, at least in terms of utterances.

5 Conclusion

We proposed and evaluated an interrogation speech manipulation detection method. A fingerprinting method optimized for speech using line spectral pairs (LSPs) was proposed. The LSP was coarsely quantized for robustness to common degradations which are not attempts at manipulation. This fingerprint was embedded as a watermark to the interrogation speech signals. While the speech is being played out for evidence, the fingerprint is extracted and matched to the fingerprint embedded as watermark. If the fingerprints match, then the speech signal can be regarded as manipulation-free. If not, some form of manipulation can be suspected.

We first optimized the threshold for the vector distance between the fingerprint extracted from the speech signal and the watermark. Then, the detection rate as well as the false negative rate was evaluated. We found that at least at the speech utterance level, the method detects all manipulated utterances, and also detects no manipulation in utterances without manipulation, i.e., 100% accurate detection.

In this paper, we have evaluated the detection of substitutions. However, we need to evaluate the detection rate of other forms of manipulations, such as deletions and insertions.

So far, we have only tested with clean speech utterances. However, actual speech, especially when recorded in an interrogation room, may include a significant amount of noise. Since we need to extract the LSP parameters from the speech signal, we also need to test with noisy speech since this may influence the LSP parameters significantly.

Also, we only discriminated manipulation of speech in units of utterances so far. However, since we do calculate the distance frame by frame, it should be possible to discriminate manipulation on a frame by frame basis. We do need to synchronize the frame position if manipulation is applied since this may alter the frame position, consequently affecting the detected position of the manipulated frames and the intact frames.

Acknowledgments. This work was supported in part by the Cooperative Research Project Program of the Research Institute of Electrical Communication, Tohoku University (H26/A14).

References

1. Advisory Panel on White House Tapes: Report on a technical investigation conducted for the U.S. District Court for the District of Columbia by the advisory panel on White House tapes. Technical report, U.S. District Court for the District of Columbia, May 1974
2. Boney, L., Tewfik, A.H., Hamdy, K.N.: Digital watermarks for audio signals. In: Proceedings of IEEE International Conference on Multimedia Computing and Systems. IEEE, Hiroshima (1996)
3. Itakura, F.: Line spectrum representation of linear prediction coefficients of speech signals. J. Acoust. Soc. Am. **57**, 535 (1975)

4. Kukucka, J.: Lights, camera, justice: the value of recording police investigations. The Huffington Post online article, July 2014. http://www.huffingtonpost.com/jeff-kukucka/lights-camera-justice-the_b_5404579.html
5. Kyodo News: Japanese police to tape all interrogations of suspects facing lay judge trials. The Japan Times online article, September 2016. http://www.japantimes.com/news/2016/09/16/crime-legal/japanese-police-tape-interrogations-suspect-facing-lay-judge-trials/
6. NII Speech Resources Consortium: ASJ continuous speech corpus for research. http://research.nii.ac.jp/src/en/ASJ-JIPDEC.html. Accessed 2 Mar 2016
7. Sugamura, N., Itakura, F.: Speech data compression by LSP analysis-synthesis technique. Trans. Inst. Electron. Inf. Commun. Eng. **J64-A**(8) (1981). (in Japanese)
8. Tousignant, L.: The secret of Nixon tape's 18-minute gap revealed. New York Post online article, August 2014. http://nypost.com/2014/08/03/after-40-years-john-dean-re-examines-nixon-tapes-18-minute-gap

Detection of Singing Mistakes from Singing Voice

Isao Miyagawa, Yuya Chiba, Takashi Nose, and Akinori Ito(✉)

Graduate School of Engineering, Tohoku University, Sendai, Japan
{miyagawa, yuya, aito}@spcom.ecei.tohoku.ac.jp,
tnose@m.ecei.tohoku.ac.jp

Abstract. We investigate a method of detecting the wrong lyrics from the singing voice. In the proposed method, we compare the input singing voice and the reference singing voice using dynamic time warping, and then observe the frame-by-frame distance to find the error location. However, the absolute value of the distance is affected by the singer individuality of the reference and input singing voice. Thus, we attempted to adapt the singer individuality into the reference singer's one by a linear transformation. The results of the experiment showed that we could detect the wrong lyrics with high accuracy when the different part of the lyrics was long. In addition, we investigated the effect of iterative linear transformation, and we could not find any benefit from the second or third linear transformations.

Keywords: Singing voice · Lyrics mistake · Dynamic time warping

1 Introduction

In recent years, the karaoke culture is popular among all generations. People go to karaoke with friends or alone to release stress. Besides the backing sounds, the scoring game is very popular as a way of enjoying karaoke.

When singing a song alone, it is difficult to notice the wrong lyrics. If the karaoke machine could point out the singing mistake, it could be a different game, and also it would contribute improving the singing skill of the singer.

There are several studies on the evaluation of the singing voice for karaoke, but most of them are to evaluate musical characteristics or voice quality, such as the accuracy of the pitch and length of the singing voice [1], singing skill including singing technique [2], singing enthusiasm [3], and so on. However, the lyrics are not treated in these studies at all, because it is very difficult to recognize the lyrics from the singing voice [4].

Two types of methods can be considered for detection of wrong lyrics. One is to match the input singing voice and the text information of lyrics. For example, we can make the model of the right lyrics using the Hidden Markov Model and calculate the probability of the input voice [5]. However, singer adaptation is necessary to evaluate the singing voice with high accuracy, which means that the singer-independent mistake detection is difficult. On the other hand, there are "guide vocal" function in some karaoke [6]. This is to play the vocal part of the song by the natural or synthetic voice with accompaniment. By using this function, we can exploit reference singing voice

J.-S. Pan et al. (eds.), *Advances in Intelligent Information Hiding and Multimedia Signal Processing*, Smart Innovation, Systems and Technologies 82,
DOI 10.1007/978-3-319-63859-1_17

with right lyrics and detect error without preparing the text of lyrics. The error detection system is not dependent on the language if we use the guide vocal function. Therefore, we aim at the detection of singing mistake with high accuracy by matching the input singing voice and reference singing voice.

2 Proposed Method

2.1 Overview of the Method

The overview of the proposed method is illustrated in Fig. 1. First, we convert the input and the reference singing voice into sequences of feature vectors, MFCC (Mel Frequency Cepstral Coefficients) in this case. Next, we compare the two MFCC sequences using the DTW (Dynamic Time Warping) based on the Euclidean distance. Finally, we detect mistakes by thresholding the frame-by-frame distances.

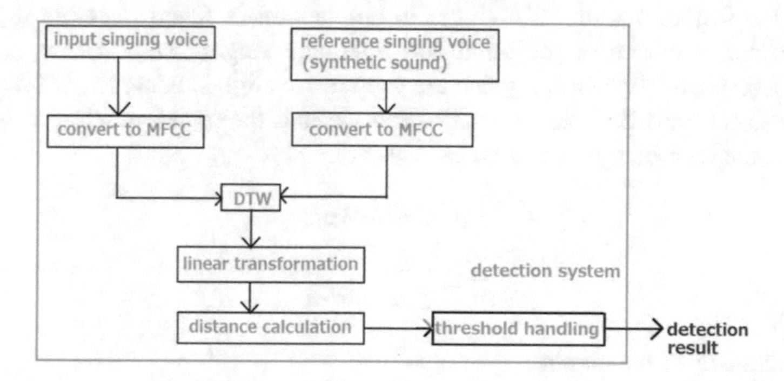

Fig. 1. Overview of the proposed method

2.2 Dynamic Time Warping

The DTW is a nonlinear matching method of two sequences [7]. We define the feature sequences of the input and reference singing voice as $\boldsymbol{x}_1 \ldots \boldsymbol{x}_n$ and $\boldsymbol{y}_1 \ldots \boldsymbol{y}_m$, respectively. Then we calculate the cumulative distance $g(i,j)$ of input voice frame $1 \ldots i$ and the reference voice frame $1 \ldots j$ as follows.

$$d(i,j) = \|\boldsymbol{x}_i - \boldsymbol{y}_i\|^2 \tag{1}$$

$$g(1,1) = d(1,1) \tag{2}$$

$$g(i,j) = d(i,j) + \min \begin{cases} d(i-1,j) + g(i-2,j-1) \\ g(i-1,j-1) \\ d(i,j-1) + g(i-1,j-2) \end{cases} \tag{3}$$

After calculating all the cumulative distances, the optimum correspondence is found by back-tracing the minimum path of the cumulative distance.

3 Linear Transformation and Smoothing

When we associate two sequences by the DTW and determine the distance, the distance of error location is larger when the singer of the input and reference singing voice is the same, but distances of the correct lyrics part is also large when the singers are different. Figures 2 and 3 show the frame-by-frame square distances of the input and reference singing voice after applying the DTW. Both of them are synthetic singing voice. Black color part indicates the wrong lyrics part. The vertical axis is the square distance, and the horizontal axis is the frame number. Figure 2 shows the distance when the singers are the same, and Fig. 3 shows that the gender of the singers is different and the key is 1 octave different. The frame-by-frame distance is large only in the error part in Fig. 2, which is clearly different from the correct part. However, we can't discriminate the correct and error parts in Fig. 3 because the frame-by-frame distance of the correct part is larger. This example suggests that we cannot detect the error location using only the DTW. Therefore, we applied a method based on the linear transformation of the feature to adapt the feature vectors of a singer to another one's feature vectors [8]. Let the feature vector sequence of the input and reference singing voice as $(\boldsymbol{x}_1 \ldots \boldsymbol{x}_n)$ and $(\boldsymbol{y}_1 \ldots \boldsymbol{y}_n)$, respectively, assuming that the vectors are aligned using the DTW. Then we apply the linear transformation $\boldsymbol{x}'_i = A\boldsymbol{x}_i$, and we find the transformation matrix A that minimizes the sum of squares of error vector $\boldsymbol{ei}$.

$$\boldsymbol{e}_i = \boldsymbol{y}_i - A\boldsymbol{x}_i \tag{4}$$

$$Z = \sum_{i=1}^{n} \|\boldsymbol{e}_i\|^2 \tag{5}$$

To minimize Z, we obtain

$$A = C_{yx} C_{xx}^{-1} \tag{6}$$

$$C_{xx} = \sum_{i=1}^{n} \boldsymbol{x}_i \boldsymbol{x}_i^T, C_{yx} = \sum_{i=1}^{n} \boldsymbol{y}_i \boldsymbol{x}_i^T \tag{7}$$

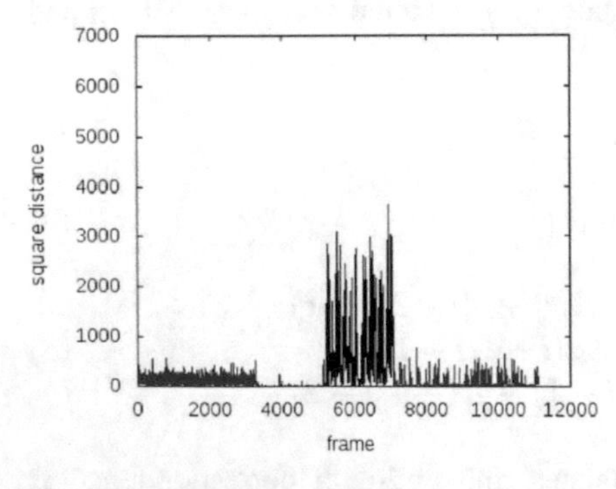

Fig. 2. Square distance, woman-woman (The same singer)

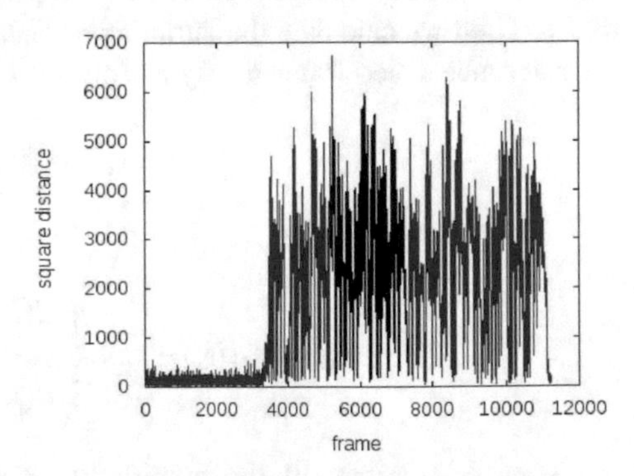

Fig. 3. Square distance, woman-man (1 octave below)

This equation assumes that $\boldsymbol{x}$ and $\boldsymbol{y}$ are singing voices with exactly same lyrics, but a lyrics error is included in a real singing. At the time of linear transformation, it is inappropriate to use the parts with lyrics errors because those parts have incorrect associations between different phonemes. Therefore, we exclude the error-like parts, which have larger distances than the pre-defined threshold.

Furthermore, we apply the linear transformation iteratively. In the first iteration, we roughly estimate the transformation matrix. Then we improve the estimation using correct lyrics part in the second iteration. By repeating DTW and linear transformation, we can reduce the distances of the correct parts while keeping the distance of the error location large.

An example of square distance after linear transformation is shown in Fig. 4. This figure corresponds to Fig. 3. The DTW and the linear transformation are iterated three times, and the threshold is set to 3000. When comparing Fig. 4 with Fig. 3, the distances of the error part around 5500th–7000th frame are larger than the other parts. From this example, it can be seen that the iteration of the DTW and linear transformation is effective. The threshold of linear transformation was determined as 1.5 time of the average square distance according to the result of the preliminary experiment.

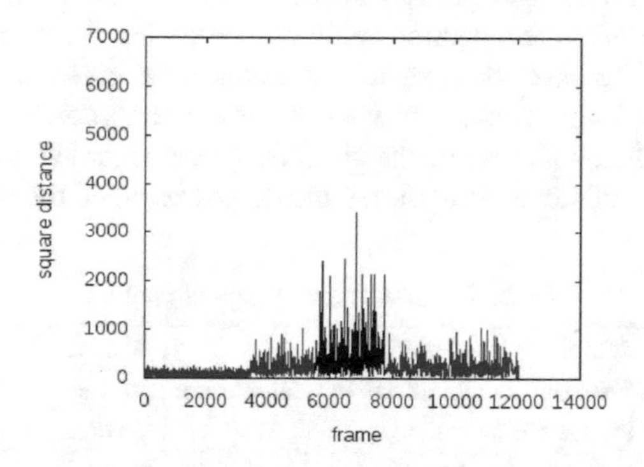

Fig. 4. Square distance after linear transformation, woman-man (1 octave below)

After transforming the vectors and calculating the square distances frame by frame, we apply the moving average filter to the sequence of the distances to remove isolated peaks. The length of the filter should be optimized in the following experiment.

4 Experiment

4.1 Detection Experiment of the Error Lyrics

We conducted an experiment to detect lyrics mistakes to investigate whether the proposed method could accurately detect the lyrics error. Singing voices of eleven male students were used as the input singing voice. We examined two songs, "*Sekaini hitotsudakeno hana*" and "*Soramo toberu hazu*". 11 male singers sang the correct lyrics.

Reference singing voices are synthesized using Vocaloid (KAITO V3, male voice). We prepared three reference singing voices: "Correct", "With different phrase", "With different words or syllables". Thereafter we call "With different phrase" and "With different words or syllables" as "large error" and "small error" voices, respectively. Here, we simulated the singing mistakes by changing lyrics of the reference singing voice. A reference voice with errors contains multiple lyrics mistakes. Eight "large error" parts and nine "small error" parts are included in the reference singing voices, 17 error parts in total. There were six "large error" parts in the "*Sekaini hitotsudakeno hana*", and two "large error" and nine "small error" parts in the "*Soramo toberu hazu*". Since we examined the voices from 11 singers, we have 187 errors to be detected.

Sound analysis conditions are shown in Table 1. Note that the temporal derivatives of MFCC (Δ and $\Delta\Delta$) were not used since it was found that those parameters degraded the detection performance from the results of the preliminary experiments. The DTW and linear transformation were iterated three times. After that, we smoothed the distance using the moving average filter.

After smoothing the distances, we detected the parts with large distances as the mistaken parts. The detection was performed by comparing the distance with the threshold, which was α times larger than the average of all of the smoothed distances. We judged the correctness of the detection result by comparing the detection result with the original error part. Because the detected mistakes were given as frame sections, we regarded the detected part as correct when the detected part and the labeled error part intersected each other. Otherwise, the detection was regarded as a false alarm. By changing the number of smoothing frames and α, we examined the change of F-measure.

Table 1. Conditions of sound analysis

Sampling frequency	16 kHz
Window length of MFCC calculation	25 ms
Frame shift	10 ms
Window function	Hamming
Order of MFCC	13 (+power)

The F-measure with respect to the smoothing frame width and α is shown in Fig. 5. The maximum value of F-measure is 0.61, which is not high enough. Therefore, we investigated the detection results of the large errors and the small errors individually. The maximum F-measure of the large errors is 0.90, whereas that of the small errors is 0.23. From these results, the small errors are found to be difficult to detect.

4.2 The Number of Iterations of the Linear Transformation

The DTW and linear transformation were iterated three times in the previous experiment but the optimum number of iterations was not clear. Therefore we conducted an experiment to change the number of iterations of the linear transformation. The result is shown in Fig. 6. From the figure, we found that the maximum value of F-measure did not change much with respect to the number of iterations. Therefore, only one linear

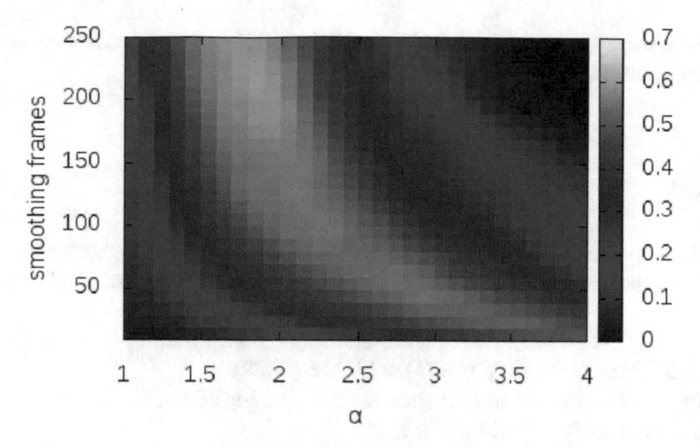

Fig. 5. The result of F-measure

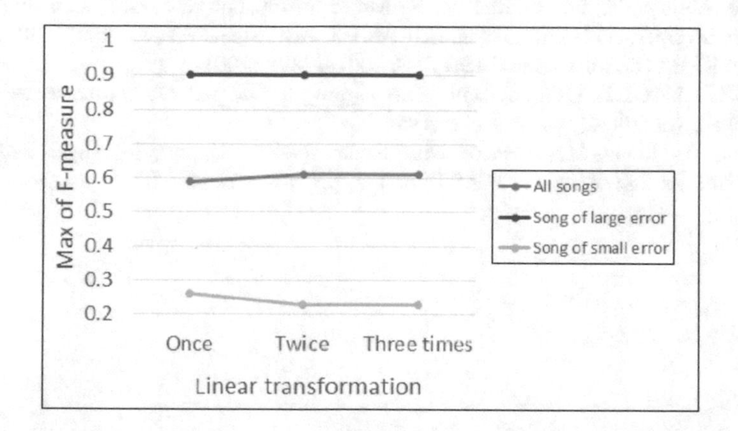

Fig. 6. The number of times the linear transformation and the F-measure

transformation was considered to be enough. Note that the transformation in this experiment was conducted between the same gender; the optimum number of iterations for cross-gender transformation should be investigated.

5 Conclusion

In this research, we investigated a method of detecting the wrong lyrics from the singing voice for karaoke. The proposed method utilizes the guide vocal as the reference, and detects the singing error parts using the DTW. The linear transformation is used to adapt the singer's individuality to the reference voice. The result of the experiment showed that we could detect the wrong lyrics with high accuracy when the wrongly sang part was long, but it was difficult to detect mistakes with one phoneme or one syllable. Also, we found that iteration of the linear transformation was not needed for singer adaptation.

References

Takeuchi, H., Hoguro, M., Umezaki, T.: A KARAOKE system singing evaluation method that more closely matches human evaluation. Trans. Inst. Electr. Eng. Jpn. C **130**(6), 1042–1053 (2010)

Nakano, T., Goto, M., Hiraga, Y.: An automatic singing skill evaluation method for unknown melodies. Inf. Process. Soc. Jpn. **48**(1), 227–236 (2007)

Daido, R., Ito, M., Makino, S., Ito, A.: Automatic evaluation of singing enthusiasm for karaoke. Comput. Speech Lang. **28**, 501–517 (2014)

Mesaros, A., Virtanen, T.: Automatic recognition of lyrics in singing. EURASIP J. Audio Speech Music Process. **2010**, article No. 4 (2014)

Suzuki, M., Hosoya, T., Ito, A., Makino, S.: Music information retrieval from a singing voice using lyrics and melody information. EURASIP J. Adv. Signal Process. **2007**, 038727 (2006)

Panasonic: KARAOKE machine, Patent JP-A-2001-42879 (2001)

Berndt, D.J., Clifford, J.: Using dynamic time warping to find patterns in time series. In: KDD Workshop, vol. 10(16), pp. 359–370 (1994)

Matsumoto, H., Inoue, H.: A piece wise linear special mapping for supervised speaker adaptation. In: Proceedings of ICASSP, vol. 1, pp. 449–452 (1992)

A Study of Audio Watermarking Method Using Non-negative Matrix Factorization for a Duet of Different Instruments

Harumi Murata[1(✉)] and Akio Ogihara[2]

[1] Department of Information Engineering, School of Engineering, Chukyo University, Toyota, Japan
murata_h@sist.chukyo-u.ac.jp
[2] Department of Informatics, Faculty of Engineering, Kindai University, Higashi-hiroshima, Japan

Abstract. In this paper, we propose an audio watermarking method using non-negative matrix factorization (NMF) for a duet of different instruments. NMF is applied to host signal and the amplitude spectrogram is factorized into the basis matrix and the activation matrix. In the proposed method, the activation coefficients of each musical instrument can be obtained by NMF, and watermarks can be embedded into a specific instrument or multiple instruments. Onset and offset times are estimated from the activation matrix, and its duration is defined as one note. However, it is possibility that onset and offset times are changed due to embedding process, and watermarks cannot be extracted correctly unless note information is known. Therefore, in this paper, onset and offset times are estimated after smoothing activation coefficients, and errors of note estimation are reduced. From this, we want to extract watermarks even if note information is unknown.

Keywords: Audio watermarking · Nonnegative matrix factorization · Duet

1 Introduction

Digital watermark is a technique to embed other digital data into digital content such as music, images, and videos. For audio signals, the sound quality of the stego signal should not deteriorate. With current methods [1], high sound quality means that the difference between the host and stego signals is small. In these methods, watermarks are embedded by operating the components of the host signal. Therefore, the noise resulting from embedding watermarks tends to be perceived as an annoying sound.

However, host signals are not always shown to users in actual systems that apply information hiding technology; in fact, it is believed that host signals are not shown in many cases. Therefore, there is no problem even if another sound,

J.-S. Pan et al. (eds.), *Advances in Intelligent Information Hiding and Multimedia Signal Processing*, Smart Innovation, Systems and Technologies 82,
DOI 10.1007/978-3-319-63859-1_18

not in the host signal, is perceived when the sound quality of the stego signal is musical. In this case, we can regard the sound quality of the stego signal as high in this paper.

Accordingly, we have proposed an audio watermarking method using nonnegative matrix factorization (NMF) [2]. NMF is an algorithm that factorizes a nonnegative matrix that is an amplitude spectrogram of a target signal into two nonnegative matrices that correspond to the spectral patterns of the target signal and the activation of each spectrum. A group of spectral patterns of the target signal and the intensity variation of each spectral pattern are represented with the basis matrix and the activation matrix, respectively. In this study, watermarks are embedded by modifying the activation coefficients which are obtained by NMF. Furthermore, in this paper, host signal is extended to a duet for different instruments. The activation coefficients of each musical instrument can be obtained, and watermarks can be embedded into a specific instrument or multiple instruments.

However, it is possibility that watermarks cannot be extracted correctly in case that note information varies between embedding and extracting processes. Hence, in [2], note information is known when watermarks are extracted. Generally, watermarks should be extracted by blind detection. Therefore, in this paper, onset and offset times are estimated after smoothing activation coefficients, and errors of note estimation are reduced.

2 Nonnegative Matrix Factorization (NMF)

NMF is one of the techniques used for separation of an audio mixture that consists of multiple instrumental sources [3]. The following equation represents the factorization of a simple NMF.

$$\boldsymbol{Y} \simeq \boldsymbol{A}\boldsymbol{B}, \tag{1}$$

where $\boldsymbol{Y}$ is an observed nonnegative matrix, which represents the time-frequency amplitude spectral components obtained via short-time Fourier transform (STFT), and $\boldsymbol{A}$ and $\boldsymbol{B}$ are nonnegative matrices. In addition, the matrix $\boldsymbol{A}$ is called basis matrix which represents spectral patterns of observed spectrogram $\boldsymbol{Y}$, and $\boldsymbol{B}$ is called the activation matrix which involve activation information for $\boldsymbol{A}$.

Moreover, the multiplicative update algorithms of standard NMF based on Euclidean distance are shown in Eqs.(2) and (3).

$$a_{m,k} = \frac{[\boldsymbol{Y}\boldsymbol{B}^{\mathrm{T}}]_{m,k}}{[\boldsymbol{A}\boldsymbol{B}\boldsymbol{B}^{\mathrm{T}}]_{m,k}} a_{m,k}, \tag{2}$$

$$b_{k,n} = \frac{[\boldsymbol{A}^{\mathrm{T}}\boldsymbol{Y}]_{k,n}}{[\boldsymbol{A}^{\mathrm{T}}\boldsymbol{A}\boldsymbol{B}]_{k,n}} b_{k,n}, \tag{3}$$

where $a_{m,k}$ and $b_{k,n}$ are the coefficients of matrices $\boldsymbol{A}$ and $\boldsymbol{B}$.

3 An Audio Watermarking Method Using NMF for a Duet of Different Instruments

The host signal is divided into frames of L samples and is transformed to the STFT domain with 50% overlap between successive frames. The amplitude spectrogram $\boldsymbol{Y}(\in \mathcal{R}^{\Omega \times T})$ is factorized into the basis matrix and the activation matrix. In this paper, we use two kinds of instrumental signals having similar timbre to sounds included in the host signal as training signals. Therefore, the basis matrices $\boldsymbol{H}(\in \mathcal{R}^{\Omega \times L})$ and $\boldsymbol{F}(\in \mathcal{R}^{\Omega \times K})$ are obtained by prior learning.

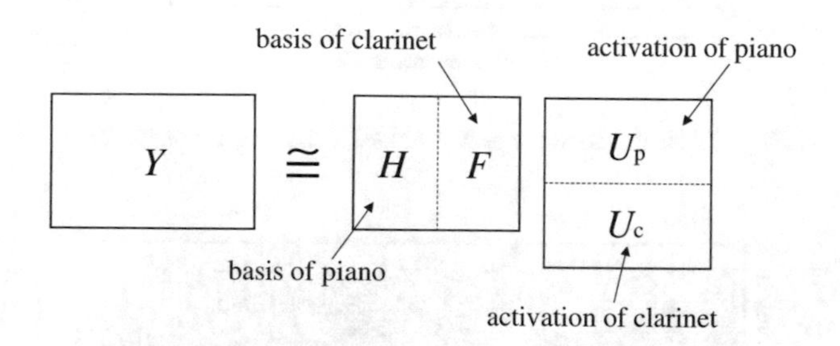

Fig. 1. An example of applying NMF to a duet of piano and clarinet.

Figure 1 shows an example of applying NMF to a duet of piano and clarinet. The amplitude spectrogram $\boldsymbol{Y}$ is factorized as follows.

$$\boldsymbol{Y} \simeq \boldsymbol{H}\boldsymbol{U}_{\mathrm{p}} + \boldsymbol{F}\boldsymbol{U}_{\mathrm{c}} \tag{4}$$

where, $\boldsymbol{H}$ and $\boldsymbol{F}$ are the basis matrices of piano and clarinet, and $\boldsymbol{U}_{\mathrm{p}}$ and $\boldsymbol{U}_{\mathrm{c}}$ are the activation matrices of them. In this paper, watermarks are embedded into instrumental sound of playing the melody. In this case, the accompaniment part is played in piano and the melody part is played in clarinet, and we embed watermarks into the clarinet's activation matrix.

Next, onset and offset times of each clarinet's note are estimated from the obtained activation matrix $\boldsymbol{U}_{\mathrm{c}}$. The duration between the estimated onset time and offset time is defined as one note, and one-bit watermark is embedded into one note. The estimated notes are regarded as root notes, and we modify the activation coefficients of root note for embedding watermarks.

It is possibility that onset and offset times are changed due to embedding process. In Fig. 2, onset time of the third note is close to that of the fourth note. The order of embedding and extracting is changed in case that onset time of the fourth note is estimated to be faster than that of the third note because the activation coefficients are modified due to the embedding process. Accordingly, watermarks cannot be extracted correctly. Therefore, watermarks are embedded and extracted in order of pitch like Fig. 3. Moreover, NMF is applied to every fixed section, and watermarks can be extracted even if the stego signal is clipped. Here, watermarks are embedded and extracted per 15 s based on the evaluation criteria for audio information hiding technologies [4].

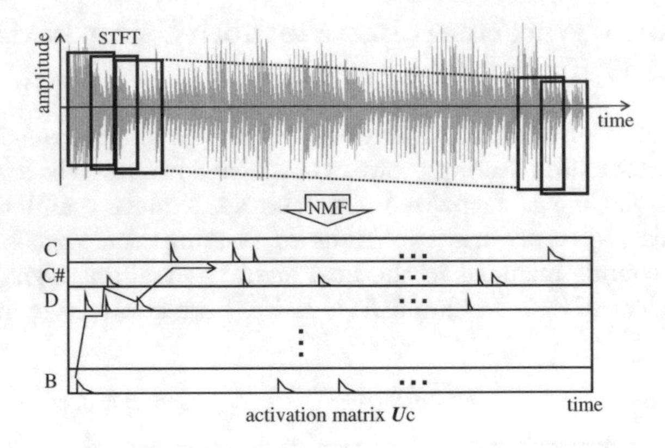

Fig. 2. Embedding watermarks in order of onset time.

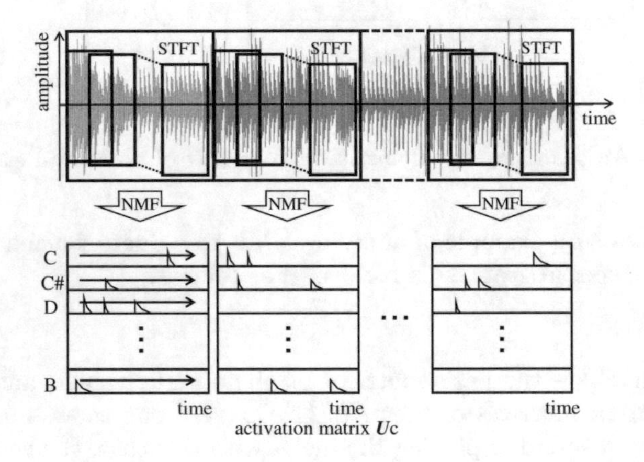

Fig. 3. Embedding watermarks in order of pitch.

3.1 Note Estimation

Because activation coefficients include a lot of unnecessary extreme values in order to estimate onset and offset times, it is necessary to smooth activation coefficients before note estimation. Therefore, the down-sampling of 1/2 times is performed with respect to activation coefficients, and spline interpolation is performed with respect to the down-sampled activation coefficients.

After smoothing activation coefficients, local maximum values of coefficients are searched. Here, if a local maximum value is more than p times global maximum value, it is regarded as note. Next, local minimum values are searched in order to estimate onset and offset times of each note. If the local minimum value adjacent to a local maximum value is regarded as onset time or offset time, its value might be not fully attenuated as shown in Fig. 4. Therefore, if the local minimum values are less than q times local maximum value, the closest

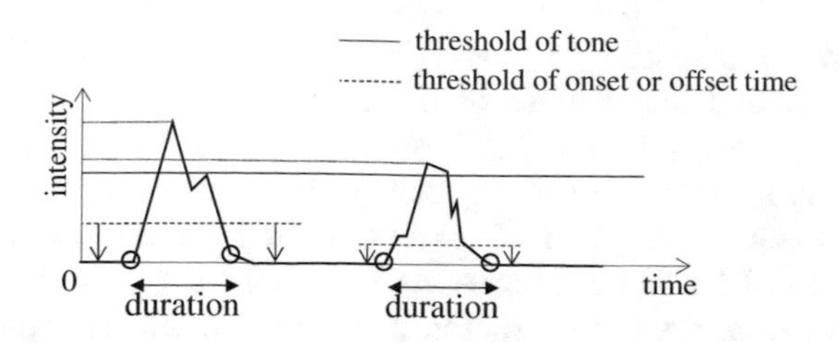

Fig. 4. Determination of onset time and offset time.

minimum values to a local maximum value are regarded as onset time and offset time. Here, p is less than 1 and q is a small fraction close to zero.

3.2 Embedding and Extracting Watermarks

Before embedding watermarks, a quantization level of activation coefficients is made coarse in order to improve the tolerance against attacks. In this method, activation coefficients are re-quantized in 10 steps.

For embedding watermarks, we use another activation matrix $\boldsymbol{U}_{\mathrm{w}}(\in \mathcal{R}^{K \times T})$ for embedding watermarks. Initially, $\boldsymbol{U}_{\mathrm{w}}$ is an empty matrix.

Embedding rules and modification of $\boldsymbol{U}_{\mathrm{w}}$

- the condition of a watermark bit “0”
  We define the condition of a watermark bit “0” if the maximum activation coefficient of root note is odd.
- the condition of a watermark bit “1”
  We define the condition of a watermark bit “1” if the maximum activation coefficient of root note is even.

If the condition of an embedding watermark bit and the maximum activation coefficient of root note is not equal, t times activation coefficients of $\boldsymbol{U}_{\mathrm{c}}$ corresponding to root note are added in $\boldsymbol{U}_{\mathrm{w}}$ corresponding to same section. Here, t is a scale factor such that the embedding condition is satisfied.

Reconstruction of Stego Signal. After embedding watermarks, the amplitude spectrogram is obtained using the basis matrix $\boldsymbol{F}$ and the activation matrix $\boldsymbol{U}_{\mathrm{w}}$.

$$\boldsymbol{Y}' = \boldsymbol{Y} + \boldsymbol{F}\boldsymbol{U}_{\mathrm{w}} \tag{5}$$

The amplitude spectrogram $\boldsymbol{Y}'$ is inversed short-time Fourier transformed and the stego signal is obtained. However, phase information of each frequency uses the same as the host signal.

Extracting Watermarks. As a same manner as the embedding process, the stego signal is divided into frames of L samples per 15 s and is transformed to the STFT domain with 50% overlap between successive frames. NMF is applied to the amplitude spectrogram $\boldsymbol{Y}'$ by using the basis matrices $\boldsymbol{H}$ and $\boldsymbol{F}$, and the activation matrices $\boldsymbol{U}'_{\mathrm{p}}$ and $\boldsymbol{U}'_{\mathrm{c}}$ are factorized. Onset and offset times are estimated by Sect. 3.1 and root notes are determined. For an instrument of the melody part, watermarks are extracted in order of pitch. If the maximum activation coefficient of $\boldsymbol{U}'_{\mathrm{c}}$ corresponding to root note is odd, a watermark bit "0" is extracted. Otherwise, a watermark bit "1" is extracted.

4 Experiments

To confirm the validity of the proposed method, we conducted an evaluation of its tolerance against various attacks based on [4] and an evaluation of sound quality. For testing, we used 10 pieces of music selected from [5] of 60 s duration at a 44.1 kHz sampling rate, 16-bit quantization and the stereo channel. These are duets of piano and clarinet. The embedding watermark bits used the payload [4]. The embedding parameters were set $L = 2048$, $p = 0.25$, $q = 10^{-4}$, respectively. Iteration number of NMF was 200 in embedding and extracting processes. Moreover, the supervised signals of basis matrices were used 36 notes of a single piano sound from octave 3 to octave 5 and 36 notes of a single clarinet sound from octave 4 to octave 6 made by Cubase Artist7.

From the experimental results, the average embedding capacities of watermarks was 4.28 bps. It is difficult to say that embedding capacity is enough, because 6 bps embedding capacity is required in [4]. However, it seems that embedding capacity can be increased by embedding watermarks into piano part.

4.1 Sound Quality Evaluation

We evaluated the objective sound quality by PEAQ. PEAQ uses some features of both host and stego signals and represents the quality comparison result as ODG. The ODG ranges from 0 to −4, with higher values indicating greater watermark transparency. Table 1 shows the ODG results, which are calculated as follows.

(A) ODGs between the original PCM host signals and stego signals. The ODG values should be more than −2.5.
(B) ODGs between the original PCM host signals and MP3-coded stego signals. The arithmetic mean of the 20 ODGs should be more than −2.0.

From the results, ODGs of (A) and (B) satisfied the criteria.

Moreover, we conducted mean opinion score (MOS) for the subjective evaluation of sound quality involving seven test subjects. Test subject listened to the stego data of propose method. A test subject scored the sound on a scaled from 1 to 5, a score of 1 being lowest quality and a score of 5 begin highest quality. The average score of all stego data was 3.98, and the subjective sound quality was high.

Table 1. The results of ODG: (A) comparing host signals to stego signals, (B) comparing host signals to degraded signals.

Music number	ODG	
	(A)	(B)
1	−0.858	−1.297
2	−1.124	−1.672
3	−0.356	−1.149
4	−0.376	−1.113
5	−0.467	−1.039
6	−0.342	−1.003
7	−0.399	−1.073
8	−0.916	−1.510
9	−0.334	−1.188
10	−0.563	−1.096
Average	−0.574	−1.214

4.2 Tolerance Against Attacks

We examined the tolerance against the following attacks.

- MP3 128 kbps (joint stereo)
- A series of attacks that mimic D/A and A/D conversions
- MP3 128 kbps (joint stereo) tandem coding
- MPEG4 HE-AAC 96 kbps
- additive Gaussian noise overall SNR 36 dB
- bandpass filterling 100 Hz – 6 kHz, −12 dB/oct.

Bit error rate (BER) was calculated by the following equation

$$\mathrm{BER} = \frac{\text{number of error bits}}{\text{number of embedding bits}} \cdot 100\ [\%] \tag{6}$$

The BER must be less than 10 % according to the criteria.

Figure 5 shows the results of BER. In case of no attack, note information was same in embedding and extracting processes, and watermarks could be extracted correctly. Moreover, tolerance against additive Gaussian noise satisfied the criteria. As the reason tolerance against other attacks did not satisfy the criteria, note was estimated incorrectly in the extracting process, and all of watermarks after error note were not extracted correctly. Therefore, it seems that note estimation should be more robustness.

In case that note information is known, tolerance against AAC, additive Gaussian noise and bandpass filtering satisfied the criteria. Moreover, BERs of 7 pieces of music were less than 10% against MP3. It seems that tolerance improvement can be expected if onset and offset times can be estimated correctly.

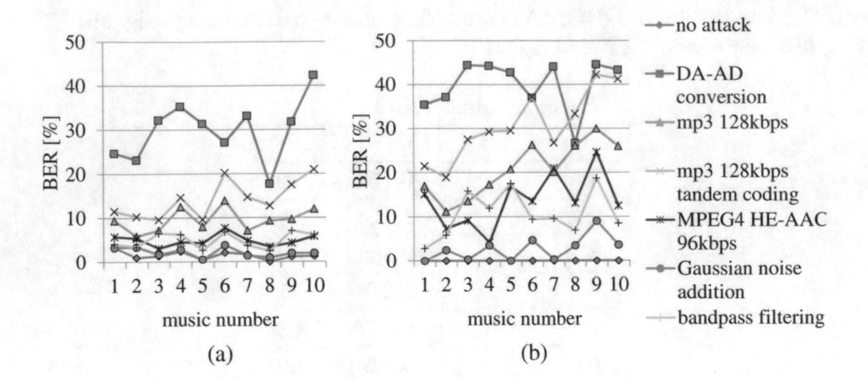

Fig. 5. The results of BER: (a) in case that note information is known, (b) in case that note information is unknown

5 Conclusions

In this paper, we proposed an audio watermarking method using NMF for a duet of different instruments. From the experimental results, sound quality of stego signals was high, but tolerance against attacks was low in case that note information was unknown. Therefore, we make note estimation more robustness and tolerance against attacks should be improved.

References

1. Lie, W.N., Chang, L.C.: Robust and high-quality time-domain audio watermarking based on low-frequency amplitude modification. IEEE Trans. Multimedia **8**(1), 46–59 (2006)
2. Murata, H., Ogihara, A.: Digital watermark for real musical instrument sounds using non-negative matrix factorization. In: International Technical Conference on Circuits/Systems, Computers and Communications, pp. 677–680 (2016)
3. Lee, D.D., Seung, H.S.: Algorithms for non-negative matrix factorization. Neural Inf. Process. Syst. **13**, 556–562 (2001)
4. IHC Evaluation Criteria and Competition, http://www.ieice.org/iss/emm/ihc/IHC_criteriaVer5.pdf, Accessed March 30 2017
5. Blancou, V., Fujii, K., Okada, T.: 40 Etudes for Clarinet and Piano. UNIVERSITY EDUCATION PRESS (2006)

A Wind Noise Detection Algorithm for Monitoring Infrasound Using Smartphone as a Sensor Device

Ryouichi Nishimura[1(✉)], Shuichi Sakamoto[2], and Yôiti Suzuki[2]

[1] Resilient ICT Research Center, National Institute of Information and Communications Technology (NICT), Sendai, Japan
ryou@nict.go.jp

[2] Research Institute of Electrical Communication, Tohoku University, Sendai, Japan

Abstract. Infrasound monitoring is promising for early warning systems to mitigate damage of disaster. However, wind noise contains the same frequency components as infrasound does, and they need to be separated. To achieve this purpose, a wind noise detection algorithm is proposed. Unlike conventional methods that typically use two microphones, the proposed method assumes that one pressure and one acoustic sensor is available. This assumption comes from a requirement that a smartphone is used as a sensor device. Wind noise is detected as anomaly detection of the microphone signal, using extreme value distribution. Comparing with the data obtained by an anemometer, it is shown that the proposed method successfully determines time periods where wind noise exists under a practical environment, depending on the condition of wind.

1 Introduction

When the sound is recorded outdoors, it is often contaminated with wind noise. A windscreen is used to alleviate bad effects of such noise but in some applications, this approach is not applicable. For example, it is difficult to find a windscreen especially designed for smartphones, mobile phones or hearing aids. This is mainly because of the limitation in shape and space of the vicinity of device. Signal processing is an alternative approach to reduce the noise when a physical approach is not applicable. An algorithm to detect wind noise is required as a pre-process of such signal processing. A typical algorithm compares samples of two microphones having different sensitivity to wind noise [8]. Other algorithms evaluate similarity of samples of two microphones because wind noise generally shows low similarity between them while meaningful sound such as speech and music shows high similarity [1]. Based on this concept, J.A. Zakis and C.M. Tan has recently proposed a novel method [9]. It counts positive and negative samples of each of the signals captured by two microphones and evaluates the similarity between them by means of χ^2-test. All of these methods assume that two closely situated microphones are available. Therefore, they would become

J.-S. Pan et al. (eds.), *Advances in Intelligent Information Hiding and Multimedia Signal Processing*, Smart Innovation, Systems and Technologies 82,
DOI 10.1007/978-3-319-63859-1_19

unavailable when only one microphone is available. This paper addresses the problem of wind noise detection for such a situation that a smartphone is used as a sensor device for infrasound monitoring.

Infrasound is sound whose frequency is lower than the human audible range. Its radiation is sometimes observed in conjunction with large geographical activities such as volcanic eruption, earthquakes, tsunamis, avalanches and so on [7]. Other than that, it is also observed in conjunction with artificial activities such as launching of rockets, subsonic aircraft, satellite or spacecraft re-entrance and so on. Therefore, observation of infrasound is promising for developing early warning systems of such event that may put our lives in peril. Infrasound inherently constitutes of low frequency components but wind noise also has dominant power in the same frequency range. Therefore, it is necessary to discriminate infrasound and wind noise when a low frequency signal is observed. It is possible to perform this process by human vision or hearing but it would be a tiresome task because monitoring should be performed 24 h a day, and the huge number of sensors will be required for infrasound monitoring. Therefore, automatic discrimination is important.

2 Wind Noise

2.1 Characteristics of Wind Noise

We think that there are two types of wind noise. One is turbulence generated as the wind is passing through an interfering object. The other is generated by pushing a microphone directly with the wind itself. We mainly focus on the latter type of wind noise in this paper. Such wind causes an abrupt and strong change in atmospheric pressure at the microphone. Therefore, we assume that the wind noise of the signal resembles an output of a step size input to the sensor. A step function is represented as

$$u(t) = \begin{cases} 1 & t \geq 0 \\ 0 & t < 0 \end{cases}, \tag{1}$$

and its Fourier transform is

$$\mathrm{F}[u(t)] = \frac{1}{j\omega} + 0.5\delta(\omega). \tag{2}$$

Eq. (2) suggests two interesting properties of the wind noise:

1. It would have a power spectrum spread over the whole frequency range.
2. All the frequency components would be in phase.

Taking these two properties and AC coupling of usual microphone into account, we can expect an abrupt increase in amplitude at the portions where wind noise happens. Consequently, it is considered that an algorithm for wind noise detection can be realized as anomaly detection in amplitude of the signal.

2.2 Anomaly Detection

There are many research works on anomaly detection in diverse research fields. For example, it has recently become popular in the research field of Internet security in order to detect malicious attacks [2,6].

A basic approach is based on entropy and its derivatives. Entropy H in statistics is defined by

$$H = \mathrm{E}[-\log p(x)], \tag{3}$$

where $p(x)$ is a probability and E is the expectation operator. When the entropy of an event is high, it means that this event unlikely happens. Because a decision of wind noise needs to be made frame by frame of the time period, it would be better to use a probability distribution directly instead of entropy. The proposed method uses cumulative distribution function instead of probability distribution because only abnormal *increase* in amplitude is interesting. Accordingly, the signal of the frame is considered as wind noise when

$$\mathrm{cdf}(b) \geq q_{\mathrm{th}} \tag{4}$$

holds, where $\mathrm{cdf}(b)$ is a cumulative distribution function, b is the maximum amplitude of the frame, and q_{th} is a threshold which is typically 0.95 or 0.99.

2.3 Acoustic Sensor (microphone)

We employed extreme value distribution to form the cumulative distribution function of Eq. (4). This is because we want to test the maximum amplitude of each frame. The type I extreme value distribution, also known as Gumbel distribution, is defined as

$$p(x|\mu, \theta) = \exp\left\{\exp\left(-\frac{x-\mu}{\theta}\right)\right\}, \tag{5}$$

where μ is a parameter that shifts the distribution and θ controls the scale. This distribution is often used as an approximation to model the maxima of a long period in various research fields such as meteorology, finance and risk management [3,4]. In this study, we approximately apply this function to the maxima for relatively short time periods.

To obtain the envelope of the signal, an analytic signal of the observed signal is first derived and then its absolute values are calculated. The envelope signal are segmented into frames of a specified length and for each time frame, the maximum value is extracted. This process is repeated for a specified period, typically a day. Fitting to the extreme value distribution is then performed. As is seen in Eq. (5), since the extreme value distribution has two parameters, the maximum likelihood estimates of these parameters are obtained by fitting. The proposed algorithm assumes that the frame where Eq. (4) with $q_{\mathrm{th}} = 0.95$ holds is regarded as presumably containing wind noise. The block diagram of the wind noise detection using acoustic sensor input is depicted in Fig. 1.

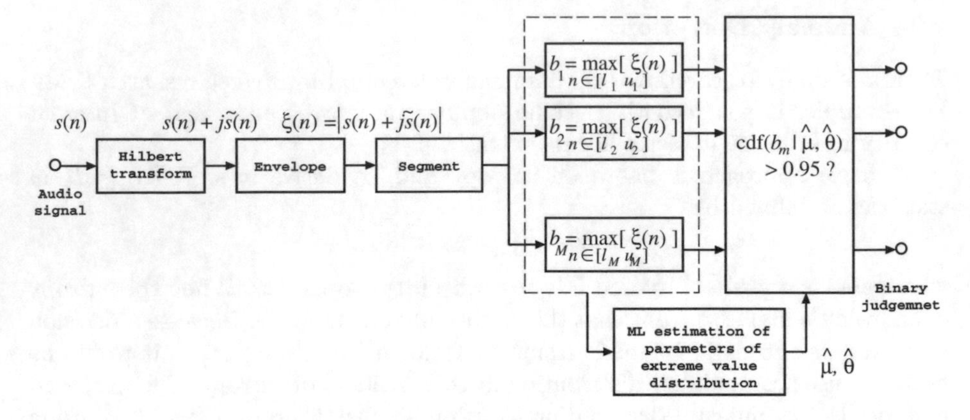

Fig. 1. Block diagram of wind noise detection with acoustic sensor. In this figure, l_i and u_i are respectively lower and upper temporal bounds of the i-th segment.

2.4 Pressure Sensor

Wind noise affects a pressure sensor signal in a similar manner, and the methodology employed for the acoustic sensor signal can also be applied to it. However, the main purpose of this sensor is to capture infrasound. Therefore, some modification was made in the algorithm. The signal is segmented with a constant frame length. For each frame, discrete Fourier transform is calculated and the power spectrum is derived. Select only frequency bins corresponding to infrasound; the frequency range of interest was set to 0.003 to 1 Hz in this study. Distribution of the power is evaluated every frequency bin within this frequency range. It is noteworthy that the power is not in decibel but in amplitude. Fitting of the data to Gamma distribution is finally performed. Gamma distribution is defined as

$$p(x|k,\theta) = x^{k-1}\frac{e^{-x/\theta}}{\Gamma(k)\theta^k}, \tag{6}$$

where $\Gamma(k)$ is a Gamma function of order k, and θ is a scale parameter. When the time period where the probability of the low-frequency power is significantly unlikely, namely anomaly, it is regarded as containing either or both of infrasound components and wind noise.

2.5 Acoustic and Pressure Sensors

Combining the acoustic and pressure sensors, we can discriminate the time period containing infrasound and that containing wind noise. A common acoustic sensor is insensitive to infrasound because of its available frequency range. On the other hand, a pressure sensor is sensitive to both infrasound and wind noise. Therefore, the frame where both the acoustic and pressure sensor signals show anomaly is regarded as wind noise. In contrast, the frame where the acoustic sensor signal shows anomaly while the pressure sensor signal does not is regarded as general sound such as speech and music. Similarly, the frame where the pressure sensor signal shows anomaly while the acoustic sensor signal does not is regarded as infrasound we are interested in.

This classification can be summarized in Table 1.

Table 1. Classification of the signal based on the positive/negative decision of anomaly using both acoustic and pressure sensors

		Acoustic	
		Positive	Negative
Pressure	Positive	Wind noise	Infrasound
	Negative	Acoustic signal	-

3 Experiment Under Actual Environment

3.1 Setup of Measurement

Performance of the propose method was evaluated under a practical situation. Recording was carried out using a custom app made for Android smartphones, which simultaneously records an audio signal captured by the microphone at a sampling rate of 48 kHz and a pressure signal captured by the built-in pressure sensor of the smartphone. Sampling rate of the pressure is not consistent because of API available to access the sensor. Therefore, the obtained signal is later re-sampled with a constant interval which is chosen so as to keep the total number of samples unchanged. The resultant sampling rate depends on the model of smartphone and it was approximately 90 Hz for the one used. A Sony Xperia Z5 Compact was steadily hanged with a mounter to a handrail on the roof of the Resilient ICT research center of NICT in Sendai, Japan. A photograph of the surrounding environment of the experiment is shown in Fig. 2.

Fig. 2. Environment of the measurement

3.2 Results

Figure 3 shows the results of the acoustic sensor signal obtained on March 21 and 22. The lower panel shows only the time periods where inclusion of wind noise is suspected by the proposed method, and the upper panel is the remaining signal.

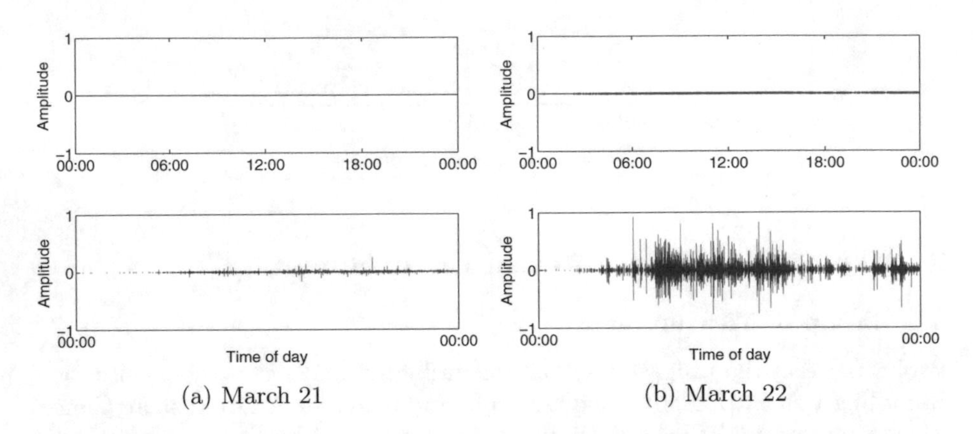

Fig. 3. Results of wind noise detection (lower panel: wave form of the extracted time periods where wind noise is suspected, upper panel: the remaining parts)

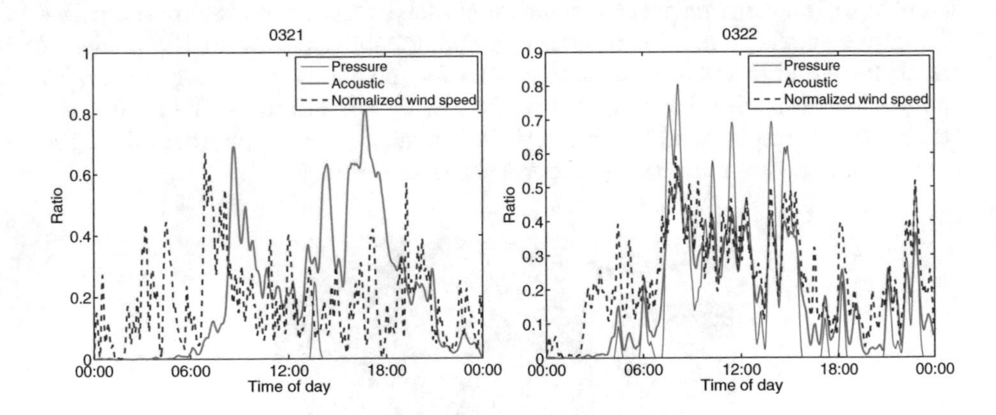

Fig. 4. Comparison with anemometer data

Existence of wind is also determined based on the wind speed measured by an anemometer placed near the smartphone [5]. Because intervals of the data are significantly different between the anemometer and the acoustic sensor, comparisons are made after converting the noise detection results into the ratio of the number of frames considered as wind noise to the total number of frames within the interval of anemometer data, 2 min. Wind speed is normalized with the maximum speed during the day under consideration. Averaged wind speeds are 0.56 and 2.39 m/s on March 21 and 22, respectively. Results are shown in Fig. 4, where smoothing by simple averaging over eight consecutive samples are performed in order to emphasize the outline of the results. Correlation coefficients between the wind detection ratio of the acoustic sensor and the normalized wind speed shown in Fig. 4 is 0.07 on March 21 and 0.88 on March 22. From the results, we can guess that the proposed method may shows better performance on a windy day than a calm day.

4 Discussion

Sensitivity to wind noise varies depending on the data used in the fitting. In the experiments, fitting was carried out day by day. Therefore, the sensitivity becomes low on windy day, and vise versa. This property is pros and cons and needs kept in mind. For example, we found that there were time periods where wind noise was not detected despite of high normalized wind speed. This may account for the reason that the correlation coefficient was quite low on March 21. Moreover, there is actually no direct relationship between wind speed and appearance of wind noise. However, it is easy to expect that more windy a day is, more often wind noise happens. In addition, as shown in Fig. 2, interfering objects exist asymmetrically around the smartphone. Therefore, the wind direction might affect this discrepancy.

5 Conclusion

A new algorithm to detect wind noise is presented, assuming a smartphone as a sensor device. The algorithm exploits both pressure and acoustic signals captured by the built-in sensors and integrates anomaly information of both the signals to detect infrasound. Comparing with the data of anemometer, it is shown through practical measurement that the proposed method can identify the time periods where wind noise is presumably included. However, the performance of the present algorithm seems sensitive to the condition of wind. Making it more stable is one of the next research topics.

Acknowledgment. The authors would like to thank to Dr. Suzuki at NICT for providing the data recorded by the anemometer. This work is partly supported by JSPS KAKENHI (17K01351).

References

1. Elko, G.: Reducing noise in audio systems. US Patent 7,171,008 (2007)
2. Eskin, E.: Anomaly detection over noisy data using learned probability distributions. In: Proceedings of the International Conference on Machine Learning, pp. 255–262. Morgan Kaufmann (2000)

3. Feng, S., Nadarajah, S., Hu, Q.: Modeling annual extreme precipitation in China using the generalized extreme value distribution. J. Meteorol. Soc. Jpn. Ser. II **85**(5), 599–613 (2007)
4. McNeil, A.J., Frey, R.: Estimation of tail-related risk measures for heteroscedastic financial time series: an extreme value approach. J. Empirical Finan. **7**(3–4), 271–300 (2000)
5. Observation system of the patch of blue sky for optical communication (OBSOC). http://sstg.nict.go.jp/OBSOC/?lang=e
6. Patcha, A., Park, J.M.: An overview of anomaly detection techniques: existing solutions and latest technological trends. Comput. Netw. **51**(12), 3448–3470 (2000)
7. Pichon, A.L., Blanc, E., Hauchecorne, A. (eds.): Infrasound Monitoring for Atmospheric Studies. Springer, New York (2010)
8. Rasmussen, K., Frederiksen, P., Rasmussen, F., Petersen, K.: Wind noise insensitive hearing aid. US Patent 7,181,030 (2007)
9. Zakis, J.A., Tan, C.M.: Robust wind noise detection. In: 2014 IEEE International Conference on Acoustics, Speech and Signal Processing (ICASSP), pp. 3655–3659, May 2014

Study on Speech Representation Based on Spikegram for Speech Fingerprints

Dung Kim Tran(✉) and Masashi Unoki(✉)

School of Information Science, Japan Advanced Institute of Science and Technology, 1-1 Asahidai, Nomi, Ishikawa 923-1292, Japan
{kimdungtran,unoki}@jaist.ac.jp

Abstract. This paper investigates the abilities of spikegrams in representing the content and voice identifications of speech signals. Current speech representation models employ block-based coding techniques to transform speech signals into spectrograms to extract suitable features for further analysis. One issue with this approach is that a speaker produces different speech signals for the same speech content; therefore, processing speech signals in a piecewise manner will result in different spectrograms, and consequently, different fingerprints will be produced for the same spoken words by the same speaker. For this reason, the consistency of speech representation models in the variations of speech is essential to obtain accurate and reliable speech fingerprints. It has been reported that sparse coding surpasses block-based coding in representing speech signals in the way that it is able to capture the underlying structures of speech signals. An over-complete representation model – known as a spikegram – can be created by using a matching pursuit algorithm and Gammatone dictionary to provide a better alternative to a spectrogram. This paper reports the ability of spikegrams in representing the speech content and voice identities of speakers, which can be used for improving the robustness of speech fingerprints.

Keywords: Speech fingerprint · Spikegram · Matching pursuit algorithm · Gammatone filterbank · Non-negative matrix factorization

1 Introduction

Applying auditory filterbank to speech signals to obtain their representation models, known as spectrograms [1–3], is one of the most commonly used methods in signal processing. The main drawback of this block-based coding approach is that it separates input signals into frames and then independently transforms all frames. Thus, the representation models are unable to convey the underlying structures and sequential dependencies of the utterances [4]. Due to the fact that a speaker produces different speech signals for the same content, different spectrograms are produced for the same content and the same speaker. Therefore, a data driven coding technique is preferable as it is able to adapt to variations

J.-S. Pan et al. (eds.), *Advances in Intelligent Information Hiding and Multimedia Signal Processing*, Smart Innovation, Systems and Technologies 82,
DOI 10.1007/978-3-319-63859-1_20

in speech signals. Recent studies have revealed that sparse coding outperforms other methods in speech signal representation [4,5] because data-driven analysis is able to capture the underlying structures of speech. Our work utilizes a matching pursuit algorithm [6,7] and Gammatone dictionary to produce over-complete representations of speech signals that are known as spikegrams.

2 Literature Review

Some of the most challenging issues in speech signal processing are time scaling and signal shifting. In reality, it is very difficult for one speaker to produce the same speech signals for a given content. Figure 1 illustrates two speech signals and their corresponding spectrograms that have the same content and have been produced by the same speaker. The recorded signals may perceptually sound similar but they are technically unaligned, their amplitudes vary, and their phonemes are scaled. Thus, the processes of dividing speech signals into blocks and independently processing these blocks are not ideal for capturing the structures and sequential relationships of speech features. For these reasons, block-based coding techniques produce different spectrograms for speech signals that have the same content and same speaker. As a result, different speech fingerprints are created because of the inconsistency of spectrograms in representing speech signals.

An experiment was carried out to evaluate the ability of block-based coding in representing speech signals. A Gammatone filterbank [8] was used in this experiment to process ten speech signals with the same content produced by two speakers. Although this is one of the most well-known techniques used in speech processing, testing the distortion between the original signals and the

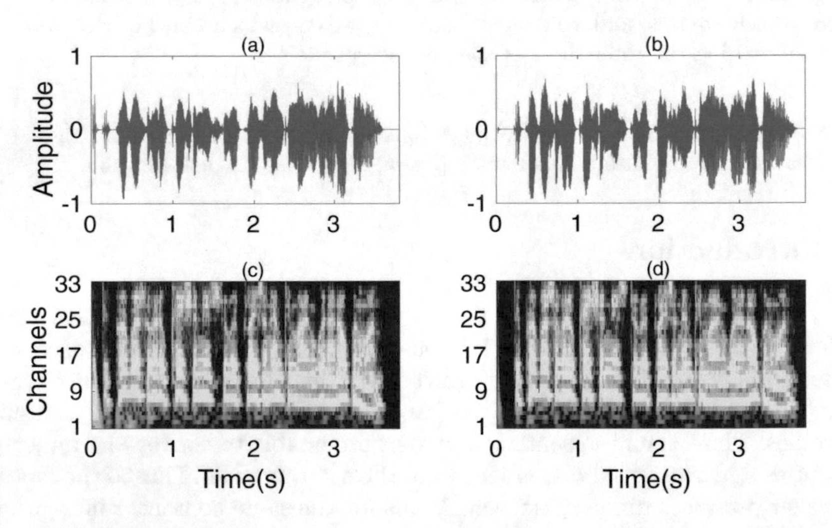

Fig. 1. Speech signals and corresponding spectrograms: (a) and (b) are speech signals of same content produced by same speaker. (c) and (d) are spectrograms corresponding to (a) and (b).

resynthesized signals could only achieve mean $\mu = 18.9$ and standard deviation $\sigma = 1.2$ in the signal to noise ratio (SNR), and mean $\mu = 3.9$ and standard deviation $\sigma = 0.1$ in the perceptual evaluation of speech quality (PESQ). Based on the work by Ellis [9], another experiment was also conducted to investigate the effectiveness of spectrograms in the process of producing speech fingerprints. The results obtained from this experiment by using the previously described data proved that different fingerprints were generated for speech signals that had the same content and the same speaker despite the fact that this method worked very well with audio fingerprints.

3 Speech Representation Based on Spikegrams

3.1 Non-negative Matrix Factorization

Non-negative matrix factorization is a method used for decomposing a matrix into two other matrices that only contain non-negative elements. The approximation is given by:

$$|X| \approx W \times B, \tag{1}$$

where $X \in R^{F \times T}, W \in R^{F \times E}$, and $B \in R^{E \times T}$. The matrices B and W are commonly known as the dictionary of bases, or the base spectrum, for the former and the weight matrix, or activations, for the latter. The speech signals were decomposed with a matching pursuit algorithm by using a dictionary of Gammatone kernels, and the weight matrix produced by the process is referred to as a spikegram in this paper.

3.2 Gammatone Dictionary

One of the main components used in creating an over-complete representation of speech signals is a dictionary of bases. The dictionary of bases we used in this research was created by using a Gammatone function because its impulse response is similar to that of the human cochlear system. Some other basis functions are the Gabor function [7] and Gammachirp. The impulse response of the Gammatone filter is given by:

$$g(t) = at^{n-1}e^{-2\pi bt}\cos(2\pi ft + \phi), \tag{2}$$

where a, t, n, b, f, and ϕ correspond to the amplitude, time, order of the filter, bandwidth of the filter, center frequency, and phase. Figure 2 illustrates three impulse responses of a Gammatone filterbank. The Gammatone filterbank was configured to use 33 channels, a center frequency of 600 Hz, and the sampling frequency of the input signal (i.e., 16,000 Hz) as a baseline for calculating the results presented in this paper. This resulted in a dictionary that had 33 Gammatone kernels.

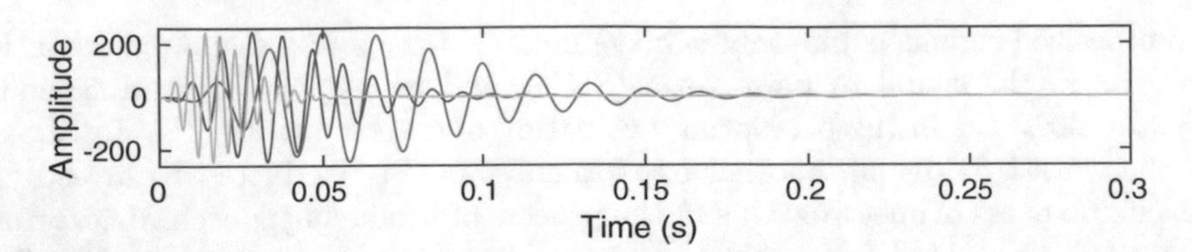

Fig. 2. Three gammatone impulse responses also known as gammatone kernels

3.3 Matching Pursuit Algorithm

An input signal in the matching pursuit technique can be decomposed into a linear combination:

$$x = \sum_{i}^{K} w_i b_i + r, \tag{3}$$

where x is the signal vector, K is the number of iterations, $w_i \in W$ and $b_i \in B$ are the weight and kernel at the i-th iteration, and r is the residual. According to this equation, minimizing the residual will maximize the approximation of the input signal and its linear combination. Therefore, the matching pursuit algorithm performs a search at each iteration for the largest inner product between the residual and a kernel, i.e.:

$$w_i = \arg\max | < r_i, b_i > |^2, \tag{4}$$

where $< r_i, b_i >$ is the inner product between the residual, r_i, and a kernel, b_i, at the i-th iteration. The residual is then updated as:

$$r_{i+1} = r_i - w_i b_i, \tag{5}$$

3.4 Creating Spikegrams

This research employed a matching pursuit algorithm [6] by using the dictionary of Gammatone kernels described in the previous section and some computational improvements [7] to create equivalent spikegrams for speech signals. The six overall computation steps are below.

1. Reverse all Gammatone kernels and then convolve signals with each kernel,
2. Find the largest peak over the convolution set,
3. Fit the signal with the kernel,
4. Subtract the signal from the kernel, record the spike, and adjust convolutions,
5. Repeat step 2, and
6. Cease when stopping criteria are met.

The matching pursuit algorithm implemented in this research used two stopping criteria; the first was when the SNR reached 40 dB and the second was the total number of iterations.

The top panel in Fig. 3 illustrates the original and the reconstructed signals from the dictionary of Gammatone kernels B and weight matrix W of a speech utterance. The residual of the approximation process is depicted in the middle panel. The bottom panel illustrates the spikegram of this speech utterance.

An experiment was conducted by using the same data as those presented in Sect. 2 to evaluate the capabilities of the matching pursuit algorithm and Gammatone dictionary in decomposing speech signals. This method achieved mean $\mu = 40.0$ and standard deviation $\sigma = 0.0$ in SNR, and mean $\mu = 4.3$ and standard deviation $\sigma = 0.2$ in PESQ when the distortion between the original signals and the resynthesized signals were tested.

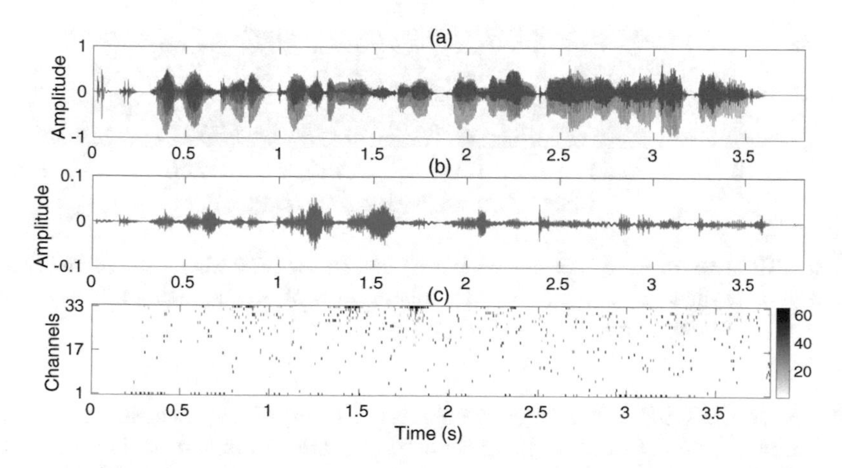

Fig. 3. Speech signal representation using spikegrams. (a) represents superimposition of speech signal in Fig. 1(a) and its approximation. (b) is residual signal after approximation. (c) is spikegram of speech signal created using matching pursuit algorithm and Gammatone filterbank.

4 Considerations

Signal shifting and phoneme scaling unaligned the features on the spikegrams due to the fact that the lengths of spikegrams varied depending on the lengths of recorded signals. As a result, it was difficult to directly compare two spikegrams. Furthermore, spike distributions and patterns were cues and these were used in identifying linguistic content and the voice identities of speakers. Therefore, it was necessary to normalize spikegrams and then encode spike distributions and patterns into a uniformed dimension to compare their features.

A local binary pattern (LBP) [10] was employed for this process. One characteristic of LBP is that it is able to demonstrate local changes in spike distributions that are important in comparing changes in content and utterers of speech. Each value on a spikegram is chosen as a center value and eight other values, i.e., four preceding and four trailing the center value, are also selected

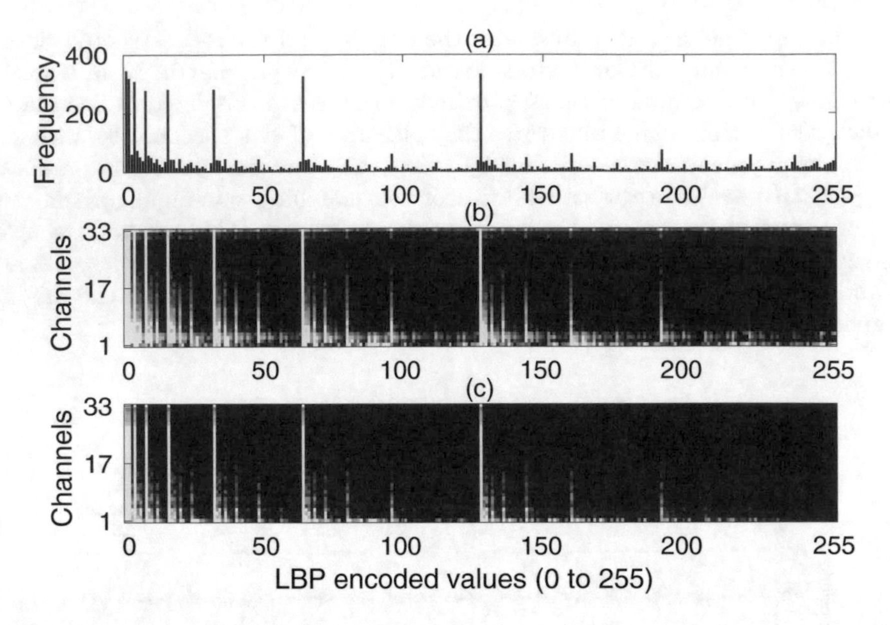

Fig. 4. LBP histograms of "kon-nichiwa". (a) illustrates LBP histogram of one channel of speech utterance. (b) and (c) are LBP histograms of 33 channels of "kon-nichiwa" produced by two speakers.

to calculate the LBP. After this process, each value of a spikegram is converted to a decimal number ranging from zero to 255 and a spikegram is transformed into an LBP matrix. Then, histograms for all channels of the LBP matrix are calculated to form a (33×255) histogram matrix, as shown in Fig. 4.

As spikegrams are calculated using a sparse coding technique, a spike in the middle of an ocean of zeros is very common. The stripes in Fig. 4 are the results of this pattern; more specifically, they are positioned at $2^0, 2^1, 2^2, 2^3, 2^4, 2^5, 2^6$, and 2^7. Figures 4(b) and (c) are histogram matrices of two speech signals of the same content produced by two different speakers. The color patterns of the upper channels of these histogram matrices are considerably different. This serves as a cue in comparing histogram matrices.

An experiment was conducted to evaluate the capabilities of spikegrams in capturing the underlying structures of speech signals, where 75 speech signals of three different types of content including /ohayo-gozaimasu/, /kon-nichiwa/, and /konbanwa/ produced by five speakers that included three males and two females were recorded. Then, the histogram matrices of equivalent spikegrams of speech signals were calculated using the previously described process.

One method that was used to evaluate the differences in spike distributions involved subtracting the histogram matrices by one another. Figure 5(a) shows the results obtained from subtracting the histogram matrices of two speech signals that had the same spoken word produced by the same speaker. The remaining subtraction is almost flat, which indicates that the spike distributions of

Table 1. Comparing Frobenius norm of LBP histograms of different speech utterances to evaluate effectiveness of spikegrams.

	Mean	Standard deviation
Same speaker and same content	$\mu = 408.3$	$\sigma = 20.3$
Same speaker and different content	$\mu = 3878.0$	$\sigma = 194.3$
Different speakers and same content	$\mu = 5072.6$	$\sigma = 878.8$
Different speakers and different content	$\mu = 5260.1$	$\sigma = 2747.2$

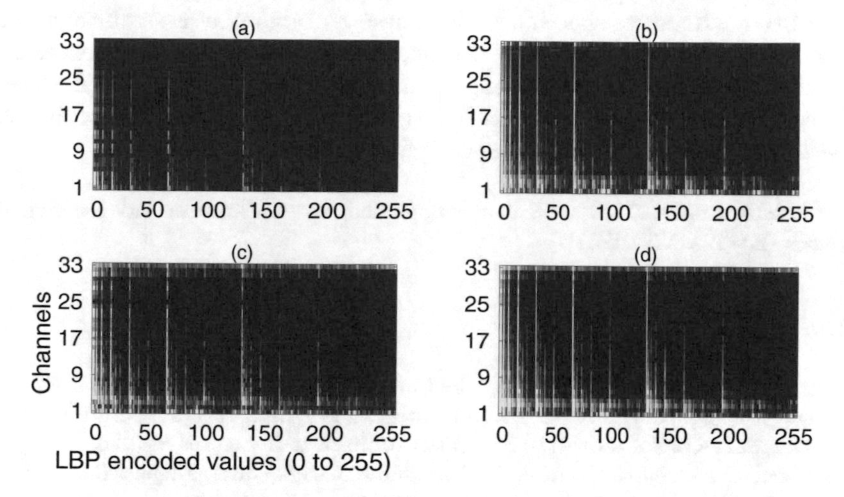

Fig. 5. Comparison of LBP histograms. Color patterns in (a) are almost flat, which indicate that two LBP histograms are similar. Color patterns in (b), (c), and (d) are spikier, which indicates that LBP histograms are different.

the original two spikegrams are very similar. Figures 5(b), (c) and (d) are the respective results from subtracting signals in the same manner that have the same content with different speakers, different types of content with the same speaker, and different types of content with different speakers. The color patterns of these figures indicate that changes in linguistic content or speakers result in very different spike distributions.

The Frobenius norm was another method used for comparing the histogram matrices. Table 1 summarizes the mean distances and standard deviations of the results. The mean distance and standard deviation obtained by comparing histogram matrices that have the same content and same speaker are about 10 times lower than the others. The other results demonstrate that changes in content or speakers notably increase the distances between histogram matrices and the farthest distances and largest variations of all of these are between different types of content and different speakers.

5 Conclusion

We found that the matching pursuit algorithm outperformed the Gammatone filterbank in decomposing speech signals by evaluating the original speech signals and the resynthesized signals with SNR and PESQ, and sparse coding provided a better representation model for speech signals. More importantly, spikegrams could be used to replace spectrograms as a better alternative in the process of creating speech fingerprints. Based on the experiments, we concluded that the spike distributions of spikegrams conveyed the linguistic content and voice identities of the utterers of speech signals. These features play a significant role in the accuracy and reliability of speech fingerprints. Further developments focus on building a method of extracting features of spikegrams that represent the linguistic content and voice identities of speakers and of creating a hashing technique to pack these features into speech fingerprints.

Acknowledgments. This work was supported by a Grant-in-Aid for Scientific Research (B) (No. 17H01761).

References

1. Cano, P., Batle, E., Kalker, T., Haitsma, J.: A review of algorithms for audio fingerprinting. In: IEEE Workshop Multimedia Signal Processing (2002)
2. Wang, A.L.-C.: An Industrial-Strength Audio Search Algorithm (2003)
3. Milano, D.: Content Control: Digital Watermarking and Fingerprinting, White Paper: Video Water Marking and Fingerprinting
4. Pichevar, R., Najaf-Zadeh, H., Thibault, L., Lahdili, H.: Auditory-inspired sparse representation of audio signals. Speech Commun. **53**(5), 643–657 (2011)
5. Mallat, S.G., Zhang, Z.: Matching pursuits with time-frequency dictionaries. IEEE Trans. Signal Process. **41**(12), 3397–3415 (1993)
6. Evan, S., Lewicki, M.S.: Efficient coding of time-relative structure using spikes. Neural Comput. **17**(1), 19–45 (2005)
7. Mineault, P.: Matching pursuit for 1D signals. https://www.mathworks.com/matlabcentral/fileexchange/32426-matching-pursuit-for-1d-signals
8. Unoki, M., Akagi, M.: A method of signal extraction from noisy signal based on auditory scene analysis. Speech Commun. **27**(3–4), 261–279 (1999)
9. Ellis, D.: Robust Landmark-Based Audio Fingerprinting. https://labrosa.ee.columbia.edu/matlab/fingerprint/
10. He, D.C., Wang, L.: Texture unit, texture spectrum, and texture analysis. IEEE Trans. Geosci. Remote Sens. **28**(4), 509–512 (1990)

Embedding Multiple Audio Data Using Information Misreading Technique

Naofumi Aoki(✉)

Graduate School of Information Science and Technology,
Hokkaido University, N14 W9 Kita-ku, Sapporo 060-0814, Japan
aoki@ime.ist.hokudai.ac.jp

Abstract. Information hiding technique is in general employed for practical applications such as steganography and watermarking. Although the mainstream of the researches on information hiding technique focuses on these practical applications, entertainment usage such as in trick arts may also be within its scope. This paper describes an idea of making audio trick art based on the concept of information misreading technique that retrieves multiple information from identical data by changing the format of data interpretation.

Keywords: Information hiding technique · Information misreading technique · Audio trick art

1 Introduction

Information hiding technique is in general employed for practical applications such as steganography and watermarking [1]. Steganography is a technique for transmitting secret information by using surface information as its cover data. On the other hand, watermarking is a technique for protecting the copyright of surface information by using secret information as its watermark. Both of them exploits the redundancy of surface information as a container for embedding secret information.

Although the mainstream of the researches on information hiding technique focuses on these practical applications, entertainment usage such as in trick arts may also be within its scope. It is true that the idea of trick arts is enjoying a sort of games exploring multiple meanings from ambiguous information. Although it is in general employed for entertainment usage, the concept of trick arts may potentially be applicable as an efficient technique to data communications transmitting multiple information by just single data.

This paper describes an idea of making trick arts for audio information. The proposed technique mixes two pieces of different audio information into just single data. The proposed technique introduces the concept of information misreading technique that retrieves multiple information from identical data by

J.-S. Pan et al. (eds.), *Advances in Intelligent Information Hiding and Multimedia Signal Processing*, Smart Innovation, Systems and Technologies 82,
DOI 10.1007/978-3-319-63859-1_21

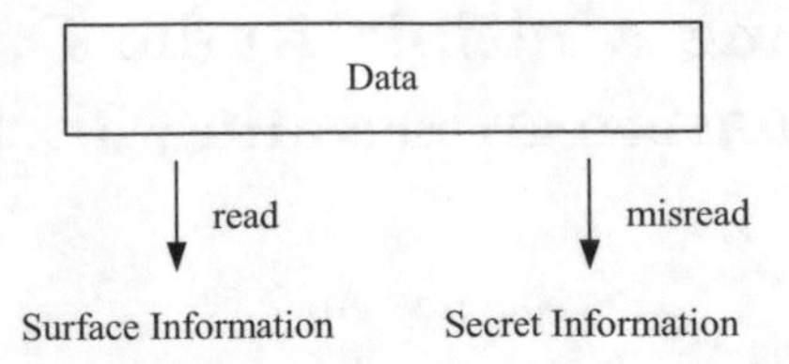

Fig. 1. Concept of information misreading technique.

changing the format of data interpretation. As shown in Fig. 1, surface information is retrieved by reading data according to an ordinary correct format, while secret information is retrieved by misreading it according to another incorrect format. This paper describes how this idea can work as audio trick art.

2 Examples of Trick Arts

Researches on the mechanisms of human cognition indicate that the meanings of ambiguous information exclusively change according to their context. Several examples of trick arts indicate that it may be possible to mix multiple information into just single data.

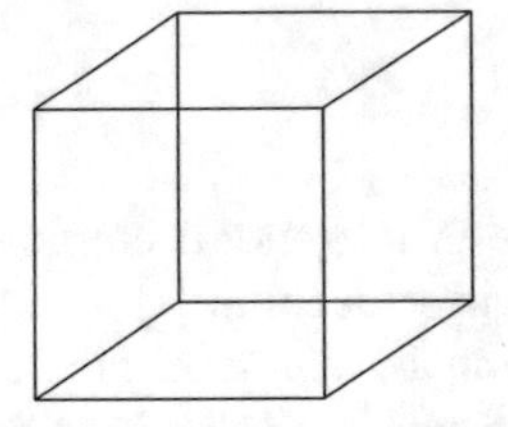

Fig. 2. Necker cube.

One of the examples is Necker cube [2]. As shown in Fig. 2, Necker cube can be interpreted in two ways. These images are derived from different assumptions how the 3-D object is reconstructed from its 2-D projection. This type of trick arts exploits information ambiguity as a container for embedding multiple information.

Another example is seen in some inverted paintings drawn by artists such as Arcimboldo. As shown in Fig. 3, two pieces of different visual information can be retrieved from the same picture by just changing its viewpoint. This indicates that the meanings of information are exclusively determined by the format how we receive the information.

Fig. 3. The Greengrocer: an example of inverted paintings drawn by Arcimboldo.

3 Audio Trick Art

This paper describes an idea of making trick arts for audio information. Digital audio data is nothing but just a sequence of binary data. Its meaning completely depends on the format of data interpretation.

Digital audio data employed in this study is defined by 44.1 kHz sampling rate and 16 bit quantization level, the standard resolution employed in the commercial audio CD (Compact Disk).

The proposed technique exploits information ambiguity of multiple bit binary data for making trick arts. As shown in Fig. 4, multiple bit binary data can be received as in four different bit and byte patterns. Conventionally, these variations are respectively adopted in different applications [3].

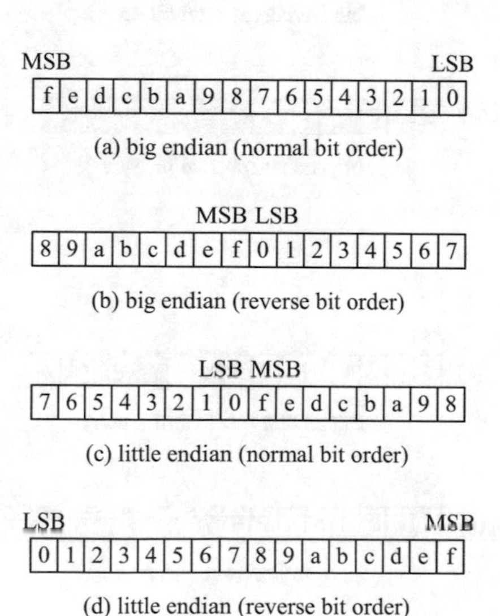

Fig. 4. Four different bit and byte patterns of 16 bit (2 byte) binary data.

Normal bit order is a format in which the least significant bit (LSB) is located at the first bit. On the other hand, when the most significant bit (MSB) is located at the first bit, it is referred to as reverse bit order. Normal bit order is employed in transmitting data for some image devices, such as facsimiles and printers. Reverse bit order is employed in transmitting data for display devices and in many data compression encoding formats.

Big endian is a format in which the low byte is located at the first byte. On the other hand, when the high byte is located at the first byte, it is referred to as little endian. Big endian is employed in Motorola's computer processor and in transmitting data by way of IP networks, while little endian is employed in Intel's computer processor.

Based on these variations of bit and byte patterns, the proposed technique mixes two pieces of different audio information into just single audio data. As shown in Figs. 5 and 6, the proposed technique replaces the low byte of audio data $s_a(n)$ with the high byte of audio data $s_b(n)$. As shown in these figures, when the original format of audio data is big endian (normal bit order), its conjugate format is little endian (reverse bit order). On the other hand, when the original format of audio data is little endian (normal bit order), its conjugate format is big endian (reverse bit order).

When the audio data processed by the proposed technique is played out according to the original format, the audio data is denoted as $s'_a(n)$ since it is perceptually similar to $s_a(n)$ even if its low byte is different from the original one.

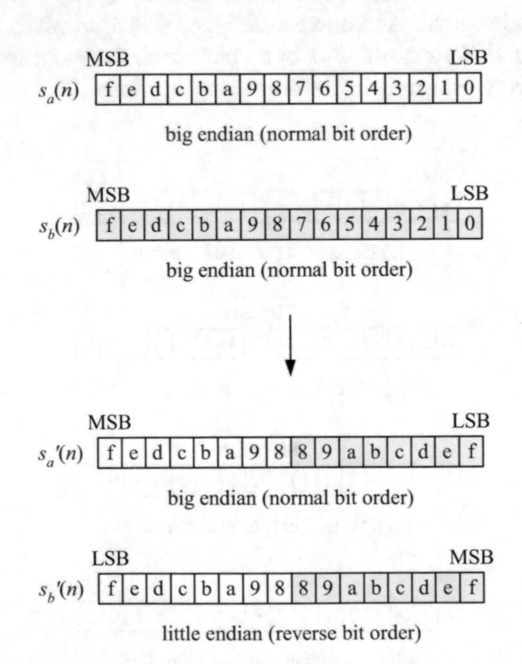

Fig. 5. Audio trick art: the original format is big endian (normal bit order), and its conjugate format is little endian (reverse bit order).

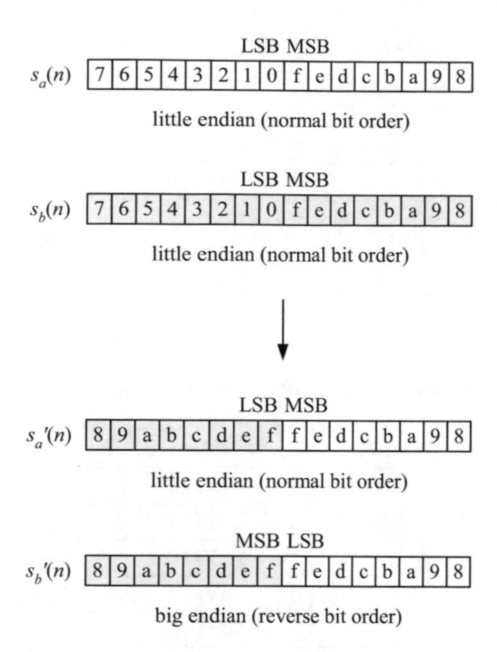

Fig. 6. Audio trick art: the original format is little endian (normal bit order), and its conjugate format is big endian (reverse bit order).

On the other hand, when the audio data processed by the proposed technique is played out according to the conjugate format, the audio data is denoted as $s'_b(n)$ since it is perceptually similar to $s_b(n)$ even if its low byte is different from the original one.

Consequently, the proposed technique may retrieve both the surface and the secret information by just changing the format of data interpretation. When the original format of the audio data is big endian (normal bit order), the secret information can be retrieved by misreading it as the data of little endian (reverse bit order). When the original format of the audio data is little endian (normal bit order), the secret information can be retrieved by misreading it as the data of big endian (reverse bit order).

4 Evaluation

As a pilot trial, the proposed technique was applied to audio data obtained from a couple of commercial audio CDs including 5 popular songs and 5 classical songs. These audio data were randomly selected.

Objective evaluation was performed in order to examine the audio quality processed by the proposed technique. In this trial, both the surface and the secret information were chosen to be the same.

PEAQ (Perceptual Evaluation of Audio Quality) was employed as the measure of the evaluation [4]. It estimates the degradation of the audio quality

Table 1. Definition of PEAQ score.

Score	Quality
0	Imperceptible
−1	Perceptible but not annoying
−2	Slightly annoying
−3	Annoying
−4	Very annoying

Table 2. Experimental results of 5 popular songs and 5 classical songs.

Song		PEAQ	DR [dB]
Popular	a	−0.203	7
	b	−0.067	7
	c	−0.159	6
	d	−0.394	7
	e	−0.001	5
Classical	f	−3.826	13
	g	−3.845	15
	h	−3.863	13
	i	−3.855	12
	j	−3.837	14

compared with the original reference data. PEAQ scores range from 0 to −4 as shown in Table 1. The higher the scores, the better the audio quality. Taking account of the characteristics of human auditory perception, PEAQ scores positively correlates with subjective evaluation.

Table 2 shows the experimental results. As well as PEAQ scores, this table also shows DR (dynamic range) of the songs. It is defined as the ratio between the peak value and the RMS (Root Mean Square) of audio data [5].

The mean and the standard deviation of the PEAQ scores calculated from the 5 popular songs were −0.165 and 0.15. On the other hand, the mean and the standard deviation of the PEAQ scores calculated from the 5 classical songs were −3.845 and 0.015. When the classical songs were played out, their secret data was slightly perceptible as background noise. On the other hand, when the popular songs were played out, their secret data was almost imperceptible. This indicates that the proposed technique in general prefers popular songs to classical songs as its audio data.

As shown in Table 2, it seems that the degradation depends on the dynamic range of the audio data. The dynamic range of popular songs is usually decreased in the mastering process by using compressor devices so that the loudness of the

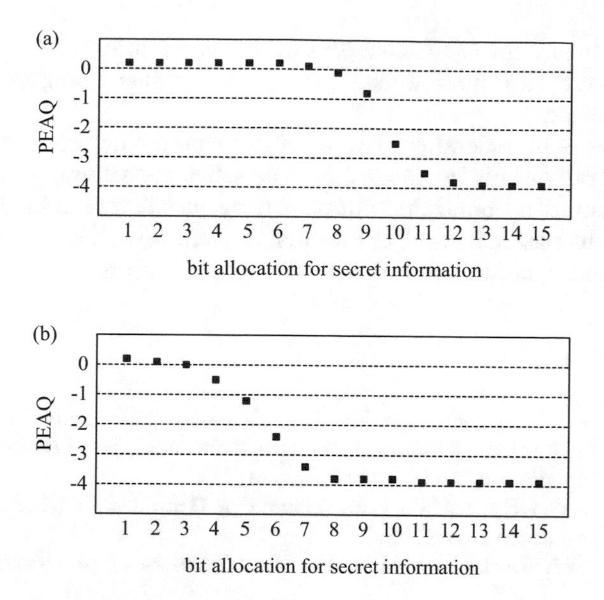

Fig. 7. Audio quality examined from (a) a popular song and (b) a classical song

audio data is increased [6]. Therefore, the degradation caused by the proposed technique becomes easy to be masked even if the low byte of audio data is completely replaced by the proposed technique. It is indicated that the proposed technique may have some advantages for the audio data of small dynamic range compared with the audio data of large dynamic range.

Figure 7 shows examples of the degradation patterns calculated from one of the popular songs and one of the classical songs. When the size of the secret information increases, the audio quality of the surface information decreases. Especially, the audio quality decreases more rapidly in the case of the classical songs. It is indicated that popular songs may contain large capacity for embedding the secret information compared with classical songs.

Optimal bit allocation for the surface and the secret information may be employed for controlling the degradation caused by the proposed technique. Allocating small resolution less than 8 bit for the audio data of small dynamic range and large resolution more than 8 bit for the audio data of large dynamic range, the degradation may potentially be balanced when these different types of audio data such as popular songs and classical songs are mixed together by the proposed technique.

5 Conclusion

This paper describes an idea of making audio trick art based on the concept of information misreading technique that retrieves multiple information from identical data by changing the format of data interpretation. The proposed technique does not require any special calculations for decoding information. Both the surface and the secret information may be retrieved by just changing the format of data interpretation.

Although it is in general employed for entertainment usage, the concept of trick arts may potentially be applicable as an efficient technique to data communications transmitting multiple information by just single data. It may potentially expand the research scope of information hiding techniques. Further investigation on making audio trick art is under consideration.

References

1. Cox, I., Miller, M., Bloom, J., Fridrich, J., Kalker, T.: Digital Watermarking and Steganography, 2nd edn. Morgan Kaufmann Publishers, San Francisco (2008)
2. http://en.wikipedia.org/wiki/Optical_illusion
3. Murray, J.D., Van Ryper, W.: Encyclopedia of Graphics File Formats. Oreilly & Associates Inc., Sebastopol (1996)
4. ITU-R, BS.1387: Method for objective measurements of perceived audio quality (2001)
5. http://www.pleasurizemusic.com
6. Katz, B.: Mastering Audio, 2nd edn. Focal Press, Burlington (2007)

A Steganography Algorithm Based on MP3 Linbits Bit of Huffman Codeword

Ru Zhang, Jianyi Liu(✉), and Feng Zhu

Beijing University of Posts and Telecommunications, Beijing, China
{zhangru, liujy}@bupt.edu.cn, 937714569@qq.com

Abstract. This paper proposes a steganography algorithm based on MP3 linbits bit of Huffman codeword by analyzing the structure of MP3 bitstream and the characteristics of the linbits of MP3 encoding. Experimental results show that the proposed algorithm not only has very high embedding capacity and keeps transparency, but also has low computational complexity and has good real-time performance for embedding and extraction.

Keywords: Compressed-domain · MP3 · Linbits · Steganography

1 Introduction

There are two main types of steganography based on audio: uncompressed domain audio steganography and compressed domain audio steganography. Nowadays, research in this field mainly focuses on uncompressed format audio, such as LSB, spread-spectrum hiding and echo hiding. With the development of the technology of digital audio compression coding, audio compression format such as MP3 has become more and more popular on the internet owing to its small storage space and high transmission speed. However, compressed format audio is obtained by removing redundant data, which has contradicts steganography since information hiding embeds information in the redundant data.

According to the embedding time, the compressed domain audio steganographic techniques can be divided into three kinds: embedding before compressed coding, embedding in compressed encoding, embedding after compressed coding. (1) The embedding algorithm before compressed coding is embedding secret information in the audio carrier before the audio signal compressed coding, mainly based on subband transform and strong robustness of low frequency coefficient of MDCT(Modified Discrete Cosine Transform) transform [1–3]. The biggest characteristic of this algorithm is that it can directly reference the classical algorithm of audio steganography in non-compressed domain, but the algorithm is mainly to modify the low frequency coefficient, and it is easy to affect the perceived quality of audio. Meanwhile the algorithm will lost some information after MP3 compressed encoding, resulting in the low correct rate of extraction. (2) The embedding algorithm in compression encoding is embedding secret information during the period of compressed coding such as the MP3Stego algorithm [4]. This algorithm usually achieves better transparency and robustness, and also can be compatible with various compressed encoding standard,

J.-S. Pan et al. (eds.), *Advances in Intelligent Information Hiding and Multimedia Signal Processing*, Smart Innovation, Systems and Technologies 82,
DOI 10.1007/978-3-319-63859-1_22

thus there is no significant effect on the compressed bit rate of the audio carrier, but it still needs to encode audio files, which has high computational complexity and poor real-time performance. (3) The embedding algorithm after compressed coding embeds the secret information in the compressed audio files directly. Kim et al. [5] has presented LSB steganography algorithm based on linbits but the limitation of linbits bit has adverse effects on the embedding capacity. Liu et al. [6] presented a steganography algorithm based on Huffman code word of count value region to replace, but the algorithm is confined to the code word of count value region, leading to its limited embedding capacity. Gao et al. [7] proposed the steganographic algorithm based on Huffman code words of big value to replace, but the result shows the statistical distribution changes significantly after steganography algorithm. Yan et al. [8] presented a steganographic method based on Huffman code replacement of multi-based way which has large capacity of hiding but poor performance for real-time. Anu Binny et al. [9] also presented a steganographic technique which embeds the text information into the audio signals using LSB technique. Sathiamoorthy et al. [10] has presented an enhanced technique for both LSB embedding and Echo hiding, based on T-codes. It was observed to be more robust for audio transmission and attacks than LSB embedding. Yang et al. [11] presented a digital watermarking scheme without changing the host audio data. Though it is very easy to embed using spatial domain and is computationally reasonable but achieving good imperceptibility and high robustness is still a challenging task.

MP3 is coded based on frame, and more than 90% of the data per frame are the Huffman codeword. This paper presents an algorithm based on MP3 linbits bit of Huffman codeword. It uses decoding block types to limit the embedding position, resulting in the consistency of the auditory and high embedding capacity.

2 Structure of MP3 Bitstream

The MP3 encoding standard has a strict limit on the format of the stream. Each frame of MP3 contains 1152 frequency coefficients which are divided into two granules. Each granule contains 576 frequency coefficients. The frequency coefficients can be divided into three parts: zero zone, the count value zone, the big value zone. Zero zone does not require encoding. The stream structure of big value zone is similar with that of count value zone and contains code word, the sign bit, and linbits composition. As shown in Fig. 1, Huffman codeword represents the code word of the coefficients. When the coefficients is larger than 15, the value over 15 is defined by binary encoding and the value of 15 can be defined by Huffman table. The part of binary encoding is defined as linbits and the maximum value of the binary encoding is $15 + 2^{linbits}$. There is no linbits for the count value zone since the coefficients the count value zone are lowers than 1. Sign represents the symbols of the coefficients. Each symbol of the coefficients is represented by a bit. When the coefficient is zero, it is unsigned.

Many code tables in MP3 Huffman codes table have the same code word with different linbits. For example, in Huffman Tables 16 to 23 and 24 to 31, these tables have the same code word but the linbits are different.

......	Huffcode	linbits_x	Sign_x	Linbits_y	Sign_y	......

(a)The bitstream structure of Big Value Zone

......	Huffcode	Sign_v	Sign_w	Sign_x	Sign_y	......

(b)The bitstream structure of Count Value Zone

Fig. 1. The structure of MP3 bitstream.

3 Steganography and Extraction Algorithm

3.1 The Algorithm of Embedding

MP3 encoding process will produce linbits when the frequency coefficient is greater than 15. If the linbits is directly embedded into the LSB, it will affect not only the audio sound quality, but also its security. Therefore the embedded process of this algorithm firstly is partial decoding of MP3, then according to the decoding rules Huffman coding table to obtain linbits bit and then the least significant bit (LSB) of the linbits bit is modified to embedding according to the improved LSB algorithm. In this paper 1–3 bits of the least significant bits are chosen for embedding and the transparency of steganography can be maintained. When Huffman table is 16 or 17, One bit of the least significant bits is set for embedding. When Huffman table is 18–31, divided into two cases: if $(linbits)_{DEC} \leq 15$ (*DEC* means a decimal number), 2 bit of the least significant bits is set for embedding; if $(linbits)_{DEC} \geq 16$, 3 bit of the least significant bits is set for embedding. Although 3 bit of the least significant bits is set for embedding, it is equivalent to the modification of 2 bit owing to the advantages of the algorithm, which maintains the transparency of steganography. The detailed steps of the extraction algorithm are:

(1) Converting secret information into 0, 1 bit streams at first, and the pretreatment is scrambling encryption which will generate the encrypted secret information, and then the length of secret information is concatenated with the encrypted secret information to create the embedding secret information.
(2) Partially decoding the MP3 carrier to get the linbits bit of the Huffman code. If the code has a linbits bit, go to step (3); otherwise going to the next code until the linbits bit is obtained.
(3) After acquiring the linbits bit, judging the number of the current decoding table and the decimal number represented by the current linbits bit, determining the number of embedding bits according to the embedding rules.
(4) According to the improved LSB algorithm, the linbits bits are embedded in the corresponding bit numbers.
(5) Repeat (2)–(4) until the MP3 file ends or the secret information is embedded at the end, getting the encrypted MP3 file.

3.2 Further Improvement of Embedding Algorithm

The position of linbits in MP3 is in the Big value Zone of every granule. However, the ranges of Big value Zone of different music are different and so are the probabilities of linbits occurrence. This leads to the variation of the steganographic capacity of different audio and then the effects on the acoustics of audio are different. Because the big value zone that occupies most energy of audio signal belongs to the low-frequency audio signals and the sound which is the easiest for human to perceive always locates in the big value zone, so the algorithm use block type to restrict location for embedding bits in the big value zone and has good performance in transparency.

During the MP3 encoding process, the operation of MDCT includes short length Windows, long length Windows, long and short length Windows and short and long length Windows. Long length Windows, long and short length Windows and short and long length Windows are long block coding and short length Windows are short block coding. Short block coding and long block coding can be combined for coding. The front part of mixed block is a long block and the rear part is a short block. In the decoding process, the sound quality can be good by reducing the number of embedded bits in factor bands of long block, short block or mixed block.

The location of modified linbits is limited by the block types and sizes of scale factor bands of every granularity decoding, as shown in Table 1. For long blocks and mixed blocks, when the scale factor is less than a specific value, the embeddable linbits position is limited and when the zoom factor is greater than a specific value, embedded position is unlimited. For short blocks, no consider the scale factor band, and embedded position is unlimited. After the above restrictions, not all linbits bits can be used as the embedding position of the secret information, but restricted to different block types with scale factor, making the capacity of audio after steganography will be reduced compared to the algorithm before the improvement, but the quality of audio is improved with limitation.

Table 1. Embed rules of different block types

Long Block	$Sfb \leq 12$	Every three embeddable location of linbits select a location for embedding
	$Sfb \geq 13$	Unlimited
Mixing block	$Sfb \leq 8$	Every three embeddable location of linbits select a location for embedding
	$Sfb \geq 9$	Unlimited
Short block	Unlimited	Unlimited

3.3 Extraction Algorithm

The extraction process only partly decodes the MP3, obtaining the number of Huffman table and the linbits bit. Then calculating the number of secret bits of the linbits by determining the size of the number of Huffman table and the decimal number of the linbits, and then extracting the corresponding linbits bits, e.g., to determine a linbits bit steganographic number is 2, if the lower two bits are "00", then directly extract "00".

The extraction algorithm is much simpler than the embedded algorithm which do not need to be encoded. The detailed steps of the extraction algorithm are:

(1) Receiving the encrypted MP3 file and key from the sending.
(2) Decoding the encrypted MP3 partly and obtaining the linbits bit of the Huffman code. If the code has a linbits bit, go to step (3); otherwise proceeding to the next code until the linbits is obtained.
(3) When obtaining the linbits position, determining the decoding table number of current and current linbits bit decimal size, then according to Table 1 embedding rule to determine the extracted bits.
(4) Then directly extract the corresponding bits from the end of the linbits.
(5) Repeat (2)–(4) to extract the bit stream of the embedded secret information, and then use the key to decrypt the secret information.

4 Experimental Results and Analysis

4.1 Experimental Environment

In order to verify the performance of the proposed Steganography algorithm, we perform a lot of simulation experiments, which can verify the real-time performance, embedding capacity and transparency of the algorithm. The experimental environment is as follows.

Hardware Environment: Intel Core i5-2400 CPU 3.10 GHz, 4.00 GB of RAM;

MP3 carriers: The experiment utilizes four MP3 sample library, each kind of sample library audio contains 100 MP3, including pop songs, rock songs, classical songs, folk songs and more. Each sample's sampling frequency is 44.1 kHz.bit rate includes 64 kbps, 128 kbps, 192 kbps, and 256 kbps. The length of sample is 3 to 5 min.

Secret information: The.txt Documents is taken as the secret documents in the experiments for encryption. The encrypted binary stream is embedded in the MP3 bit stream.

4.2 Embedded Capacity Analysis

In this paper, we measure the embedded capacity by the ability of unit time audio signal to hide the number of the secret information's bits. The bit/second (bps) is defined as the ratio of the maximum capacity of the secret information to the length of the audio. Table 2 shows the maximum average steganography capacity of these MP3 carriers.

As can be seen from Table 2, in bad situation the steganography capacity of the algorithm reaches 600bps and the steganography capacity of MP3Stego algorithm is only 76 bps, and compared with the algorithm in reference [12], the steganography capacity of our algorithm increases about 10%, which indicates that the embedding capacity increases significantly.

Table 2. Average embedding capacity of different audio (bps)

Music sample	Linbits	Reference [12]	MP3Stego
Library 1	764.7	692.9	76.5
Library 2	3288.2	3044.1	79.6
Library 3	573.4	491.3	76.3
Library 4	6531.6	5438.7	78.4

4.3 Imperceptibility

Select a MP3 audio of library 1, 128 kbps, 44.1 kHz. The experimental sample is fully embedded and then analyze the waveform in time domain before and after steganography, as shown in Fig. 2. Figure 2(a) shows the audio carrier signal waveform, Fig. 2(b) shows the audio stego signal waveform and Fig. 2(c) shows the difference between the audio carrier signal waveform and the audio stego signal waveform. It can be seen from Fig. 2 that the difference between the audio carrier signal waveform and the audio stego signal waveform is so small before and after steganography. And the obvious difference showing in figure is always in the original audio where the amplitude is relatively large and the jump points are also relatively more. These areas usually won't affect the original audio signal quality, and the human beings cannot perceive this subtle change.

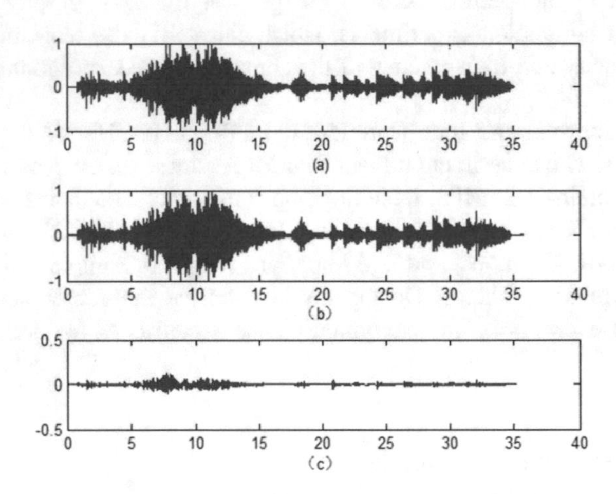

Fig. 2. Comparison of the waveforms before and after embedding

4.4 Security Analysis

Due to some statistic characteristics of the audio will inevitably be changed after embedding the secret information into audio (such as the MP3 block length distribution), therefore the security analysis of the algorithm is the evaluation of its non-detectability, which is to calculate the similarity of the probability distribution of the audio before and after the secret information embedding. The algorithm of this paper is to modify the linbits bit of the Huffman codeword, and the length of the position is not changed, that is, the structure of the modified MP3 audio stream does not change, and audio size also will not change. However, the modification of the linbits bit will change the frequency coefficient by the Huffman codeword decoding, that will have an impact on the distribution of MDCT coefficients. Therefore we test the change of the MDCT coefficient distribution of the stego audio and the host audio to evaluate the non-detectability of the algorithm. Select a MP3, 128 kbps, 44.1 kHz. The experimental sample is fully embedded, then the Euclidean distance d_{Euc} is used to compare the distribution of the MDCT coefficients of the audio.

$$d_{Euc} = \sqrt{\sum_{i=1}^{N} |P_i - Q_i|^2} \tag{1}$$

Among them, N is the dimension, P_i is the value of the P point and Q_i is the value of Q point in the i dimension. The smaller the calculated d_{Euc} value, the smaller the difference between P and Q. Table 3 shows the different samples of the d_{Euc} value of the different samples. As seen from the table, the distribution of MDCT coefficients of audio before and after the secret information is not changed obviously.

Table 3. The d_{Euc} value of different samples

Music sample	d_{Euc} value
Library 1	0.0019
Library 2	0.0012
Library 3	0.0021
Library 4	0.0034

5 Conclusion

This paper presents a steganography algorithm based on the characteristics of producing linbits during MP3 Coding to directly modify MP3 bitstream. The algorithm achieves linbits steganography through locating the linbits position and then embedding the linbits bit according to the algorithm's embedding rule. In order to improve the transparency of the algorithm, this paper further improves the algorithm. By judging the different block types of decoding and the limitation of the embedding of different scaling factors, the algorithm further guarantees the audio quality of the audio after steganography. The algorithm requires only partial decoding of MP3 and has low complexity and low computational overhead. Meanwhile the linbits Huffman code which changed by the algorithm represented by the frequency coefficients are relatively large, and the frequency is not fixed, which ensuring transparency and safety of the algorithm.

Acknowledgements. This work was supported by The National Key Research and Development Program of China under Grant 2016YFB0800404, the NSF of China (U1636112, U1636212).

References

1. Wang, M.L., Lin, H.X., Lee, M.T.: Robust audio watermarking based on MDCT coefficients. In: 2012 Sixth International Conference on Genetic and Evolutionary Computing, Kitakushu, pp. 372–375 (2012)
2. Bellaaj, M., Ouni, K.: A robust audio watermarking technique operates in MDCT domain based on perceptual measures. Int. J. Adv. Comput. Sci. Appl. **7**(6), 169–178 (2016)
3. Youssef, S.M.: HFSA-AW: a hybrid fuzzy self-adaptive audio watermarking. In: 2013 1st International Conference on Communications, Signal Processing, and their Applications (ICCSPA), Sharjah, pp. 1–6 (2013)
4. Petitcolas, F.: MP3Stego. Computer Laboratory, Cambridge, August 1998
5. Kim, D.-H., Yang, S.-J., Chung, J.-H.: Additive data insertion into MP3 bitstream using linbitss characteristics. In: Proceedings on ICASSP 2004, Montreal, Canada, vol. 220, pp. 181–184 (2004)
6. Liu, X., Guo, L.: High capacity audio steganography in MP3 bitstreams. Comput. Simul. **24**, 110–113 (2007)
7. Gao, H.: The MP3 steganography algorithm based on Huffman coding. Acta Scientiarum Naturalium Universitatis Sunyatseni **46**, 32–35 (2007)
8. Yan, D., Wang, R., Zhang, L.G.: A high capacity MP3 steganography based on Huffman coding. J. Sichuan Univ. **48**, 1281–1286 (2011)
9. Binny, A., Koilakuntla, M.: Hiding secret information using LSB based audio steganography. In: International Conference on IEEE Soft Computing and Machine Intelligence (ISCMI), pp. 56–59 (2014)
10. Manoharan, S., Mitra, S.: Message recovery enhancements to LSB embedding and echo hiding based on T-Codes. In: IEEE Pacific Rim Conference on IEEE Communications Computers and Signal Processing, pp. 215–220 (2009)
11. Yang, B., Wu, P., Jing, Y., Mao, J.: Lossless and secure watermarking scheme in MP3 audio by modifying redundant bit in the frames. In: 2013 6th International Conference on Information Management, Innovation Management and Industrial Engineering, Xi'an, pp. 154–157 (2013)
12. Kim, D.-H., Yang, S.-J., Chung, J.-H.: Additive data insertion into MP3 bitstream using linbits characteristics. In: Proceedings on ICASSP 2004, Montreal, Canada, pp. 181–184 (2004)

An AFK-SVD Sparse Representation Approach for Speech Signal Processing

Fenglian Li[1(✉)], Xueying Zhang[1], Hongle Zhang[1], and Yu-Chu Tian[1,2]

[1] College of Information Engineering, Taiyuan University of Technology, No.79, West Yingze Street, Taiyuan 030024, Shanxi, China
ghllfl@163.com, tyzhangxy@163.com

[2] School of Electrical Engineering and Computer Science, Queensland University of Technology, GPO Box 2434, Brisbane, QLD 4001, Australia
y.tian@qut.edu.au

Abstract. Sparse representation is a common issue in many signal processing problems. In speech signal processing, how to sparsely represent a speech signal by dictionary learning for improving transmission efficiency has attracted considerable attention in recent years. K-SVD algorithm for dictionary learning is a typical method. But it requires to know the dictionary size prior to dictionary training. A suitable dictionary size can effectively avoid the problem of under-representation or over-representation, which affects the quality of reconstruction speech significantly. To tackle this problem, an Adaptive dictionary size Feedback filtering K-SVD (AFK-SVD) approach is presented in this paper for dictionary leaning. The proposed method first selects the dictionary size adaptively based on the speech signal feasure prior to dictionary learning, and then filters out the noise caused by over-representation. The approach has two unique features: (1) a learning model is constructed based on the training set specifically for adaptive determination of a range of the dictionary size; and (2) a two-level feedback filter measure is developed for removal of speech distortion caused by over-representation. The speech signals from TIMIT speech data sets are used to demonstrate the presented AFK-SVD approach. Experimental results showed that, in comparison with K-SVD, the proposed AFK-SVD method can improve the quality of the reconstructed speech signal in PESQ by 0.8 and SNR by 3 - 7 dB in average.

Keywords: AFK-SVD · Speech signal sparse representation · Dictionary learning · Dictionary size · Feedback filter

F. Li — Authors would like to acknowledge the National Natural Science foundation of China (NSFC)(No.61371193) for its financial support for this research. Financial support from the Science and Technology Department of Shanxi Provincial Government under the international collaboration grant scheme (No.2015081007) and Special Talents Projects Grant Scheme (grant no.: 201605D211021) are also appreciated.

J.-S. Pan et al. (eds.), *Advances in Intelligent Information Hiding and Multimedia Signal Processing*, Smart Innovation, Systems and Technologies 82,
DOI 10.1007/978-3-319-63859-1_23

1 Introduction

Sparse representation is an effective way to capture signal features. It is benefit for saving storage space and transmission resources. Speech signal is typically sparse in both time and frequency domains. Representing speech signal by sparse approximation can reduce the transmission bandwidth requirements in speech codecs. Sparse representation assumes that a test speech signal is a linear combination of a few atoms in a dictionary. The atoms in the dictionary can be obtained through dictionary learning from training samples.

There have been quite a few dictionary learning methods for getting the atoms. The K-Singular Value Decomposition (K-SVD) algorithm is a typical method for designing over-complete dictionaries through sparse representation [1]. It minimizes the reconstruction error of each atom during updating dictionary atoms. However, the K-SVD algorithm requires the dictionary size or the total number of atoms to be determined in advance. The dictionary size is an important factor in sparse representation and can affect the reconstruction error. The larger the dictionary size, the smaller the reconstruction error [2]. However, in K-SVD, the dictionary size can not selected automatically based on the signal. This affects the quality of the dictionary learning. Generally, the compression rate can be improved if the dictionary size is set an ideal value. Beyond this ideal size, the compression performance maybe deteriorates [3]. In [4], the dictionary size was analyzed and constructed from online kernel sparsification. The results showed that under certain technical conditions, the dictionary size could grow sub-linearly to the number of input data. Although there have been some studies to improve the efficiency of K-SVD method [5,6], how to select the dictionary size is still a fundamental issue yet to be solved.

For solving this problem, an Adaptive dictionary size Feedback filtering K-SVD (AFK-SVD) approach is presented in this paper. The approach gives a model for adaptive determination of the dictionary size, and a two-level feedback filter measure for removal of speech distortion caused by over-representation. The approach allows the dictionary size to grow based upon the complexity of input signals [7]. Experiments results demonstrated that the proposed method can improve the reconstruction quality of the speech signal.

The paper is organized as follows. Section 2 presents our AFK-SVD approach. Section 3 evaluates the performance of the approach. Finally, Sect. 4 concludes the paper.

2 The AFK-SVD Approach

The realization of the K-SVD method is to minimize its objective function by using sparse coding and updating dictionary repeatedly. The dictionary size needs to be decided in advance, and once determined it will not be changed during the dictionary learning procedure. If the dictionary size is not appropriately selected, under-representation or over-representation will happen, which will affect the quality of the reconstructed speech signals significantly. Under-representation may lead to an increased approximation error. This is because the reconstruction algorithm cannot pursuit a suitable atom if the atom quantity is less than normal. Over-representation, may also result in degradation of the quality of the reconstructed speech signals and is even less tolerable than under-representation. It is similar to adding a "*cila*" noise to the speech signal. In fact, it is a series of random pulses superimposed on the signal waveform in the time domain. Therefore, it is necessary to denoise the signals by a feedback filtering to the reconstructed speech signals. For finding a suitable dictionary size adaptively according to the speech signal characteristics and weakening the influence of noise, an Adaptive dictionary size feedback Filtering K-SVD (AFK-SVD) approach is presented in the following.

2.1 Adaptive Dictionary Size Model

Let x_i be the ith column vector of sparse coefficient matrix X. The l_0 norm of x_i is $\|x_i\|_0$, which is the number of non-zero elements in $x(i)$. Then, the sparsity L of x_i is expressed as:

$$L = \|x_i\|_0 \tag{1}$$

The sparse ratio (SR) is defined as:

$$SR = n/L \tag{2}$$

where n is the number of samples. SR characterizes the degree of sparsity of the samples. The higher the SR is, the fewer the non-zero samples is (e.g., the more sparse the samples). Our experimental results have shown that SR and the dictionary size are negatively correlated. That is, the higher the SR, the fewer the atoms.

Some speech signal sentences in TIMIT database are used for finding the most suitable dictionary size under different SR values. Statistical analysis showed that with the increase of SR, the most suitable dictionary size (DS) decreases exponentially. Based on this, the adaptive dictionary size model is constructed to predict the dictionary size. This prediction provides a reference range for the dictionary size. Figure 1 gives the relationship of SR and the dictionary size DS by curve fitting. It is noticed that the dictionary size is not positively correlated with the speech quality. Because for a fixed SR value, a larger dictionary size does not mean better quality of the reconstructed speech signals.

Let x and y denote SR and the optimum dictionary size DS, respectively. Their relationship can be expressed as:

$$y = 424.422\exp(-0.031x) \tag{3}$$

2.2 Two-Level Feedback Filtering

The dictionary size selection model determines the dictionary size adaptively prior to dictionary learning. But experimental results have shown that random noises still exist in the reconstructed speech signals even though the optimum dictionary size is used. During the reconstruction, it is inevitable to generate

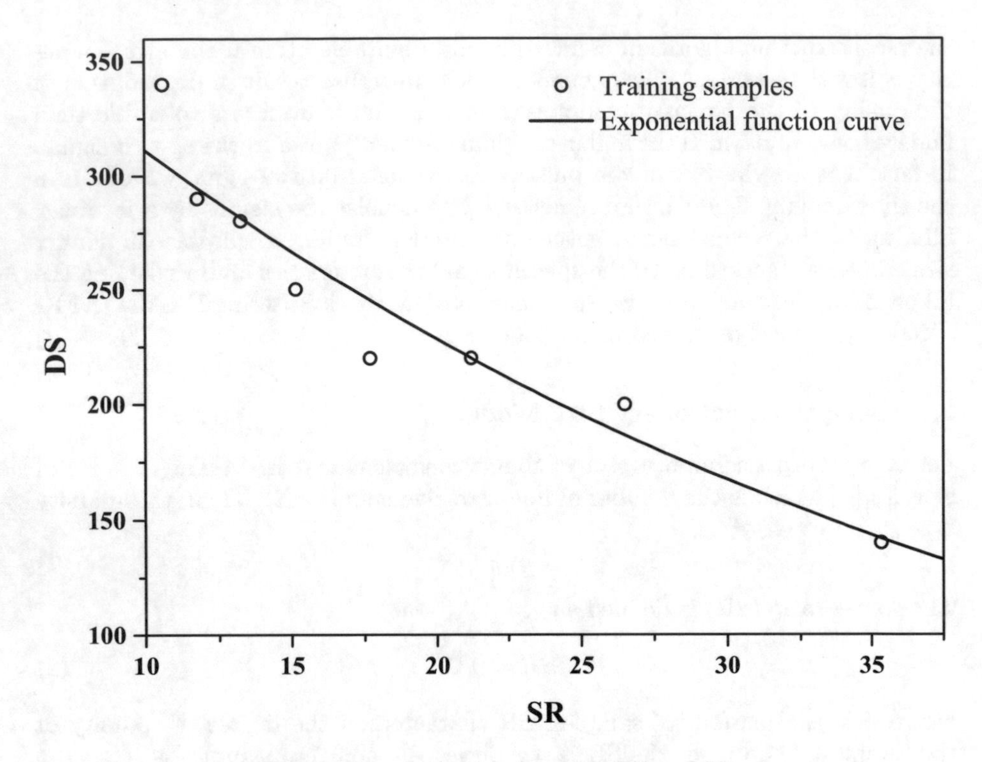

Fig. 1. The relationship between DS and SR.

some random noises from the reconstruction algorithm. So a two-level feedback filtering measure is used to filter out noises. In detail, it is implemented by smooth filtering the distortion using median smooth and linear smooth filters.

The procedure of the two-level feedback filtering is as follows:

- Find the distortion position by comparing the reconstructed speech with the clean speech signals;
- Filter the speech signal by mean smooth filtering in its neighbourhood; this will reduce the difference between the distortion signal and its neighbourhood signal;
- Compare the smoothed signal with the clean speech signal again to find the remaining distortion point.
- To the remaining distortion point, get its neighbourhood again and filter the signal by linear smooth filtering.

2.3 AFK-SVD Algorithm

Our AFK-SVD algorithm uses the adaptive dictionary size model to select the dictionary size adaptively according to the SR value. Then, the Orthogonal Matching Pursuit (OMP) algorithm is used for sparse decomposition. During dictionary updating, the K-SVD algorithm is employed to find the minimized objective function. Finally, the two-level feedback filtering measure is used to filter out noises and thus to smooth out the distortion points. The following is the detailed procedure of the AFK-SVD algorithm:

(1) Set sparsity L, calculate SR according to Eq. (2). Use the model to predict size K of initial dictionary. Get the initial dictionary $D^{(0)} \in R^{n \times K}$ by selecting K samples randomly from input samples.
(2) Sparse coding. From $D^{(0)}$, solve under-determined equation using OMP algorithm, get coefficient matrix $\mathbf{X}^{(0)}$.
(3) Solve approximation error E_k.
(4) Get the corresponding column of E_k that corresponds to non-zero coefficient column of coefficient matrix $\mathbf{X}^{(0)}$, and then combine these columns to form E_k^R.
(5) Get $E_k^R = U\Delta V^T$ by the K-SVD method.
(6) Update dictionary and sparse coefficient by using the first column of U to replace d_k and the first row of ΔV^T to replace x_T^k, respectively. It means $d_k = U(:,1)$,$x_T^k = \Delta(1,1)V(1,:)$.
(7) Repeat step (3) $\sim$ (7) until to the iteration times.
(8) The first level feedback filtering. $Y^{'}$ is the sparse approximation of input signal Y. T_1 is a threshold. $\Delta Y = Y^{'} - Y$. Find the sample which satisfies $|\Delta Y| > T_1$. To the neighbourhood of $Y^{'}$, carry out a median smoothing to get $Y^{''}$.
(9) The second level feedback filtering. T_2 is another threshold. $\Delta Y' = Y^{''} - Y'$. Find the sample which satisfies $|\Delta Y'| > T_2$. To the neighbourhood of $Y^{''}$, carry out a linear smoothing for getting final denoised speech signal.

3 Performance Evaluation

To evaluate the AFK-SVD approach, some male and female speech signals are used from TIMIT dataset. The window length is 512 samples, e.g., a time interval of 64 ms. The sampling frequency is 8 kHz.

Experiment 1: The relationship between sparse ratio SR and dictionary size DS. Experiments are conducted to find the relationship between SR and DS. For a fixed SR value, gradually increase DS to find the quality changes of the reconstructed speech signal. Repeating this process for a series of fixed SR values. Table 1 gives the relationship of SR, DS and the quality of the reconstructed speech signal. In the Table 1, SNR (Signal to Noise Ratio) and PESQ (Perceptual evaluation of speech quality) are used as criteria to evaluate the speech quality.

It is seen from Table 1 that with the increase of SR, the required optimum dictionary size(DS) gradually decreases. This implies that SR and DS are negatively correlated. The reason is that the more sparse the speech signal is, the less the useful information is carried after the speech signal is represented sparsely. Eventually, the atoms used to express the signal become fewer.

A similar conclusion can be get about the speech quality. With an increased loss of the speech information, both SNR and PESQ decrease. Moreover, for a fixed SR, more atoms do not mean better speech quality. For example, when SR = 12, it is seen from Table 1 that when DS increases from 270 to 290, the SNR increases from 23.64 to 25.18, PESQ increases from 3.4 to 3.6. However, with dictionary size further increases from 290 to 310, both SNR and PESQ decrease quickly: SNR from 25.18 down to 20.48, and PESQ from 3.6 down to 2.96, respectively. This is because a smaller dictionary size causes under-representation of the speech signal, while a bigger dictionary size maybe lead to over-representation. With the increase of the dictionary size, the error gradually decreases until the optimum. Then, with the further increase of the dictionary size, over-representation appears, leading to discontinuity in the reconstructed speech signal.

Figure 2 gives changes of SNR and PESQ with the increase of dictionary size when SR is equal to 12. It is seen that the initial values of SNR and PESQ are low. With the dictionary size increases, both SNR and PESQ are improved. When the dictionary size is 350, both SNR and PESQ get the maximum: SNR = 27 dB, PESQ = 3.7. Then, with the further increase in the dictionary size, the speech quality deteriorates quickly. When dictionary size is 380, the SNR becomes lower than 15 dB and the PESQ is lower than 3.0. This means that for a given SR, there is an optimal dictionary size that gives optimum speech quality.

Experiment 2: Remove the distortion from over-representation by using feedback filtering. Compared to the distortion caused by under-representation, the speech distortion resulting by over-representation is less tolerance. In this paper, the two-level feedback filtering method is used to remove the distortion. The first level filtering is the median smoothing filter. The input is a total of 21 samples, which are sampled before and after the discontinuity are selected. The second level filtering is the linear smoothing filter. It further smooths out the speech signal that has already been filtered by the first level filtering. Again, a total of 21 samples are used as input, which are sampled before and after the discontinuity are selected.

Table 2 gives the speech quality comparison for K-SVD and AFK-SVD in over-representation conditions. It is seen from the table that for a given SR, both SNR and PESQ of the AFK-SVD are much better than those of the K-SVD. This implies that AFK-SVD has better performances than K-SVD.

Table 1. The relation of SR,DS and speech quality

SR	12			15			21			36		
DS	270	290	310	230	250	270	190	210	230	130	150	170
SNR	23.64	25.18	20.48	20.22	22.39	15.87	17.41	18.90	14.30	14.15	15.12	12.77
PESQ	3.4	3.6	2.96	3.05	3.35	2.48	2.48	2.84	2.99	2.47	2.66	2.15

Table 2. Speech quality comparisons for K-SVD and AFK-SVD

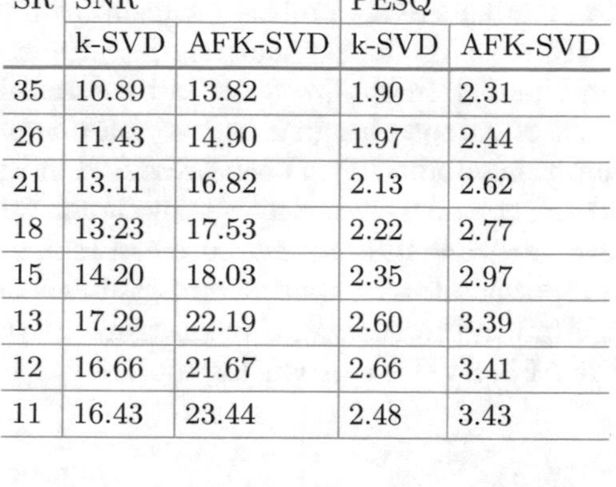

SR	SNR		PESQ	
	k-SVD	AFK-SVD	k-SVD	AFK-SVD
35	10.89	13.82	1.90	2.31
26	11.43	14.90	1.97	2.44
21	13.11	16.82	2.13	2.62
18	13.23	17.53	2.22	2.77
15	14.20	18.03	2.35	2.97
13	17.29	22.19	2.60	3.39
12	16.66	21.67	2.66	3.41
11	16.43	23.44	2.48	3.43

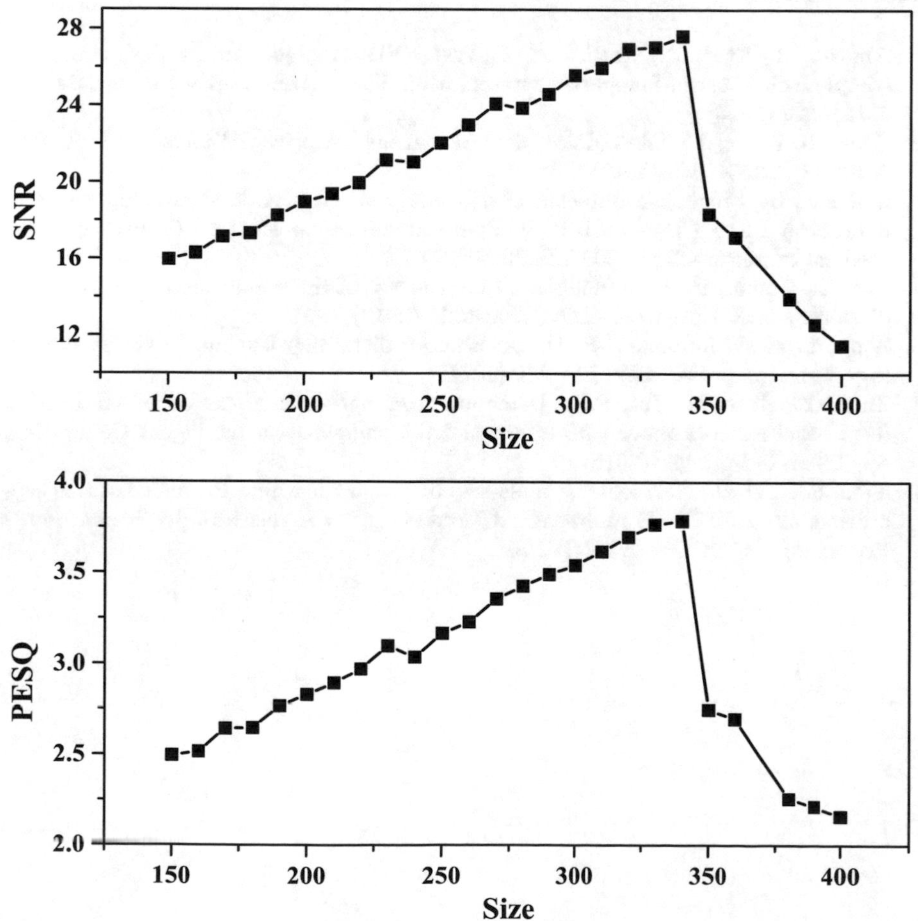

Fig. 2. SNR (upper plot) and PESQ (lower plot) vs DS.

4 Conclusion

Sparse representation can effectively improve transmission efficiency of speech signal. K-SVD algorithm for dictionary learning is a typical method in signal processing. However, the dictionary size need to be determined in advance. It limits the improvement of the speech signal reconstruction quality. A novel AFK-SVD sparse representation approach has been presented in this paper to address this issue. The approach can adaptive select the dictionary size based on speech signal, and a two-level feedback filtering measure can remove the speech distortion caused by over-representation. Experimental results showed that the AFK-SVD approach can improve the quality of reconstructed speech signals significantly. Therefore, the AFK-SVD is a promising tool for speech signal processing.

References

1. Aharon, M., Elad, M., Bruckstein, A.: *rmk*-SVD: an algorithm for designing overcomplete dictionaries for sparse representation. IEEE Trans. Signal Process. **54**(11), 4311–4322 (2006)
2. Zhou, J., Wang, J.: Fabric defect detection using adaptive dictionaries. Text. Res. J. **83**(17), 1846–1859 (2013)
3. Bierman, R., Singh, R.: Influence of dictionary size on the lossless compression of microarray images Twentieth IEEE International Symposium on Computer-Based Medical Systems: CBMS 2007. IEEE (2007)
4. Sun, Y., Gomez, F., Schmidhuber, J.: On the size of the online kernel sparsification dictionary. arXiv preprint arXiv: 1206.4623 (2012)
5. Zhou, Y., et al.: Immune K-SVD algorithm for dictionary learning in speech denoising. Neurocomputing **137**, 223–233 (2014)
6. Zhou, Y., Zhao, H., Lie, P.X.: Detection from speech analysis based on K–SVD deep belief network model. In: International Conference on Intelligent Computing, pp. 189–196. Springer (2015)
7. Tjoa, S.K., et al.: Harmonic variable-size dictionary learning for music source separation. In: 2010 IEEE International Conference on Acoustics Speech and Signal Processing (ICASSP). IEEE (2010)

An Automatic Detection Method for Morse Signal Based on Machine Learning

Zhihao Wei[1,2,3], Kebin Jia[1,2,3(✉)], and Zhonghua Sun[1,2,3]

[1] College of Electronic Information and Control Engineering, Beijing University of Technology, Beijing, China
kebinj@bjut.edu.cn
[2] Beijing Laboratory of Advanced Information Networks, Beijing, China
[3] Beijing Advanced Innovation Center for Future Internet Technology, Beijing, China

Abstract. In this paper, an automatic detection for time-frequency map of Morse signal is proposed base on machine learning. Firstly, a preprocessing method based on energy accumulation is proposed, and the signal region is determined by nonlinear transformation. Secondly, the feature extraction of different types of signal time-frequency maps is carried out based on the graphics. Finally, a signal detection classifier is built based on the feature matrix. Experiments show that the classifier constructed in this paper has the generalization ability and can detect the Morse signal in the broadband shortwave channel, which improve the accuracy of Morse signal detection.

Keywords: Morse signal · Machine learning · Feature extraction · Classifier

1 Introduction

Short wave telegraph is one of the main communication methods that Morse code is transmitted through shortwave channel, and it has the characteristics of narrow signal bandwidth, simple equipment, high mobility and survivability, it is also the main working mode of shortwave tactical communication in the background of strong noise interference. The traditional Morse newsletter adopts the listening mode of the artificial duty mode, which has numbers of shortcomings. Firstly, due to the complexity of the signal type of the communication environment and the uncertainty of the noise, the sending and receiving personnel need rich experience in signal detection, which lead to high cost in personnel training. Secondly, due to the limited concentration of people's attention, person who keep detecting long time Morse signal will lead to higher instability of signal detection accuracy. So, an automatic detection for Morse signal become the need of signal detection [1, 2]. New methods have been used to recognize Morse signal, such as voice recognition [3], signal analysis in both time and frequency domain [4], which expand the coverage of the research.

In this paper, the automatic detection method of Morse signal based on machine learning is proposed. The non-linear transformation is used to preprocess the wideband signal to obtain a high-contrast time-frequency signal matrix. The signal localization method based on energy accumulation is used to local the signal region. And the machine

J.-S. Pan et al. (eds.), *Advances in Intelligent Information Hiding and Multimedia Signal Processing*, Smart Innovation, Systems and Technologies 82,
DOI 10.1007/978-3-319-63859-1_24

learning method is used to construct the classifier for Morse signal detection, and finally experiments shows the performance of the automatic detection method for Morse signal.

2 Pretreatment

2.1 Time-Frequency Image Preprocessing Based on Energy Accumulation

In the shortwave communication transmission environment, the visibility of the time-frequency signal matrix is poor due to the influence of channel attenuation and noise interference. In order to reduce the influence of noise on the signal localization, the time-frequency map is reduced dimensioned by the method of accumulating the energy in the Y-axis coordinate direction of the statistical time-frequency diagram to obtain the one-dimensional energy matrix Y, thus highlighting the energy accumulation of the signal, energy accumulating function is as follows,

$$Y_j = \sum\nolimits_{i=x_0}^{i=x_{max}} I(x_i, j) \tag{1}$$

Where the Y_j represents the j position's value in the one-dimensional energy matrix Y, and x_0 and x_{max} represent the coordinate zero point and the maximum coordinate value of the X-axis of the signal time-frequency diagram.

2.2 Signal Region Selection Based on Nonlinear Transform

Image enhancement is one of the important aspects of signal time-frequency processing [5]. Since the noise fluctuation in the time-frequency matrix is large, the energy accumulation value of the high-noise part approaches and even submerges the energy accumulation value of the non-noisy signal. Therefore, the one-dimensional energy matrix Y obtained based on Sect. 2.1 is nonlinearly transformed, in order to highlight the energy accumulation value of the signal part. In this paper, we use the Formula (2) to nonlinearly transform the one-dimensional energy matrix Y_j to obtain the enhanced one-dimensional energy matrix D.

$$Dj = Dk_1 * Dk_2 * (Dk_1 > 0) * (Dk_2 > 0) \tag{2}$$

Where Dj is the j position's value of the enhanced one-dimensional energy matrix D, and $Dk_{\mathrm{m}}(m = 1, 2, 3, 4)$ is corresponding to one-dimensional energy matrix Y respectively, and Y_j is the center where energy are different between each locations, as shown in Fig. 1.

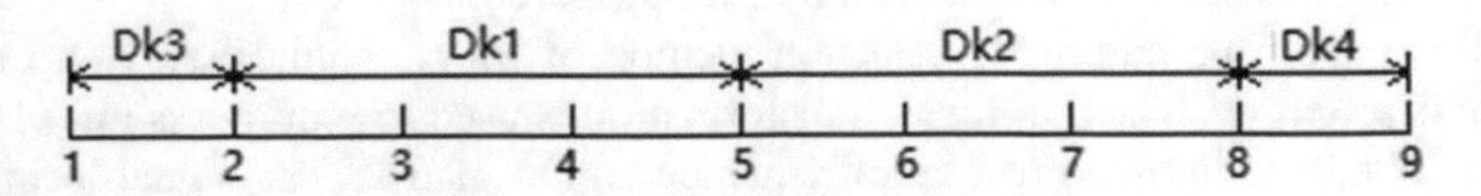

Fig. 1. Variable of nonlinearly transform.

Then, based on the mean value of the one-dimensional energy matrix D which is $\bar{D}$, and traverse the matrix D according to the Formula (3) to obtain the one-dimensional matrix S for identifying the signal position in the matrix D.

$$L_j = \begin{cases} 0, & Dj \leq \overline{D} \\ 1, & Dj > \overline{D} \end{cases} \tag{3}$$

The position of the number 1 in the matrix S corresponds to the Y-axis position of the signal of the time-frequency signal matrix I. Thus, the positioning of the signal in the time-frequency diagram is achieved.

3 Morse Signal Detection Classifier

3.1 Characteristic Matrix Base on Feature Extraction

The classifier is one of the machine learning methods for image recognition. Generally, the classifier for image recognition is constructed and trained by the template image sample set [6]. However, due to the various kinds of signals in the shortwave communication channel which is the environment studied in this paper, including FM, AM, Morse, FSK, voice signal, etc., the signal bandwidth and length is different between each other. Thus the signal size of the corresponding time-frequency map is not the same, more difficult to size normalization and template image sample set to establish. Therefore, it is necessary for the signal feature extraction method which can calculate the rules of the time-frequency map of different sizes according to a series of the same characteristics, and then the signal characteristic matrix of the same data structure is constructed to facilitate the classifier construction.

The feature extraction of data is actually the process of describing the most obvious features of the data with less data [7]. Because the time-frequency spectrum of the signal studied in this paper is a two-dimensional grayscale, which is finally used to detect the Morse signal for the purpose, so the main consideration to extract the signal time-frequency graphics features, and in the X-axis, Y-axis two features The Morse signal is selected in the classification of the frequency map corresponding to the highest correlation characteristics.

By comparing the correlation of the graphical features, the intermittent characteristic in the X–axis of the time-frequency map and the distribution characteristic in the Y-axis of the time-frequency are chosen as the main features. The intermittent characteristic is described as below. The discontinuous characteristic σ is calculated by the Formula (4).

$$\sigma = \sum_{j=1}^{M} \sqrt{\frac{1}{N} \sum_{i=1}^{N} (I_i - \mu_j)^2} \tag{4}$$

Where M and N are the maximum coordinate values of the time-frequency diagram I along the X and Y axes, I_i is the gray value at I position along the X-axis of the time-frequency diagram I, and μ_j is the average gray value at j position's value in the time-frequency diagram I along the Y-axis.

3.2 Classifier Construction

The construction of the classifier includes two main steps: classifier model selection and parameter setting, and classifier training based on training sample set [8]. The decision tree classifier is selected as the classifier model, and the GINI metric algorithm is chosen as the splitting principle of the spanning tree, which the maximum classification tree is set at 20.

Then, according to the method of Sect. 3.1, the training sample set is established based on the time-frequency map of broadband short-wave signal, and the classifier is trained based on the training sample set to obtain the classifier which can realize Morse signal detection.

4 The Experimental Results and Analysis

The study in this paper is based on the time-frequency map of broadband short-wave signal. According to the method in Sect. 2, the signal is located from the broadband time-frequency diagram. The Morse signal and the non-Morse signal are partially listed below (Figs. 2 and 3).

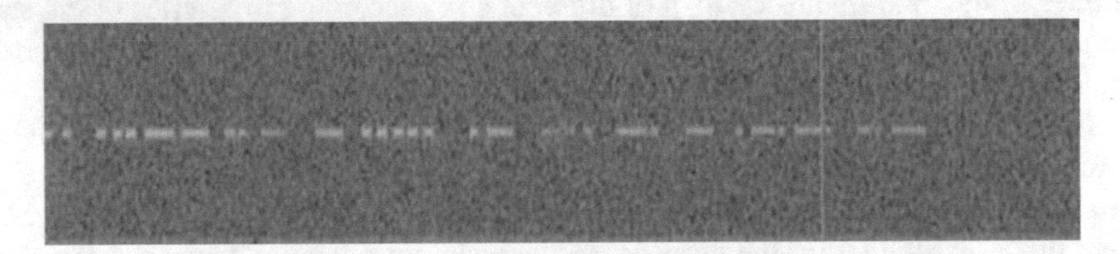

Fig. 2. Morse signal waveform.

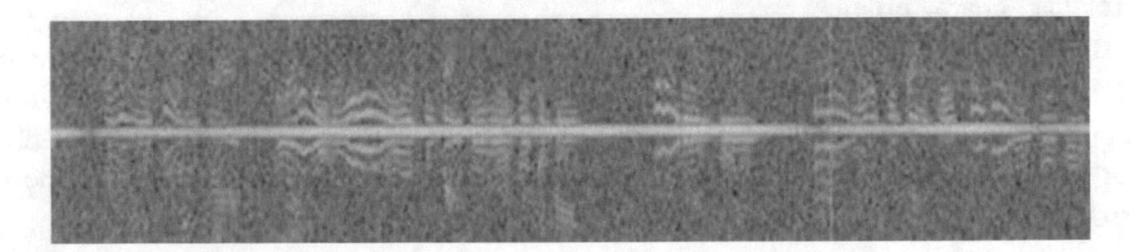

Fig. 3. Non-Morse signal waveform.

The signal time-frequency graph is constructed by the time-frequency diagram of the signal obtained from the signal in the time-frequency data Data1 of the broadband short-wave signal, and the Morse and non-Morse signal types are marked by the Flag. As shown in Table 1.

Table 1. Dataset of signal

Flag	Type	Quantity
1	Morse	533
0	Sweep	20
0	Speech	77
0	Multitone	32
0	MPSK & MQAM	54
0	FM	43
0	CW	61
0	AM	214
0	8FSK	22
0	4FSK	19
0	2FSK	38

The feature set of 1113 samples is constructed by feature extraction of the time-frequency graph according to the method in Sect. 3.1.

Then, according to the classifier construction method in Sect. 3.2, the classifier is constructed based on the feature matrix, and the performance of the classifier is tested by experiments.

The performance of the decision tree classifier are shown by Figs. 4 and 5, which include the ROC curve and truth value matrix. In the ROC curve of the decision tree classifier, the area under the curve named AUC is 0.97, and the false alarm rate of the Morse signal is 5.9% from the truth matrix.

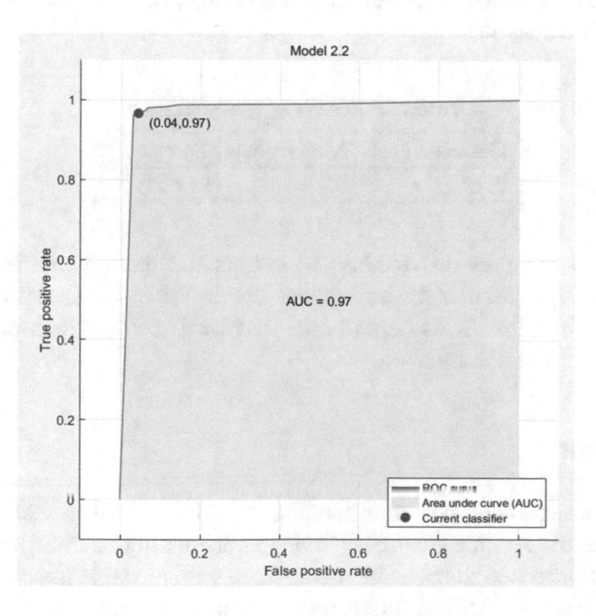

Fig. 4. The ROC curve of the classifier.

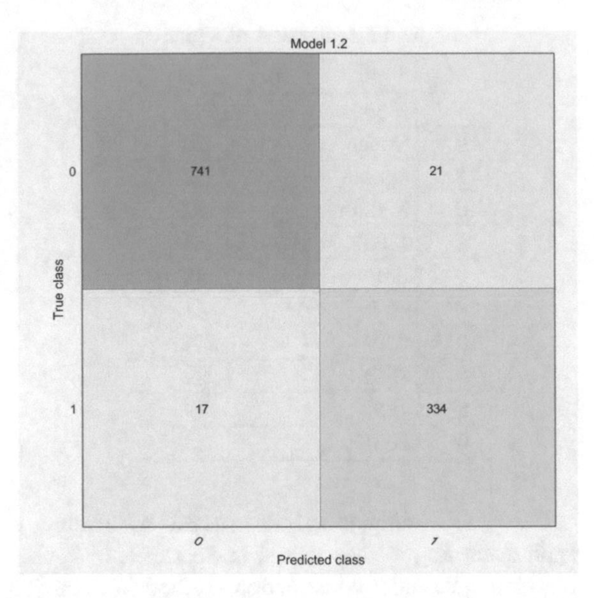

Fig. 5. The truth value matrix.

By analyzing the performance data of the decision tree classifier, it is concluded that the decision tree classifier can automatic recognize the Morse signal.

The generalization ability of the decision tree classifier is then further tested. Another broadband shortwave time-frequency data Data2 is selected to test the decision tree classifier. The Data2 dataset consists of 492 signals, of which 325 Morse signals and 167 non-Morse signals.

Table 2. Results of generalization

Correct signals	Miss signals	Correct Rate
298	27	91.69%

According to the previous method to extract the feature matrix, and input the feature matrix to the classifier for testing, the test results of the classifier generalization ability is shown in Table 2. Its correct rate reached 91.69%, which proves the generalization ability of the classifier.

5 Conclusion

In this paper, an automatic detection method of Morse signal in wideband shortwave signal is proposed. A time-frequency map preprocessing method based on energy accumulation is proposed for time-frequency map with unstable noise, and the location of the signal region is realized based on the nonlinear transformation. Then, due to difference between template sizes of time-frequency maps, a graphic description method based on graphic features is proposed, and the feature matrix is established based on the signal time-frequency map. Finally, a classifier for Morse signal detection is constructed by using the established feature matrix. The ability of the Morse automatic detection classifier based on the decision tree is proved by experiments. Also, the generalization ability of the classifier is proved by the experiments using unfamiliar data.

Acknowledgments. This paper is supported by the Project for the Key Project of Beijing Municipal Education Commission under Grant No. KZ201610005007, Beijing Postdoctoral Research Foundation under Grant No.2015ZZ-23, China Postdoctoral Research Foundation under Grant No. 2016T90022, 2015M580029, Computational Intelligence and Intelligent System of Beijing Key Laboratory Research Foundation under Grant No.002000546615004, and The National Natural Science Foundation of China under Grant No.61672064.

References

Singh, A., Thakur, N., Sharma, A.: A review of supervised machine learning algorithms. In: 2016 3rd International Conference on Computing for Sustainable Global Development, pp. 1310–1315 (2016)

Zahradnik, P., Simak, B.: Implementation of Morse decoder on the TMS320C6748 DSP development kit. In: European Embedded Design in Education and Research Conference, pp. 128–131 (2014)

Li, C.X., Zhao, D.F., Li, Q.: Auto recognizing Morse message using speech recognizing technology. Inf. Technol. **2**, 51–52 (2006)

Ma, W., Zhang, J.X., Wang, H.B.: Automatic decoding system of Morse code. Netw. Inf. Technol. **26**(6), 51–55 (2007)

Gonzalez, R.C., Woods, R.E.: Digital image processing. Prentice Hall Int. **28**(4), 484–486 (2002)

Christopher, B., Michael, G.: Machine learning classifiers in glaucoma. Optom. Vis. Sci. **85**(6), 396–405 (2008)

Murphey, Y.L., Luo, Y.: Feature extraction for a multiple pattern classification neural network system. In: International Conference on Pattern Recognition, pp. 220–223 (2002)

Rahim, N.A., Paulraj, M.P., Adom, A.H.: Adaptive boosting with SVM classifier for moving vehicle classification. Procedia Eng. **53**(7), 411–419 (2013)

Intelligent Distribution Systems and Applications

Capacity Reduction of Distribution Transformer by Harmonic Effect

Yen-Ming Tseng[1], Li-Shan Chen[2](✉), Jeng-Shyang Pan[1], Hsi-Shan Huang[1], and Lee Ku[3]

[1] Fujian Provincial Key Laboratory of Big Data Mining and Application, School of Information Science and Engineering, Fujian University of Technology, Fuzhou, China
swkl200@qq.com, jengshyangpan@gmail.com, hw5400@hotmail.com

[2] School of Management, Fujian University of Technology, Fuzhou, China
sun56@ms8.hinet.net

[3] School of Design, Fujian University of Technology, Fuzhou, China
91621@fjut.edu.cn

Abstract. This article actual picks up the load, the harmonics and the temperature with distribution transformer of the high voltage commercial building by two underground 22.8 kV feeders supplied. Train the ANN using the MATLAB training data to form the ANN until converged. Construct the recall sets by total voltage harmonic distortion V_{thd} and total current harmonic distortion I_{thd} and so on. That investigates the V_{thd} or I_{thd} mapping to the core temperature of the distribution transformer to fit by artificial neural networks and two order polynomials to reach load capacity reduction rate calculation. It is setting I_{thd} is 6 P.U., V_{thd} is 1 P.U, Q is 0.4 P.U, P is 0.95 P.U for first case, its capacity reduction rate is 26.40%. Second case, P is 0.9 P.U, the capacity reduction rate is 26.62%. Third case, P is 0.85 P.U, its reduction rate is 24.65%, and the forth-case P is 0.8 P.U, its reduction rate is 23.08%.

Keywords: Artificial neural network · Total voltage harmonic distortion · Total current harmonic distortion · Temperature · Distribution transformer · Capacity reduction rate

1 Introduction

Transformers play an important role in transmission and distribution systems from power plant and by the way have some transformer with step up or step down and final to the end users. Available for ultra-high commercial building supply voltage may be 69 kV or 22.8 kV, which depend on the step-down transformer to 220 V for livelihood or factory using. In general, for designing phase only considered the load capacity enough or not and not evaluate the total current harmonic distortion I_{thd} and total voltage harmonic distortion V_{thd} will made the temperature raising to affect the distribution system operation. In many research for transformer such as fault diagnosis, which are about for oil-filled [1–3] transformers and gas-insulated transformers

J.-S. Pan et al. (eds.), *Advances in Intelligent Information Hiding and Multimedia Signal Processing*, Smart Innovation, Systems and Technologies 82,
DOI 10.1007/978-3-319-63859-1_25

operation of the transformer, has a significant contribution to the operation. Transformer fault analysis has many different method such as using fuzzy principle [4, 5], clustering method [6], the use of microprocessor hardware architecture [7], neural network and network structure [8–10], wavelet theory [11, 12] and signal decomposition [13] and other methods that made transformer protection to improve. However, there are must according to the specifications of various types of transformers, manufacturing test reports, operating records and related standards to investigate the transformer status. Nevertheless, the former discussion of the rated capacity in accordance with the temperature rise does not contain the harmonic voltage and current caused capacity decline that made the proposal method and algorithms to develop to ensure the operation security.

2 Loadings Data of Transformer

Actual to the ultra-high commercial building that supply side by underground 22.8 kV distribution system that measure the data such as loading, harmonic, temperature and other relevant information of transformer by installation in the measuring instrument at demand side of transformer. It is divided into three types of workdays, Saturday and Sunday in one week. Usually, the data of transformer for the five days is similar to that shown in Fig. 1, so the average load curve for each time period can represents the workday daily load curve. According to the load curve of the working day and the power supply three-stage time-of-use price due to demand period and distinguish between peak period (9:00 to 17:30), half-peak period (8:00 to 8:30 to 6:00 pm 7:00), and off-peak period (the remaining time) such as Fig. 2. Because the transformer almost burning out in the peak period of working day, so for the peak period to be discussed.

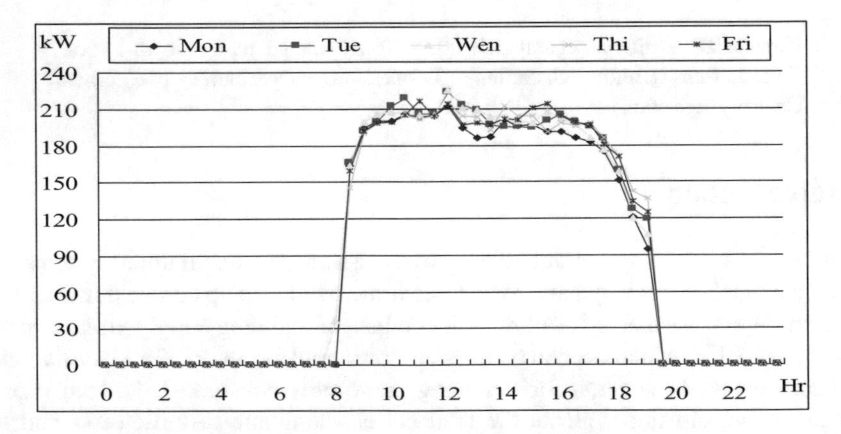

Fig. 1. Workday real power curve thought Monday to Friday

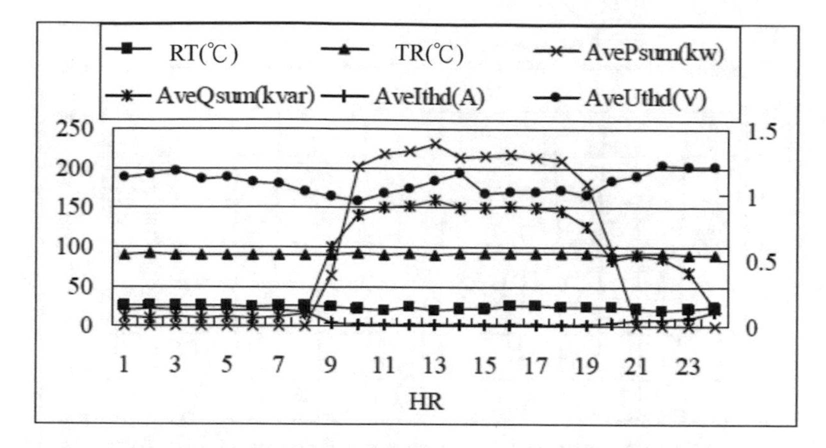

Fig. 2. Harmonic related parameter curve of working day

2.1 The Relationship Between the Load and Harmonics

The real power P and the total voltage harmonic distortion rate curve are shown in Fig. 3 which time interval is for a week from 0:00 am on Wednesday to 11:00 midnight on Tuesday. As well as the reactive power, Q and the total voltage harmonic distortion curve shown in Fig. 4, the figure for the information from Wednesday to midnight on Wednesday morning to 11:00. Theoretically, it can be known that the higher the load make the higher the harmonics will be. However, because this measurement object has been installed harmonic absorber and the workday - peak period and half-peak period when the load is larger filter device inputs. Such as shown in Fig. 3 the real power P is getting bigger and bigger, the total current harmonic distortion rate has not followed to increase, but the real power is low when the total current harmonic distortion rate will increase. Because when the filter close and the filter will begin to absorb the total

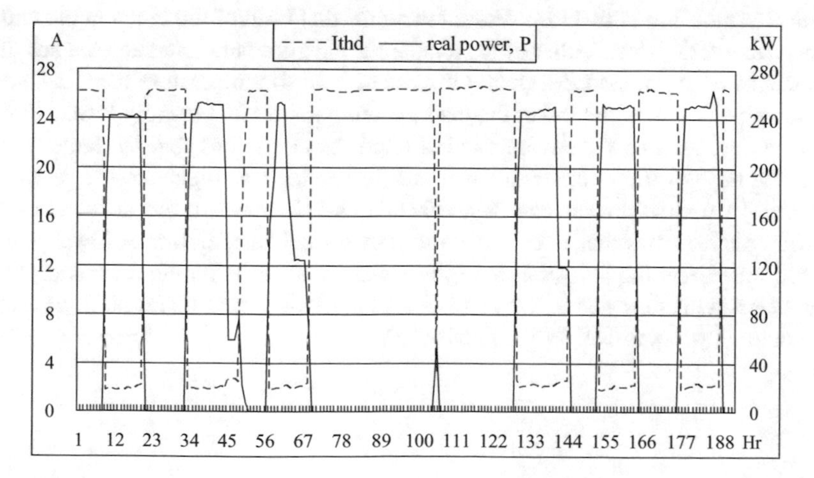

Fig. 3. Real power P and the total current harmonic distortion curve of the week

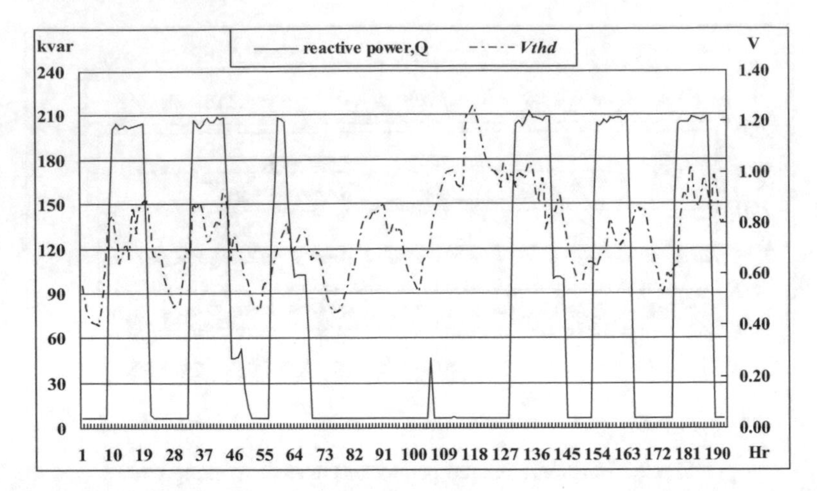

Fig. 4. Reactive power Q and the total voltage harmonic distortion curve of the week

current harmonic distortion, so made the higher the load, the total current harmonic distortion rate decreased. In this study is according to the actual operation situation to approach the temperature of transformer. Figure 4 is the reactive power Q and total harmonic voltage distortion curve which in the workday although somewhat proportional relationship, but no load on Saturday and Sunday and *Vthd* has increased, can be said to be disproportionate.

2.2 The Relationship Between the Load and Temperature of Transformer

Take a week of transformer actually measured data to form temperature and load curve for hourly unit as shown in Fig. 5 and the data period from Wednesday to midnight on Wednesday morning until 11:00. From the figure that both of the temperature and load are for a time series data. Either of the temperature, equipment rated and the actual load is associated with the load the greater the use of lead to temperature increases and has the same phenomenon, the lower the load the temperature is relatively lower. Form the view of a day, assume the device rated is fixed for conditions that the larger the load during the day so the temperature rise and lower load at night, so the temperature decreases. Of course, lower load on weekends and Sundays so the temperature dropped. The graph shows that after 115 h the transformer temperature is lower than 1 to 67 h. So over time, the temperature of the transformer or harmonic filter with the load size changes with time series, we can see that the load P and distribution transformer temperature is proportional to the relationship.

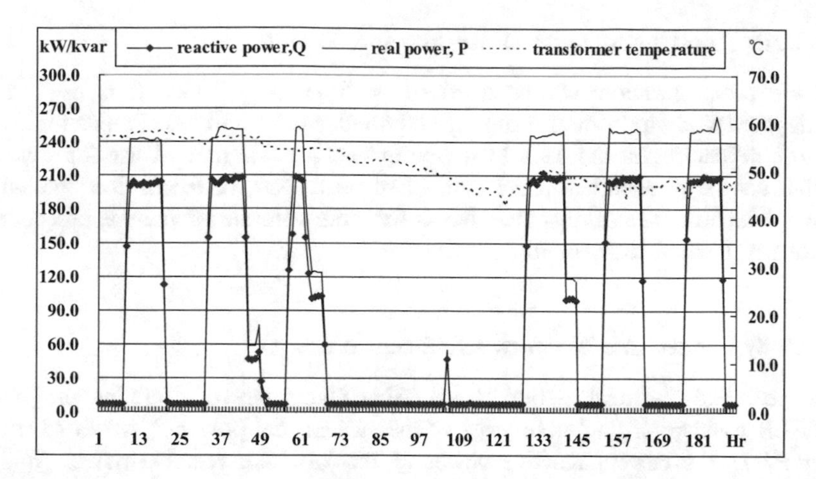

Fig. 5. Real power P, reactive power Q and the transformer temperature curve of the week

3 Artificial Neural Networks

3.1 Artificial Neural Network Structure

Figure 6 shows a neural network model consisting of multiple neuron models and their links. The structure model consists of input layers consisting of several processing units, hidden layers of several layers and neurons or neurons that are also an important part of the non-linear mapping. It is also an important part of the non-linear mapping of the output layer of the neuron, the function of the hidden layer is the interaction of the processing unit and the intrinsic structural capacity of the problem.

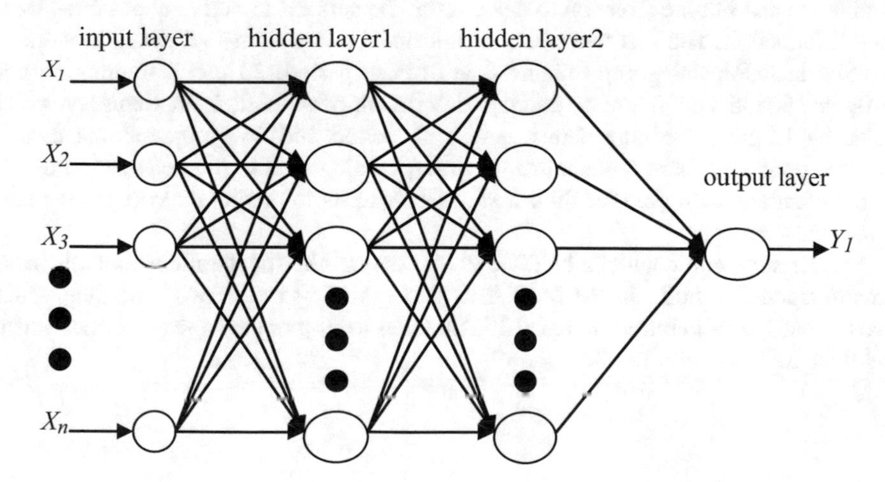

Fig. 6. Artificial neural network model

3.2 Daily Pattern and Network Architecture Design

With the power provider charge is based on time-of-use rules. It is divided into weekdays (Monday to Friday), Saturday and Sunday three daily load types for a week. Which is depending on the daily load type of load and the time-of-use for a working day that can be divided into time interval of peak load, half-peak load period and off-peak load time interval that classifies is for more accurate estimate approach of the transformer junction temperature.

3.3 Daily Pattern and Network Architecture Design

In the peak and half-peak period which using neural network training program by MATALB package. There are consist of the current real power *P*, the last hour real power *P(−1)*, the current reactive power *Q*, the last hour reactive power *Q(−1)*,the current total current harmonic distortion *Ithd*, the last hour total current harmonic distortion *Ithd*(−1), the current total voltage harmonic distortion *Vthd*, the last hour total voltage harmonic distortion *Vthd(−1)*, room temperature *Td*, the last hour room temperature *Td(−1)*, the last hour core temperature of transformer *Ti(−1)* for the input data and the transformer core temperature *Ti* for the output expected value are consist of the training data to train the ANN network to converged. With the same parameter of the training data except the expected value to feed into the input layer of ANN can obtain or reach the core temperature of transformer or called forecasting temperature of transformer.

3.4 Training Set and Recall Set

Choose about four weeks of the highest temperature of the year which make bad data elimination and outline isolation to delete from the data set to derive about three week good information. The first two weeks of information as training set which in working day that including 50 group training data in peak period, 20 group training data in half-peak period and 0 group training data in off-peak period; in Saturday which including 12 group training data in half-peak period and 36 group training data in off-peak period; in Sunday including 48 group training data in off-peak period that to form the training sets. And the third week as a recall set to predict the core temperature of the transformer.

The iterative approach can be divided into the number of iterations and allowable error iterations, or both. In the MATALB package of neural network training which give the number of iterations to reach 30,000 times to stop or the expected error value is less than 10^{-7} can stop the iteration.

4 Load Capacity Reduction Rate of Transformer

In peak period of working day which set the reactive power Q is 0.4 P.U. and then changes its load so that the actual power P from 0.8 P.U. each increase step for 0.05 P.U. until to 0.95 P.U.. The power factor at 0.92, 0.91, 0.90, 0.89 by neural network method predict the transformer core temperature, furthermore to fit by second order polynomial method.

The Distribution transformer insulation temperature set at design and production which allowable temperature value T_{rate}, due to temperature factors generated by the load capacity reduction rate of transformer as shown in Eq. (1).

$$TCRR = \frac{\Delta T_{iron}}{T_{rate}} \times 100\% \tag{1}$$

Where

TCRR: Load capacity reduction rate of transformer
ΔT_{iron}: Temperature changes
T_{rate}: Allowable temperature value

4.1 Impact Factor of Temperature of Transformer at P.F. = 0.92

P.F. is 0.92 (when real power is 0.95 P.U. and reactive power is 0.4 P.U.), the harmonic current I_{thd} and temperature of transformer curve as Fig. 7 which including second order polynomial regression fitting curve. It can be seen that the higher the I_{thd} value of the transformer, the greater the response value of the transformer temperature. So that I_{thd} will cause the distribution transformer temperature rise.

Figure 8 is P.F. = 0.92 (when real power is P = 0.95 P.U. and reactive power Q is 0.4 P.U.) the harmonic voltage and temperature of transformer curve which including second order polynomial regression fitting curve. The input response of the V_{thd} the

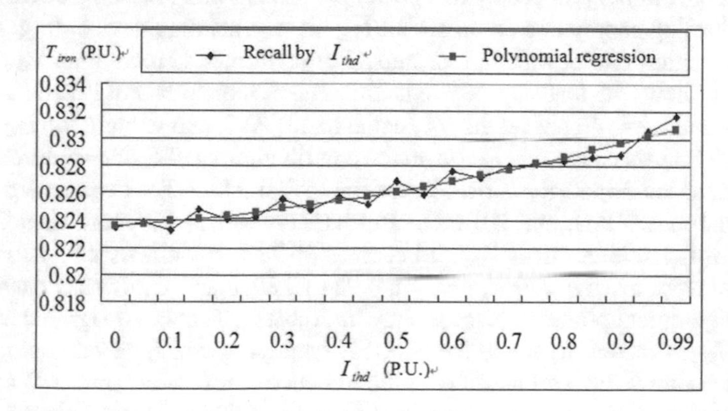

Fig. 7. P.F. is 0.92 the harmonic current I_{thd} and temperature of transformer curve

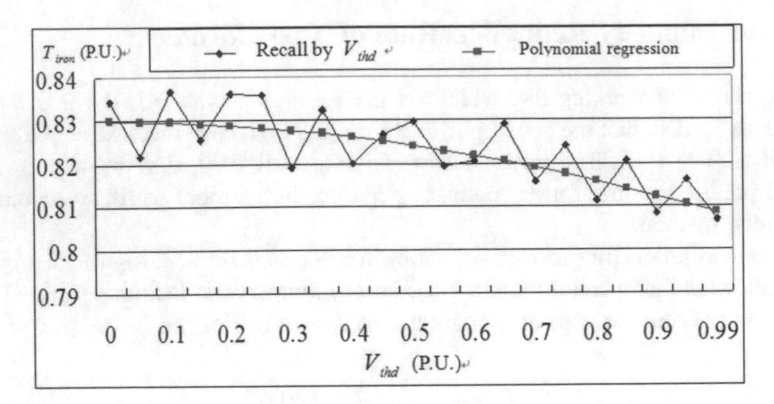

Fig. 8. P.F. is 0.92 the harmonic current V_{thd} and temperature of transformer curve

greater the predicted distribution transformer temperature value is smaller, so V_{thd} and distribution transformer temperature into Inverse.

Where I_{thd} with T_{iron} and V_{thd} with T_{iron} relationship curves have the same curve shape and trend can also be obtained by quadratic polynomial regression at PF = 0.92, PF = 0.91, PF = 0.90, PF = 0.89. Due to the I_{thd} larger by load or switching element effect will make the temperature razing of transformer. In contrast; when the load V_{thd} greater to feed in input layer of neural network and output response value of output layer is also a small value, it can be said V_{thd} Will cause the temperature of the distribution transformer to drop. According to the statistical data I_{thd} affected by the warming effect than the V_{thd} will be much more than the effect of temperature.

4.2 Capacity Reduction Rates of Transformer Are Compared at Different Loads

The real power is proportional with I_{thd} of load and reactive power is proportional relationship with V_{thd} of load. Although the distribution network has reactive power compensation, but I_{thd} produce by real power of loadings and switching devices for real power of loading supply and so on are all I_{thd} main generating device. In general, I_{thd} for load reduction rate of transformer impact or contribute is more than V_{thd}.

Table 1 Shows the load capacity reduction rate of transformer at P.F. equal to 0.92, 0.91, 0.90, and 0.89. Where set the V_{thd} equal to 1.0 P.U., and varying the I_{thd} form 1.0 P.U. to 10 P.U. because the I_{thd} is main contribution of the temperature rising of transformer to investigate the *TERR*. When the P.F. at 0.89, TERR equal zero mapping to I_{thd} equal to 1.0 P.U. and 2.0 P.U.. But TERR lager at I_{thd} more than 2 and the influence of I_{thd} is larger than V_{thd}, it becomes 36.78% when it rises at 10 P.U.

Form view of average *TERR*, PF form 0.92 to 0.89, reactive power Q fixed equal to 0.4 and varying real power to discuss show in Table 1. That has a significantly effect, when the I_{thd} is increasing made the *TERR* is increasing which called coping effect.

Figure 9 shows the load capacity reduction curve for reactive power Q setting 0.4 P.U., when the actual power is greater that copy the *TERR* is larger. Under 6.0 P.U of

Table 1. Load capacity reduction rate of transformer at different P.F.

Ithd (P.U.)	TCRR (%)				
	P.F. = 0.92	P.F. = 0.91	P.F. = 0.90	P.F. = 0.89	Average
1	0	0	0	0	0
2	0	0.88	0.11	0	0.25
3	1.65	3.42	2.35	1.78	2.30
4	4.76	6.66	5.33	4.56	5.33
5	8.79	10.6	9.05	8.08	9.13
6	13.74	15.24	13.51	12.34	13.71
7	19.61	20.58	18.71	17.34	19.06
8	26.4	26.62	24.65	23.08	25.19
9	34.11	33.36	31.33	29.56	32.09
10	42.74	40.8	38.75	36.78	39.77

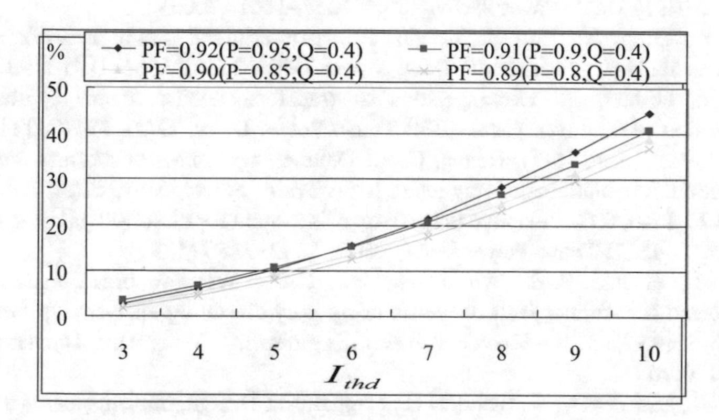

Fig. 9. Load capacity reduction curve of transformer

I_{thd} which *TERR* is not different, but when it is more than 6 P.U, it is obviously larger than other data and 42.74% at 10 P.U.

5 Conclusions

Distribution transformer factory design only real power and reactive power to form apparent power for the rated capacity, but the actual operation in transformer operation is more than the total current harmonic distortion (I_{thd}) and voltage total harmonic distortion (V_{thd}) two factors to affect the temperature rise of transformer. If the new distribution transformer does not add the current total harmonic distortion (I_{thd}) and voltage total harmonic distortion (V_{thd}) two factors on the distribution transformer temperature rise to monitoring. Due to the current total harmonic distortion (I_{thd}) is too large may lead to the distribution transformer load reduction rate bigger so that the transformer life shortened or burned. The new configuration to take into account the current total harmonic distortion (I_{thd}) and voltage total harmonic distortion (V_{thd}) on the distribution transformer equipment operating capacity reduction rate, you can avoid such a situation, so that the distribution transformer can in the safe range of effective operation. The real power is proportional with I_{thd} of load and reactive power is proportional relationship with V_{thd} of load. The higher the power factor is copied the greater the P, Q ratio and larger the I_{thd} mad the *TERR* greater and safe operation loading lower which will damage to the distribution transformer. For safety reasons, the distribution transformer must monitor the value of I_{thd} to guarantee that operation in safe lamination.

References

1. Huang, Y.-C.: A new data mining approach to dissolved gas analysis of oil-insulated power appatatus. IEEE Trans. Power Deliv. **18**(4), 1257–1261 (2003)
2. Zhang, Y., Ding, X., Liu, Y., Griffin, P.J.: An artificial neural network approach to transformer fault diagnosis. IEEE Trans. Power Deliv. **11**(4), 1836–1841 (1996)
3. Jalbert, J., Gilbert, R.: Decomposition of transformer oils: a new approach for the determination of dissolved gases. IEEE Trans. Power Deliv. **12**(2), 754–760 (1997)
4. Huang, Y.-C., Yang, H.-T., Huang, C.-L.: Developing a new transformer fault diagnosis system through evolutionary fuzzy logic. IEEE Trans. Power Deliv. **12**(2), 761–767 (1997)
5. Yang, H.T., Liao, C.C.: Adaptive fuzzy diagnosis system for dissolved gas analysis of power transformers. IEEE Trans. Power Deliv. **14**(2), 1342–1350 (1999)
6. Lin, C.-H., Huang, P.-Z., Wu, C.-H., He, C.-Z.: Window based assistant tool for oil-immersed transformer fault diagnosis using grey clustering analysis. In: Proceedings of the 27th Symposium on Electrical Power Engineering, Tsing Hua University, Taiwan, PD2.20 (2006)
7. Sidhu, T.S., Sachdev, M.S., Wood, H.C., Nagpal, M.: Design, implementation and testing of a micro-processor-based high-speed relay for detecting transformer winding faults. IEEE Trans. Power Deliv. **7**(1), 108–117 (1992)
8. Bastard, P., Meunier, M., Regal, H.: Neural-network-based algorithm for power transformer differential relays Generation. IEE Proc. Transm. Distrib. **142**(4), 386–392 (1995)
9. Perez, G., Flechsig, A.J., Meador, J.L., Obradovic, Z.: Training an artificial neural network to discriminate between magnetizing inrush and internal
10. Zaman, M.R., Rahman, M.A.: Experimental testing of the artificial neural network based protection of power transformers. IEEE Trans. Power Deliv. **13**(2), 510–517 (1998)
11. Wilkinson, W.A., Cox, M.D.: Discrete wavelet analysis of power system transients. IEEE Trans. Power Syst. **11**(4), 2038–2044 (1996)
12. Mao, P.L., Aggarwal, R.K., Bo, Z.Q.: Analysis of transient phenomena in power transformers using wavelet transforms. In: 34th University Power Engineering Conference, United Kingdom (1999)
13. Rahman, M.A., Jeyasurya, B.: A state-of-the-art review of transformer protection algorithms. IEEE Trans. Power Deliv. **3**(2), 534–544 (1988)

Base on Transmission Line Model to Investigate the Power Margins of Main Transformers

Yen-Ming Tseng[1], Rong-Ching Wu[2(✉)], Jeng-Shyang Pan[1], En-Chih Chang[2], and Peijiang Li[1,3]

[1] Fujian Provincial Key Laboratory of Big Data Mining and Application/School of Information Science and Engineering, Fujian University of Technology, Fuzhou, China
swkl200@qq.com, 596905210@qq.com, jengshyangpan@gmail.com

[2] Department of Electrical Engineering, I-Shou University, Kaohsiung, Taiwan
rcwu@isu.edu.tw

[3] School of Information Science and Engineering, Fujian University of Technology, Fuzhou, China

Abstract. Demand type of power is divided into residential, industrial and commercial three types. Feeders loading are combination with the three type demands to form residential-oriented, industrial-oriented and commercial-oriented. Different demand-oriented loading of feeders affect the load flow with different line model such as lumped model, π model and distributive model. In this paper, chosen a main transformer at Wei-Wu substation which loadings is summation by the connective feeder's loadings. By using the PSS/E software to find the maximum power and power margin under different power delivery. With the testing result, the power margins is relative with the various categories of main transformer assumes tendency of the inverse ratio.

Keywords: Main transformer · Load curve · Lumped model · π model · Distributive model · Power margin

1 Introduction

The main demand from the electric power form both Economic development and daily life which be asked for safe, reliable and stable supply of electricity providing that is based on the economic development and national industrial objectives pursued [1, 2] with the diversification of user comfort needs make the load demand substantial growth. Distribution system for the last part of the power system, and is connected to the user, so the quality of the distribution system operation [3, 4] the merits of not only affects the quality of power supply, more direct impact on users. For the quality of power supply to the power supply, voltage stability is most important and the equivalent load model affects the voltage change. Feeder loading is summation by connective user demands and main transformer loading is summation by connective feeder loadings, simultaneously. In this paper, main transformer, explore the relationship between maximum

J.-S. Pan et al. (eds.), *Advances in Intelligent Information Hiding and Multimedia Signal Processing*, Smart Innovation, Systems and Technologies 82,
DOI 10.1007/978-3-319-63859-1_26

power and power boundary change among the three different equivalent models of transmission line. Perform and execute the steps of this research are as follows:

Step 1: It is selected a main transformers as voltage 11.4 kV of Wei-Wu substations that posse's different load characteristic.
Step 2: Collect the line constant and load data of feeders is connected to main transformer.
Step 3: Put the Step2 data into PSS/E power flow software and solving the load flow under its original state including the voltage magnitudes and phase angles.
Step 4: Calculate the maximum transmission power of each feeder under the different load characteristics.
Step 5: Approach the maximum transmission powers of main transformer.
Step 6: Statistic the power margins of main transformers.

2 Transmission Model and Principle

Select the distribution system is composed of three main transformers at Wei-Wu substation whose load curves can be connect to the feeder by feeders' loadings summation, simultaneously. Load curve of feeder base on different type power consumption and statistic the each type percentage of total loading of feeder to divide into four type feeders of residential-oriented, agricultural-oriented, commercial-oriented and industrial-oriented. Therefore, according to the transmission line model such as equivalent lumped model, equivalent π model and distribution model and feeder loadings connected to main transformers to form these distribution systems to delivery power. In the distribution operation, reduce line losses improve power factor, by the three-phase load flow analysis systems required nodes voltage and phase angle [5–7] and to explore the main transformer power stability and power margins.

2.1 Equivalent Lumped Model for Transmission Line

For short-term transmission line, this only considers the series resistance and inductance to form the impedance such as the Fig. 1. the real and reactive power flow as Eqs. (1) and (2).

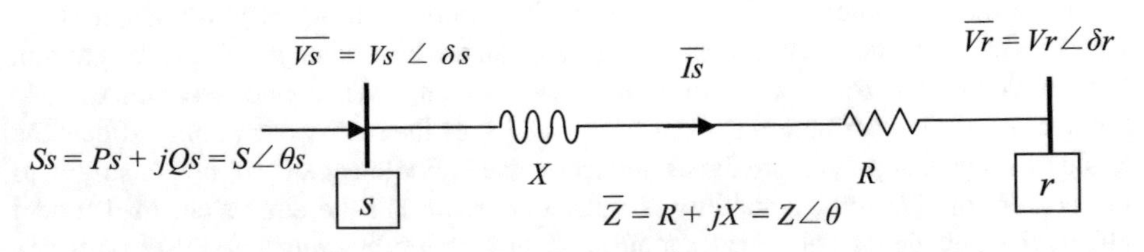

Fig. 1. Equivalent lumped model

$$Ps = \frac{Vs^2}{Z} cos\theta - \frac{VsVr}{Z} cos(\delta s - \delta r + \theta) \tag{1}$$

$$Qs = \frac{Vs^2}{Z} sin\theta - \frac{VsVr}{Z} sin(\delta s - \delta r + \theta) \tag{2}$$

Where

Vs: voltage magnitude of supply side
Vr: Voltage magnitude of demand side
Z: Impedance magnitude of transmission line
θ: Impedance angle
δs: Phase angle of supply side
δr: Phase angle of supply side
Ps: Real power of supply side
Qs: Reactive power of supply side

2.2 Equivalent π Model for Transmission Line

For middle term, short transmission line, which not only considering the series resistance and considering inductance also shunt at supply side and demand side. Equivalent π model of the distribution system are generally ignored skin effect. The admittance of supply side $\overline{Ys} = gs + jbs$ and the gs and bs are conductance and susceptance. Similarly, the admittance of demand side $\overline{Yr} = gr + jbr$ Where $gs << bs$ and $gr << br$ that gs and gr are omitted count. Therefore, $\overline{Ys} = jbs$, $\overline{Yr} = jbr$ quite are capacitive or inductive element, as shown in Fig. 2 is the equivalent π model distribution line, the real and reactive power flow as Eqs. (3) and (4).

$$Ps = \frac{Vs^2}{Z} cos\theta - \frac{VsVr}{Z} cos(\delta s - \delta r + \theta) \tag{3}$$

$$Qs = \frac{Vs^2}{Z} sin\theta - \frac{VsVr}{Z} sin(\delta s - \delta r + \theta) + YsVs^2 \tag{4}$$

Where Ys: admittance of supply side

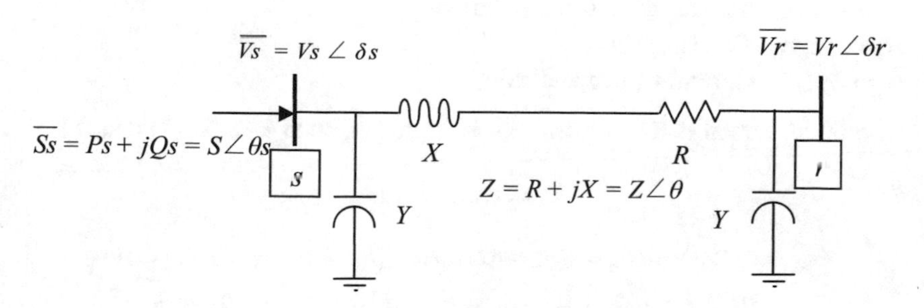

Fig. 2. The equivalent π model distribution line

2.3 Equivalent Distribution Model for Transmission Line

The exact relationship of voltages and currents between supply side and demand side if the transmission distance is belong the middle-term range that transmission line take the equivalent model π has the more accuracy. If the transmission line is long term, which based on the distance divide into N series parts of middle-term π equivalent model to reach the equivalent distribution models. By using two-port transmission parameters principle of theory, which product N small equivalent series Transmissive matrix becomes a whole transmission matrix between the supply and demand sides. Figure 3 is an equivalent distribution model of distribution lines. Equivalent π model and equivalent lumped model are applied the two-port network principle of theory of transmission parameters derived equivalent distribution model.

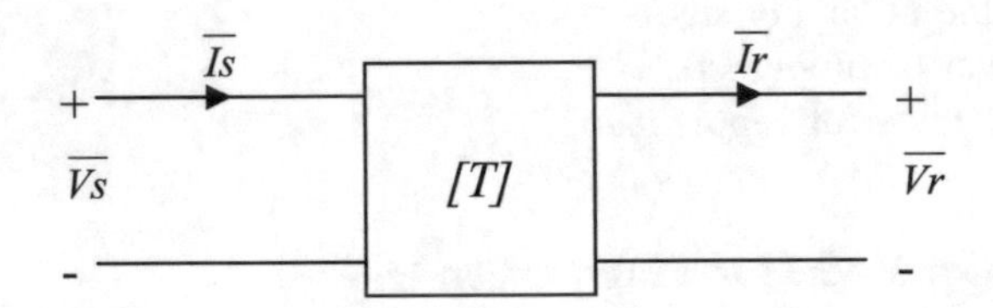

Fig. 3. Equivalent distribution model of transmission line

According to transmission network based on the double parameter definitions available transmission parameter matrix equation between the input and output, wherein A, B, C, D is the constant of proportionality, called transmission parameters such as Eqs. (5) and (6) and Eq. (5) can be expressed by the Eq. (6).

$$\begin{bmatrix} \overline{V_S} \\ \overline{I_S} \end{bmatrix} = \begin{bmatrix} \overline{A\,B} \\ \overline{C\,D} \end{bmatrix} \begin{bmatrix} \overline{V_r} \\ \overline{I_r} \end{bmatrix} \tag{5}$$

$$\left[\overline{SVI}\right] = \left[\overline{T}\right]\left[\overline{DVI}\right] \tag{6}$$

Where

$\overline{A} = A\angle\theta_A$: voltage gain
$\overline{B} = B\angle\theta_B$: transfer impedance (ohm)
$\overline{C} = C\angle\theta_C$: transfer admittance (mho)
$\overline{D} = D\angle\theta_D$: Current gain
$\overline{Ir} = Ir\angle\theta r$: Current of demand side

The feeder actual delivery power Ps is approach, such as Eqs. (7) and (8).

$$Ps = real\left[\overline{Vs} \times \overline{Is^*}\right] \tag{7}$$

$$\begin{aligned} Ps =& ACVr^2 cos(\theta_A - \theta_C) + ADV_r I_r cos(\theta_A - \theta_r + \theta_D + \delta_r - \delta) + \\ & BCV_r I_r cos(\theta_B + \theta_r - \theta_C - \delta_r + \delta) + BDI_r^2 cos(\theta_B - \theta_D) \end{aligned} \tag{8}$$

Where δ: added variable

Let voltage magnitude of supply side and demand side, the Max $[ADV_rI_rcos(\theta_A - \theta_r + \theta_D + \delta_r - \delta) + BCV_rI_rcos(\theta_B + \theta_r - \theta_C - \delta_r + \delta)]$, that exist $\delta = \delta_{max}$. By using δ_{max} substituted δ in Eq. (8) to become Eq. (9) and b Ps become to the $P_{S_{max}}$ maximum power delivery for feeder.

$$P_{S_{max}} = ACVr^2cos(\theta_A - \theta_C) + ADV_rI_rcos(\theta_A - \theta_r + \theta_D + \delta_r - \delta_{max}) + BCV_rI_rcos(\theta_B + \theta_r - \theta_C - \delta_r + \delta_{max}) + BDI_r^2cos(\theta_B - \theta_D) \tag{9}$$

3 Apply the PSS/E Power System Simulation Software

PSS/E (Power System Simulator for Engineering) simulation of power system analysis package, the operation of the power system simulation software package at PC Window environment, applied load flow analysis. Whether the trend of convergence calculation itself there are problems using the PSS/E power system simulation software to calculate the load flow [8–10]. Feeder configuration is radial type so that Gauss-Seidel Iteration method to find the parameter of load flow.

4 Main Transformer Solution Results

In this study, for each type load characteristic of the main transformer of the distribution system to explore the relationship between the maximum power and power to change the boundary between the three different equivalent transmission line models.

4.1 Discuss the Hourly Maximum Power of Feeder

The formula of Power feeder boundary FPB% is shown in Eq. (10) below. Where are the maximum power feeder $P_{S\,max}$ and actual power P_S of feeder with the equivalent distribution model? Voltage fixed size, the phase angle as the power change, in the steady state power boundary is larger, the higher the power delivery changes. Conversely, the smaller the power boundary changes in power delivery lower.

$$FPB\% = \frac{P_{S\,max} - P_S}{P_{S\,max}} \times 100\% \tag{10}$$

For feeders of power boundaries approach the main transformer maximum power $P_{T\,max}$ and minimum power $P_{T\,min}$ by feeder and main transformer connectivity such as Eqs. (11–13). Moreover, main transformer power boundary $MTPM\%$ as Eq. (14).

$$P_{T\,max} = \sum_{i=1}^{FN} P_{max,i} \tag{11}$$

$$P_T = \sum_{i=1}^{FN} P_{s,i} \tag{12}$$

$$P_{Tmin} = \sum_{i=1}^{FN} P_{min\,x,\,i} \tag{13}$$

Where *FN*: connected to main transformer of feeder number.

$$MTPM\% = \frac{P_{T\,max} - P_T}{P_{T\,max}} \times 100\% \tag{14}$$

4.2 Load Analysis of Main Transformer

Select Wei-Wu secondary substation, which contains three main transformers and associated connecting feeders to construct a small distribution system. Each of the main transformer can be connected to the feeder load statistics derived load and load curve. Figure 4 represents the load curve of main transformer, according to their curved shape that is a mix of residential and commercial consumption characteristics to form feeders load pattern, which contains two curves of real power P and reactive power Q. The real power P about the difference between the maximum and minimum is 10000 kW and reactive power Q about the difference between maximum and minimum is 3900 kVAR.

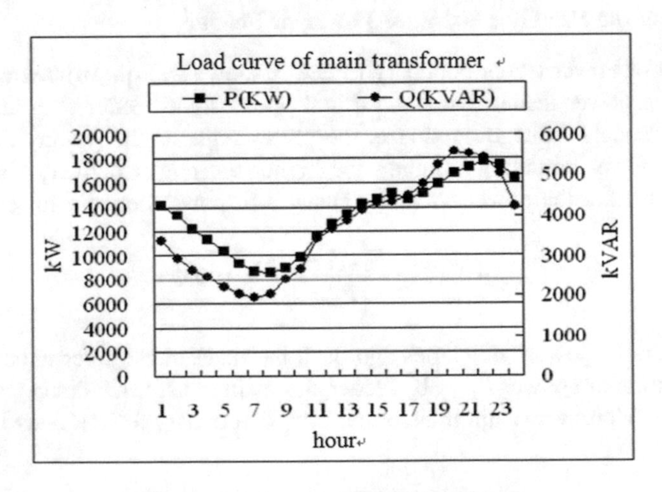

Fig. 4. Load curve of main transformer

4.3 Comparison of Three Kind's Equivalent Model of Transmission Line with Power Margin

Tables 1, 2 and 3 are three equivalent transmission line model of the main transformers power margins comparison table of Wei-Wu substation main transformers.

4.4 Load Curve of Main Transformer Base on Equivalent Lumped Model

Figure 5 represents the main transformer in accordance with its curved shape determined and load type statistic to determine what load type dominated the main transformer load characteristic. Which residential consumer dominates the load curve characteristic of main transformer form the first hour to the 9th hour. Moreover, the 10th hour to the 17th hour is another load type dominated by industrial customer, and the final time interval form the 18th hour to 24th hour is tendency patterns by commercial type load. The Fig. 7 of load curve has three plots with the real power, maximum power and power margin. It is the real power of main transformer difference between the maximum and minimum power is about 5631 kW and transmission maximum power of main transformer difference between maximum and minimum is about 5616 kW, and final curve is power margin curve that between maximum and minimum difference is about 1.46%. In the Figure the real power curve and the maximum power curve are almost overlap.

Table 1. The comparison of power margin of transformer in lumped model of feeders

Item	Actual power (MW)	Maximum power (MW)	Power margin (%)
Maximum	10.64	10.70	1.78
Minimum	5.01	5.08	0.31
Average	7.93	8.00	0.96

Table 2. The comparison power margin of transformer in π model of feeders

Item	Actual power (MW)	Maximum power (MW)	Power margin (%)
Maximum	10.08	10.89	1.84
Minimum	5.20	5.27	0.35
Average	8.12	8.19	1.01

4.5 Power Curve of Main Transformer with Equivalent π Model

Figure 6 represents load curve of main transformer according to the curve shape and consumer power demand that are residential, industrial and commercial load combination characteristic. Which residential consumer dominates the load curve characteristic of main transformer form the first hour to the 9th hour. Moreover, the 10th hour to the 17th hour is another load type dominated by industrial customer, and the final time interval form the 18th hour to 24th hour is tendency patterns by commercial type load.

The Fig. 7 of load curve has three plots with the real power, maximum power and power margin. It is the real power of main transformer difference between the maximum and minimum power is about 5629 kW and transmission maximum power of main transformer difference between maximum and minimum is about 5615 kW, and final curve is power margin curve, which between maximum and minimum difference is about 1.49%. In the Figure the real power curve and the maximum power curve are almost overlap.

Table 3. The comparison power margin of transformer in distribution model of feeders

Item	Actual power (MW)	Maximum power (MW)	Power margin (%)
Maximum	10.70	10.75	1.73
Minimum	5.07	5.14	0.31
Average	7.99	8.06	0.94

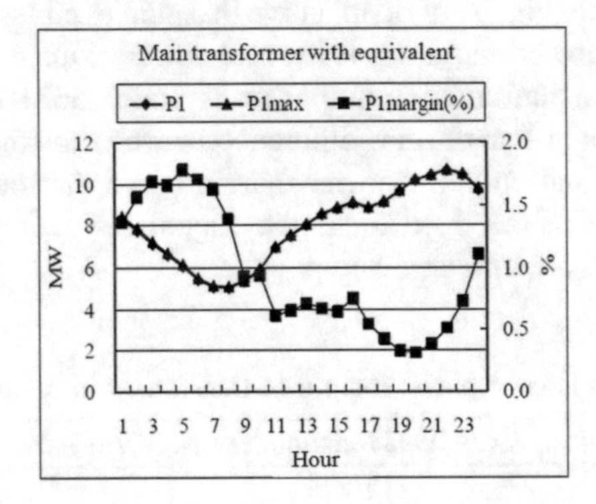

Fig. 5. Load curve of main transformer with equivalent lumped model

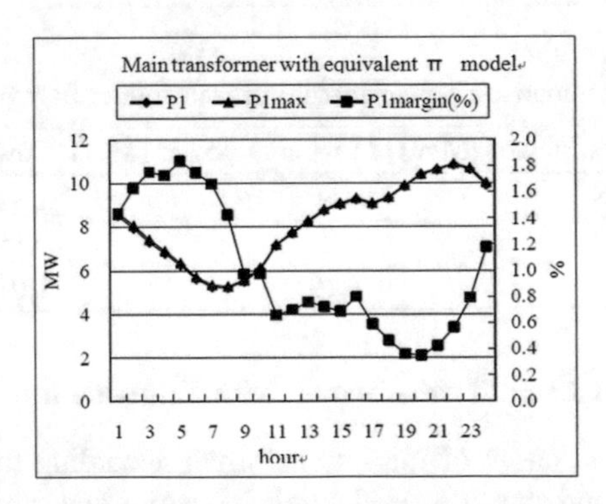

Fig. 6. Load curve of main transformer with equivalent π model.

4.6 Power Curve of Main Transformer with Equivalent Distribution Model

The Fig. 7 of load curve has three plots with the real power, maximum power and power margin. It is the real power of main transformer difference between the maximum and minimum power is about 5624 kW and transmission maximum power of main transformer difference between maximum and minimum is about 5609 kW, and final curve is power margin curve, which between maximum and minimum difference is about 1.44%. In the Figure the real power curve and the maximum power curve are almost overlap.

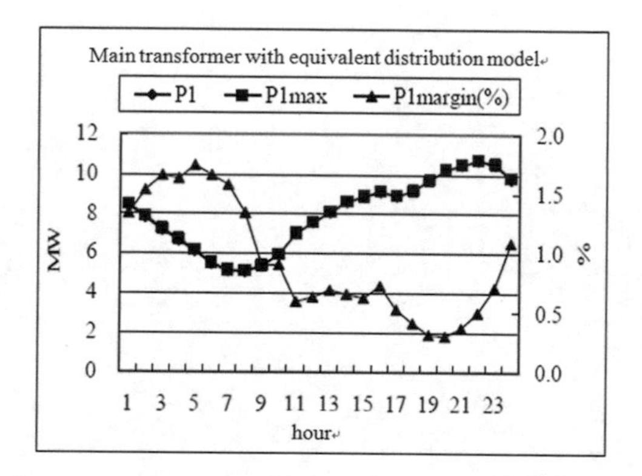

Fig. 7. Load curve of main transformer with equivalent distribution model.

5 Conclusion

It aim at substation Wei-Wu which including two main transformers with different load characteristic which base on the constant voltage magnitude on supply side and demand side approach the power margins of main transformer by feeder model of equivalent lumped, π and distribution models.

With three-phase load flow analysis by PSS/E to derive the main transformer #1 power margin that maximum value with equivalent π model is 1.84%, middle value with lumped model is 1.78%, and minimum value with distribution model is 1.74%.

It compare between power delivery and power margin by both main transformers which main transformer has more power about 10.89 MW delivery. Therefore, the trend power margin relationship of between power delivery and power margin of the main transformer is inverse ratio was also found.

References

1. Saadat, H.: Power System Analysis. McGraw-Hill, Boston (1999)
2. Pansini, J.: Guide to Electrical Power Distribution Systems. Pennwell (1996)

3. Bollen, M.H.J.: Understanding Power Quality Problems. IEEE Press (2000)
4. Heydt, G.T.: Power quality engineering. IEEE Power Eng. Rev. **21**, 5–7 (2001)
5. Chen, T.H., Chang, Y.L.: Integrated models of distribution transformers and their loads for three-phase power flow analysis. IEEE Trans. Power Deliv. **11**(1), 507–513 (1996)
6. Nelson, R., Bian, J., Williams, S.L.: Transmission series power flow control. IEEE Trans. Power Deliv. **10**(1), 504–510 (1995)
7. Chen, T.H., Chen, M.S., Hwang, K.J., Kotas, P., Chebli, E.A.: Distribution system power flow analysis-a rigid approach. IEEE Trans. Power Deliv. **6**(3), 1146–1152 (1991)
8. PSS/E-29 Program User Manual
9. PSS/E-29 Program Operation Manual
10. PSS/E-29 Program Application Guide

Development of Optical Fiber Stress Sensor Based on OTDR

Hsi-Shan Huang[1(✉)], Jeng-Shyang Pan[1], Yen-Ming Tseng[1], Weidong Fang[1], and Ruey-Ming Shih[2]

[1] School of Information Science and Engineering, Fujian University of Technology, Fuzhou, China
hw5400@hotmail.com, jengshyangpan@gmail.com, swkl200@qq.com, wdfang@126.com

[2] School of Design, Fujian University of Technology, Fuzhou, China
sunrise-design@live.com

Abstract. In this paper, an OTDR-wide full dispersion fiber stress sensor is used to sensing the signal of strain, and the geometrical bending of the optical fiber can be generated by the strain of stress, and the variation characteristic of the optical conduction modal. When stress is produced in single-mode fibers, the optical energy of a part of the optical fiber is transformed into a radiant mode, so that an OTDR can accurately identify the location of the signal source, and the OTDR uses the time-domain analysis of the optical pulse wave technique, which can simultaneously sensing the accurate energy loss and position of the source of stress signals [1].

The single-mode optical fiber is used to simulate two different practical environments. One is the soil, the other is the sand, and uses two different material protection casing, one is the plastic casing, the other is the silicone casing, carries on the real measure, and compares the different light wave pulse width, the measured result to the different average time to the stress influence. Finally, the material elasticity and deformation characteristics of two kinds of protective casing are tested. The experimental results are that the selection of silicone protective casing is ideal, while the length of 150 m multi-point monitoring, the light pulse width of the choice of 100 ns, so the results obtained in the experiment can prove the ratio of stress to elastic modulus of relationship. On the other hand, because the quality of sand and soil of different problems, so the hardness of different, the same weight trample on the specific stresses of the rendering, in different substrate materials reflect different stress changes, because the sand soft and then greatly enhance the sensitivity of stress [2].

Keywords: Full dispersion · Stress · Strain

1 Introduction

Optical fiber communication technology from the laboratory to industry and rapid growth, become a annual output of more than 100 billion yuan, today's information era of one of the pillar industries. Accompanied by the fiber industry chain of another branch of the fiber-optic sensing technology industry, experienced by sporadic research

J.-S. Pan et al. (eds.), *Advances in Intelligent Information Hiding and Multimedia Signal Processing*, Smart Innovation, Systems and Technologies 82,
DOI 10.1007/978-3-319-63859-1_27

into the development, from the military into civilian, by a single point of detection to the distributed network monitoring, in recent years is striding forward to the industrial takeoff [8].

The working principle of fiber optic sensors is to use the parameters of measurement to modulate the optical parameters transmitted within the fiber, and to detect the modulated optical signal, thus obtaining a device for measuring the value. When the propagation of light waves in the fiber is characterized by the characteristic parameters of light waves, such as intensity. wavelength, amplitude, phase, polarization and modal distributions are directly or indirectly changed due to external factors such as temperature, pressure, acceleration, voltage, current, displacement, vibration, rotation, bending, strain, and chemical quantities and biochemical quantities [3].

For this sense of measurement, optical fibers have incomparable advantages, because the pipeline is a thousand kilometers, the traditional electronic monitoring is not only easy to generate electric spark, it does not conform to the safety requirements of the petrochemical product transmission environment, and can only do the random point measurement, not to the whole monitoring, and the measurement of density and higher, the need to set up the measurement facilities will also increase the cost [9].

2 Structure and Theoretic Analysis of Sensing System

Research is a security defense system, the main sensing object is the human stepping into a region caused by the vibration, the sensor mainly has two parties to do, the first is the vibration sensing, the second is positioning vibration position. Because I am using a fully distributed sensing system, and there is no special detection point, the circuit itself is the sensor itself, then face a positioning problem, which will be applied to the light-time domain reflection technology, in the practical application, is to use the ordinary optical fiber single-mode fiber, because in some literature has the use of other special optical fibers, the effect is more remarkable, but the main point of this paper is that the use of ordinary optical fiber, no need to make special, reduce cost [10].

Around one weeks after the building, access to OTDR instrumentation, in a specific detection time period, as long as someone stepped into the field, that is, through the range of optical fiber surround, the vibration caused by the stampede will be perceived by the instrument, and positioning, and the system can achieve multi-point simultaneous positioning [4], that is, there are two or more than two people stepping into the machine can also be sensing, but there is a resolution of the limitation of precision, which is related to the power of the light pulse wave, the focus will be discussed later, distributed sensors and other types of sensors the biggest difference is the positioning of the two words, because there is no special sensing point, so the entire Ray route can be sensing points, then inevitably face positioning and positioning accuracy. The way to solve the positioning problem is to use pulsed light waves to detect the scattering light intensity of light waves. Originally such a strong light should be a certain rule with the distance, once a certain physical form changes, so long time must appear in the scattering light body change, because is the pulse wave, can calculate the start time, so can be reached the point of time to occur. This is the OTDR technology, if we determine the resolution of the system, that is when two different points are subjected to vibration, the shortest distance is how much can be detected, in the ideal state of the instrument, this is related to the time of pulse. Because if at the moment of the event, the two events are in the same pulse wave through the distance, can not be sensing [5]. The formula is as follows: dL = t*c/n (n refractive index of fiber)

When the pulse light arrives at L, the portion of the energy in the L ~ L+dL segment is returned to the instrument and the time required is the preceding t. (The distance of the dL is the distance that the light pulses can take, i.e. the resolution of the sensor can be measured) assuming that the light pulse is 5 ns, then the resolution is 1 meters, that is, the distance between 1 meters above the two vibration points, in the time Domain Analyzer can be distinguished. Because the time to walk on the light road is different. Figure 1 Principle system of distributed sensors.

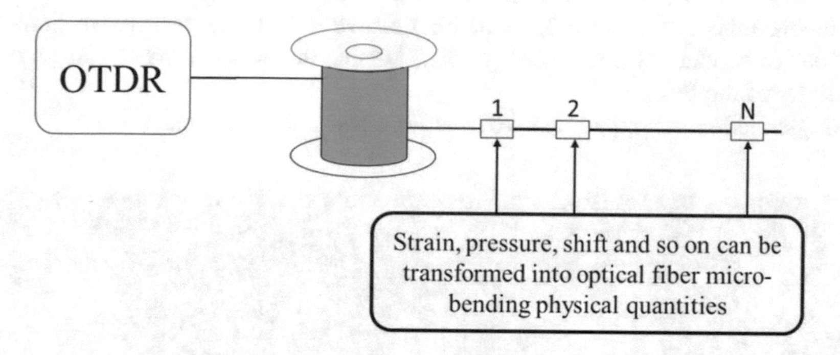

Fig. 1. Principle system of distributed sensors

3 Experimental Structure and Measurement Results

Function one is the engine is launched and power supplying: When the key is turned to ACC, the input voltage will begin to be detected via the voltage detection chip. The voltage detection chip keeps operating. When the key is turned to ON, the engine is launched, and the voltage will rise until 14 V. The current will flow through the coil. So, the magnetic switching operates to become a short-circuit switch, and it will begin to supply power to external devices [6].

Function two is battery life detection: When the key is turned to ACC, the input voltage will begin to be detected via the voltage detection chip. The voltage detection chip keeps operating. If the voltage is lower than 11 V, the buzzer will begin to operate to make a noise. The voltage regulator can regulate the voltage into 5 V for the buzzer to operate. The range of function two is from 9 V to 11 V. The driver should consider whether change their battery at this range. When the voltage lowers than 9 V, the engine isn't launched. The driver has to change their battery.

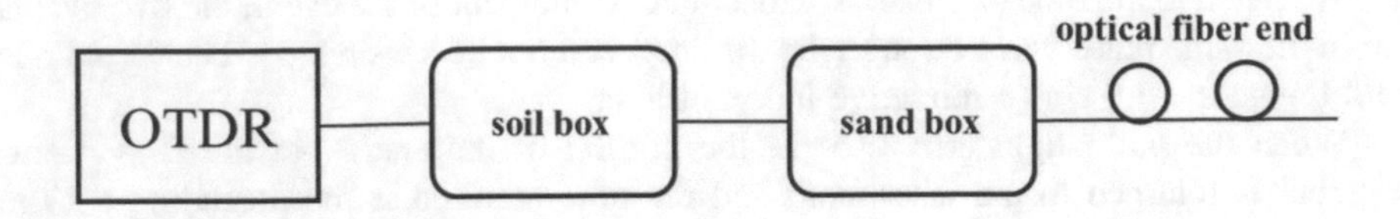

In the simulation environment, this experiment prepared two kinds of simulation environment, one is common soil, one is ordinary sand, the actual simulation of human normal weight trampling vibration, and observing the reaction of the instrument. Because it is a real-world simulation, the measurements at this time must be measured in real time, the average of zero [7]. The experiment is divided into 3 measurements: 1. is a single-mode optical fiber direct measurement, 2. is a single-mode fiber sleeve on the silicone tube, 3. It is a single-mode fiber sleeve to increase the average strength of the fiber to produce a more homogeneous strain, in the experiment can know the sensitivity of the best.

Single-mode fiber experiment 100 ns: First test point soil box

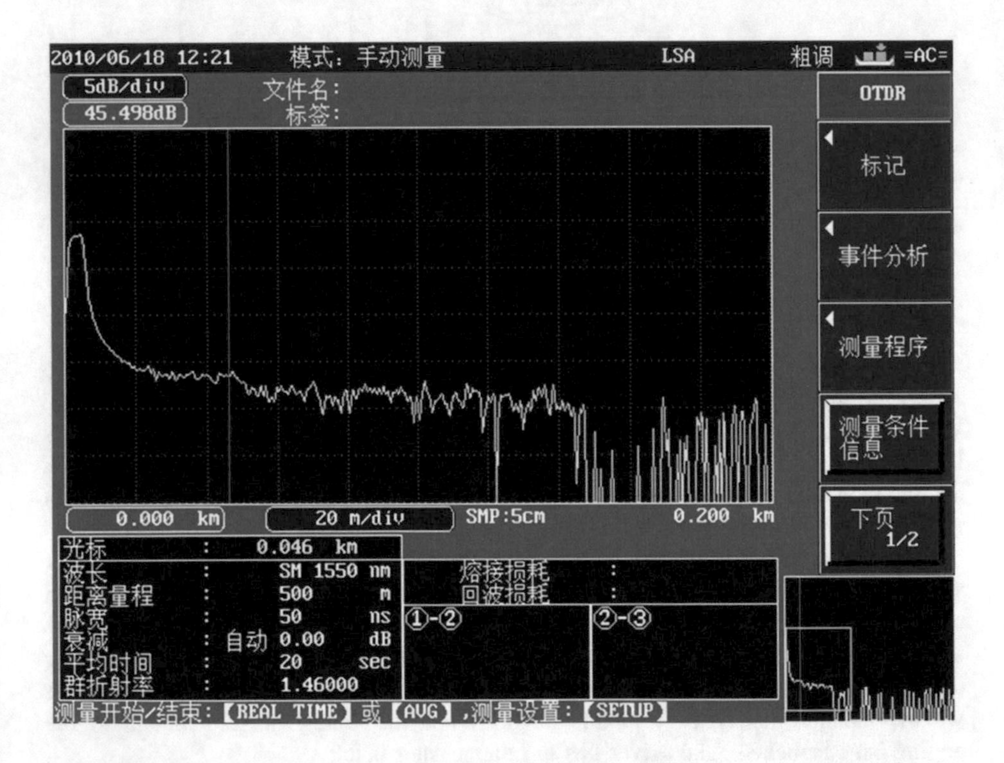

It is found that the local curves in 46 m have a downward gradient, although the back signal is doped in the noise, it is still suitable for the actual measurement.

Single-mode fiber experiment 100 ns:
Second test point sand box

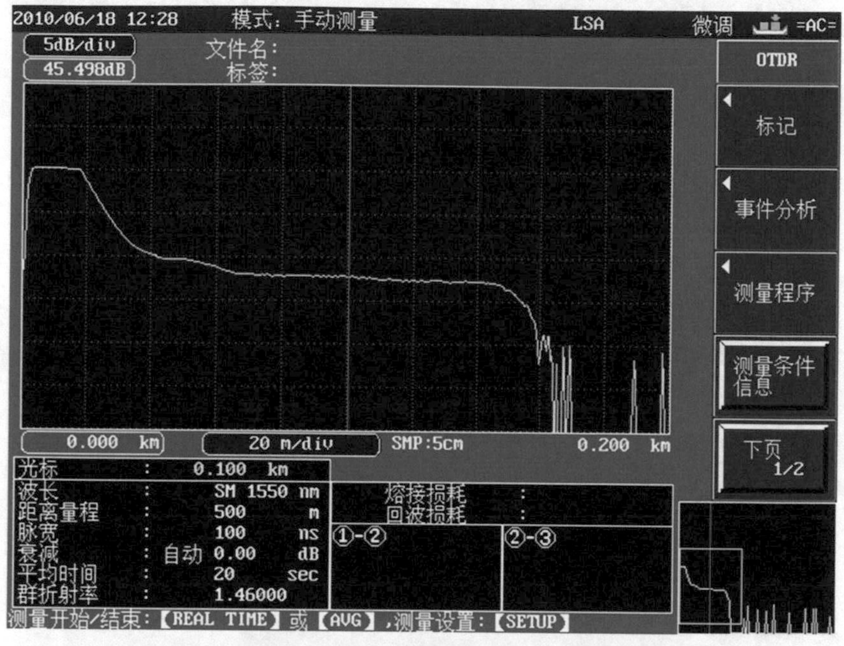

Single-mode fiber sleeve Silicone tube 100 ns: First test point soil box

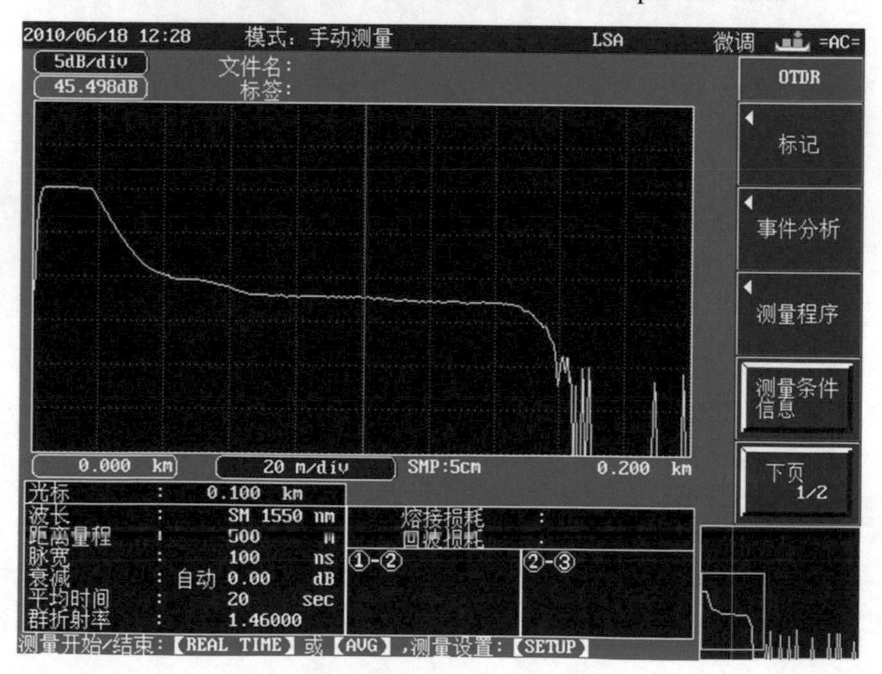

Single-mode fiber sleeve Silicone tube 100 ns: Second test First test sand box

Single-mode fiber sleeve plastic pipe 100 ns: First test point soil box

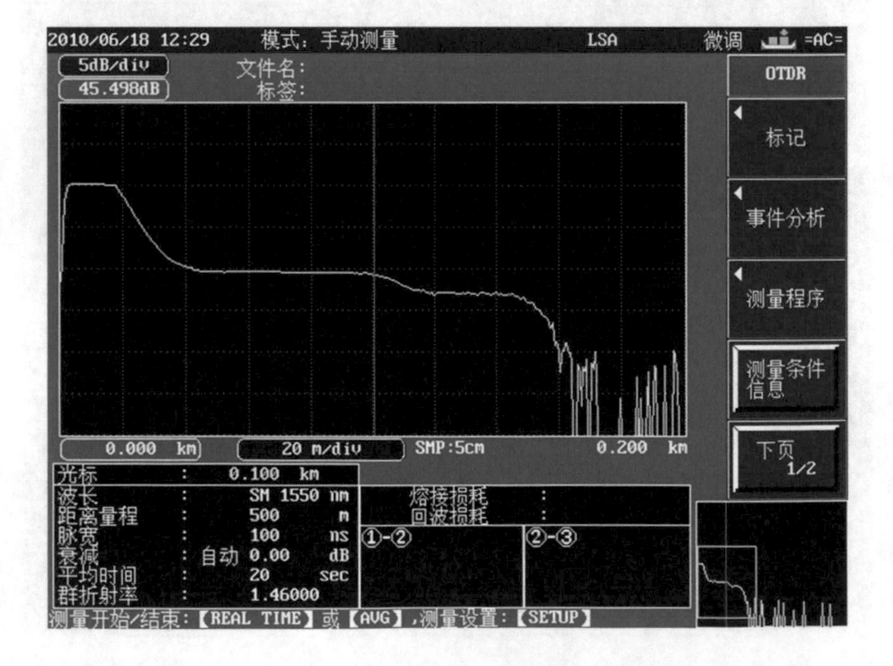

Single-mode fiber sleeve plastic pipe 100 ns: Second test point sand box

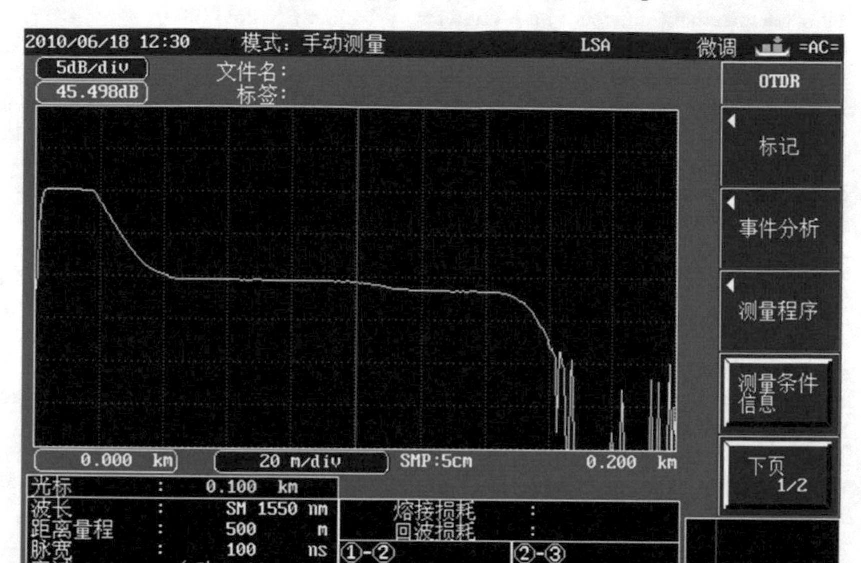

4 Conclusion

Two kinds of casing for tensile testing, under the same tensile force can make larger deformation, in the sand environment silicone casing phenomenon will be more obvious, and the difference between the two is bigger than in the soil environment, it is because the sand itself under the same stress deformation more obvious, so through the above experimental measurements, found that the sensitivity optimization is the silicone casing in the 100 ns measurement for the best. Judging from the whole experiment, it should be said that the whole measurement system from the theoretical model, and the simple measurement of the laboratory is set up, but the distance to become commercial products may also have a distance, may also need some supporting facilities to cooperate with its work, to achieve the true security of the function.

References

1. Guo, Z.S., Feng, J., Wang, H.: Cryogenic temperature characteristics of the fiber Bragg grating sensors. Cryogenics **52**(10), 457–460 (2012)
2. Guo, L.Y., Ho, I.J., Hou, Y.Y., Yang, C.H., Wu, W.L., Chen, S.K.: Comparison of plantar pressure distribution between different speed and incline during treadmill jogging. J. Sports Sci. Med. **9**(1), 154–160 (2010)

3. Mita, A., Yokoi, I.: Fiber Bragg grating accelerometer for structural health monitoring. In: Fifth International Conference on Motion and Vibration Control (2000)
4. Jung, J., Nam, H., Lee, B., Byun, J.O., Kim, N.S.: Fiber bragg grating temperature sensor with controllable sensitivity. Appl. Opt. **38**, 2752–2754 (1999)
5. Wu, Q., Hatta, A.M., Wang, P., Semenova, Y., Farrell, G.: Use of a bent single SMS fiber structure for simultaneous measurement of displacement and temperature sensing. IEEE Photonics Technol. Lett. **23**(2), 130–132 (2011)
6. Shen, C., Zhong, C.: Novel temperature-insensitive fiber Bragg grating sensor for displacement measurement. Sensor Actuators A Phys. **170**(1–2), 51–54 (2011)
7. Fu, H., Shu, X., Suo, R., Zhang, L., He, S., Bennion, I.: Transversal-load sensor by using local pressure on a chirped fiber bragg grating. IEEE Sensors J. **10**(6), 1140–1141 (2010)
8. Guru Prasad, A.S., Omkar, S.N., Vikranth, H.N., Anil, V., Chethana, K., Asokan, S.: Design and development of Fiber Bragg Grating sensing plate for plantar strain measurement and postural stability analysis. Measurement **47**, 789–793 (2014)
9. Wang, Q., Zhang, L., Sun, C., Yu, Q.: Multiplexed fiber-optic pressure and temperature sensor system for down-hole measurement. IEEE Sensors J. **8**(11), 1879–1883 (2008)
10. Dziuda, L., Skibniewski, F.W., Krej, M., Lewandowski, J.: Monitoring respiration and cardiac activity using fiber bragg grating-based sensor. Biomed. Eng. **59**(7), 1934–1942 (2012)

Application of AlN-Coated Heat Sink to Improve the Thermal Management of Light-Emitting Diode Assembles

Guo-Fu Lian[1], Ming-Der Jean[1](✉), and Tzu-Hsuan Chien[2]

[1] School of Mechanical and Automotive Engineering, Fujian University of Technology, Fuzhou, China
gflian@mail.ustc.edu.cn, mdjean@foxmail.com
[2] Department of Medical Laboratory Science and Biotechnology, Kaohsiung Medical University, Kaohsiung, Taiwan
13459421275@139.com

Abstract. This study reports the use of aluminum nitride(AlN) ceramics as a heat dissipated interface material to enhance the thermal performance of light-emitting diode (LED) packages. AlN ceramics are coated on copper/aluminum substrates by electrostatic spraying. The 18 orthogonal arrays of the LEDs with a signal-to-noise ratio are adopted and the impact of the effect of the spraying parameters on the AlN coatings is examined. Using an optimized design, the thermal performance of the LED is improved and good heat resistance characteristics are achieved. A confirmation run is used to compare results with the overall experimental tests. The results show that using AlN ceramic coating on a heat sink yields a good the thermal performance.

Keywords: Aluminum nitride · Thermal performance · Optimization · Light-emitting diode

1 Introduction

Light emitting diodes (LEDs) are growing rapidly because they are used in solid lighting devices. LEDs are more efficient than traditional light sources, including reducing overall product costs and increasing efficiency and efficiency, so they are widely used in a wide variety of applications such as signs and displays, back-lighting of LCD displays, traffic signals, and for architectural displays. Around 80% of the heat generated at the LED junction when forward biased is converted to heat rather than light [1–3]. Therefore, thermal management, which ensures the rapid dissipation of the heat that is generated in the LED chip through Cu/Al heat sinks, is crucial for LED modules. To meet out the demands in electronic industry, highly conductive materials with reasonable production cost are preferred. Various solutions have been reported on thermal properties of AlN [4–7]. Very few studies have reported the use of electrostatically sprayed AlN coatings as a thermal interface material [7–12]. In this study, AlN films are optimally deposited on Cu/Al heat sinks by electrostatic spraying. The thermal performance of a 1.2 W LED fixed on Cu/Al heat sinks with an AlN coating is conducted.

J.-S. Pan et al. (eds.), *Advances in Intelligent Information Hiding and Multimedia Signal Processing*, Smart Innovation, Systems and Technologies 82,
DOI 10.1007/978-3-319-63859-1_28

2 Experimental Details

2.1 Aluminum Nitride Coating Synthesis

AlN coatings were deposited on Cu/Al substrates (55 mm in diameter and 2/3/5 mm in thickness) using an electrostatic spraying system. The fabricated LED packaging system is composed of the LED module, an aluminum substrate, AlN coatings and a heat sink that uses both Al and Cu substrates, a 1.2 W LED light source with a metal core printed parallel circuit board and a starter. AlN is a unique ceramic material that has outstanding properties, such as high thermal conductivity (280 $Wm^{-1}K^{-1}$), low electrical resistivity and chemical stability.

2.2 Thermal Analysis

Since almost no thermal radiation or convection is observed in this study, the heat conduction rate for a component is verified and the mathematical formulation is given by Fourier's Law of heat conduction [13] as follows:

$$\dot{q} = q/A = -k\frac{dT}{dx} \tag{1}$$

where, $\dot{q}$ is the thermal conductive heat flux of a flatplate (W/m2), k is the thermal conductivity (W/m·°C), $\frac{\partial T}{\partial x}$ is the temperature gradient in the direction of the heat flow and $q = \frac{dQ}{dt}$ and A respectively represent the heat transfer rate and the heat transfer surface area (m^2). Thermal resistance, which is measured in °C/W, is the ability of a material to resist heat flow from a heat source to a sink. In other words, the resistance affects the temperature difference between the junction and the surface of the chip. The following equation is the formula that is used to calculate the resistance: Eq. (2)

$$R_{ja} = (\frac{T_j - T_a}{Q}) \tag{2}$$

where Rja is the thermal resistance from a device junction to the atmosphere, Tj is the device junction temperature under steady state operation, T_a is the ambient temperature (25° C) and Q is the heat power of the starter in the device. The thermal resistance of each LED in this array is the sum of the thermal resistance of the single LED package, and the thermal coupling resistance as follows:

$$R_{ja} = R_{js} + R_{sb} + R_{ba} \tag{3}$$

2.3 Optimal Design

In this study, 18 tests were used to allow a system attic approach for the optimization of various control factors with respect to thermal performance and quality. The optimization method, which uses orthogonal arrays, reduces the time and cost for the experiments. There are three categories for the quality of the characteristics in the analysis of the S/N ratios: the lower and the better, the higher and the better and the nominal and the better. These were used to measure the deviation in the quality of the characteristics from the desired values. In this study, the logarithmic transformation for a small-the-better characteristic, such as the T_j temperature of the AlN coated Al substrate, is used. The η value based on the loss functions is

$$\eta = -10Log\left[\frac{1}{n}\left(\sum_{i=1}^{n} y_i^2\right)\right] \quad (4)$$

Where y_i is the ith trial and n is the number of repeated tests for the ith trial. It is noted that the η value for each level of control factor is computed using the experimental data.

3 Results and Discussion

3.1 The Thermal Performance of the AlN Coatings

In this paper, the AlN coatings were used as an interface material for a 1.2 W LED package that was attached to a parallel circuit board. During the thermal test, the LED was driven using a 700 mA current when it reached a steady state. It is seen that when the driving currents are shut down after 60 min. Junction temperature of the LEDsis significantly low over time. This shows that the temperature of the LED packages is obviously improved when an AlN coated Al heat sink is driven at 700 mA, which AlN coated Al heat material is more favorable in reducing the total thermal resistance of the given 1.2 W LED package and hence supports for efficient heat dissipation. The central portion of the Cu/Al substrate exhibits the highest Tj value for AlN coatings. This may be due to solid-air interface defects in the crystal structure, which increase the thermal barrier and hence influences the heat conductance.

3.2 Thermal Resistance Analysis

In this study, the thermal performance of a 1.2 W LED on an AlN-coated Al heat sink that is used as a heat dissipated interface material and the influence of the parameters on the Tj and thermal resistance (R_{ja}) values are reported. The respective Ave. R_{ja} values with an AlN-coated Al heat sink are significantly decreased as compared to the AlN-coated Cu heat sink.

3.3 Thermal Resistance Analysis

To optimize the process to give the best thermal properties, the signal-to-noise(S/N) ratio was employed. In small-the-better case, from Eq. (4), the main effects of eight control factors are obtained. The S/N ratio is optimal at its maximum value, which the optimal levels for each control factor that LED assembly arrangement is 1.2 W were analyzed according to the η value effect determined as Al substrate plate, temperature of 100°C, thickness of 5 mm, spraying time of 30 s, reciprocating speed of 50 mm/s, spray flow rate of 4 cc/s, resin/AlN of 1:4, and baking time of 40 min due to their high S/N ratios.

4 Conclusion

In this study, the thermal performance of a 1.2 W LED that uses an electro-sprayed AlN-coated Al substrate as a heat sink and the effect of parameters on the Tj and Rja values are reported. An L18 orthogonal array is used for the LED aluminum plate and the effect of the deposition parameters on the coatings' thermal properties is determined. These results demonstrate that the temperature of the LED packages is significantly reduced if there is an AlN-coated Al substrate and a current of 700 mA. The Cu/Al substrate samples with AlN coatings also have better thermal performance when there is an optimal design for the electrostatic spraying process. The results demonstrate that junction temperature of the LED packages is effectively reduced, where there is an AlN-coated Al heat sink and a current of 700 mA. Overall, the Cu/Al substrate samples with AlN coatings can give better thermal performance when optimal design is conducted by electrostatic spraying.

Acknowledgments. This research was financially supported by the Natural Science Foundation of Fujian Province (grant number 2015J01181)

References

1. Shanmugan, S., Mutharasu, D.A., Haslan, H.: A study on AlN thin film as thermal interface material for high power LED. Int. J. Electron. Comput. Sci. Eng. **2**(1), 296–300 (2014)
2. Kuo, P.K.G., Auner, W., Wu, Z.L.: Microstructure and thermal conductivity of epitaxial AlN thin films. Thin Solid Films **253**, 223–227 (1994)
3. Lee, J.W., Cuomo, J.J., Cho, Y.S., Keusseyan, R.L.: Aluminum nitride thin films on an LTCC substrate. J. Amer. Ceram. Soci **88**(7), 1977–1980 (2005)
4. Park, S.H.: Robust Design and Analysis for Quality Engineering. Chapman & Hall, London (1996)
5. Pan, T.S., Zhang, Y., Huang, J., Zeng, B., Hong, D.H., Wang, S.L., Zeng, H.Z., Gao, M., Huang, W., Lin, Y.: Enhanced thermal conductivity of polycrystalline aluminum nitride thin films by optimizing the interface structure. J. Appl. Phys. **112** (2012)
6. Heo, Y.J., Kim, H.T., Kim, K.J., Nahm, S., Yoon, Y.J., Kim, J.H.: Enhanced heat transfer by room temperature deposition of AlN film on aluminum for a light emitting diode package. Appl. Therm. Eng. **50**, 799–804 (2013)
7. Lu, X.Y., Hua, T.C., Wang, Y.P.: Thermal analysis of high power LED package with heat pipe heat sink. Microelectron. J. **42**, 1257–1262 (2011)
8. Christensen, A., Graham, S.: Thermal effects in packaging high power light emitting diode arrays. Appl. Therm. Eng. **29**, 364–371 (2009)

9. Yung, K.C., Liem, H., Choy, H.S., Lun, W.K.: Thermal performance of high brightness LED array package on PCB. Int. Commun. Heat Mass Transfer **37**, 1266–1272 (2010)
10. Cheng, T., Luo, X.B., Huang, S., Liu, S.: Thermal analysis and optimization of multiple LED packaging based on a general solution. Int. J. Therm. Sci. **49**, 196–201 (2010)
11. Kim, Y.P., Kim, Y.S., Ko, S.C.: Thermal characteristics and fabrication of silicon sub-mount based LED package. Microelectron. Reliab. **56**, 53–60 (2016)
12. King, S.Y., Tseng, J., Zhao, J.: Design of AlN-based micro-channel heat sink in direct bond copper for power electronics packaging. Appl. Therm. Eng. **52**, 120–129 (2013)
13. Yang, K.S., Chung, C.H., Tu, C.W., Wong, C.C., Yang, T.Y., Lee, M.T.: Thermal spreading resistance characteristics of a high power lightemitting diode module. Appl. Therm. Eng. **70**, 361–368 (2014)

A Fuzzy Neural Network on the Internet Addiction for University Students in China

Chien-Hua Wang[1], Jich-Yan Tsai[2(✉)], I-Hsiang Lin[3], and Chin-Tzong Pang[4]

[1] School of Management, FuJian University of Technology, Fujian Province, Fuzhou 350118, China
thuck@saturn.yzu.edu.tw

[2] Department of Information Management, University of Kang Ning, Taipei 114, Taiwan
jytsai@ukn.edu.tw

[3] School of Economics and Management, Xiamen University of Technology, Xiamen, China
2016000008@xmut.edu.cn

[4] Department of Information Management and Innovation Center for Big Data and Digital Convergence, Yuan Ze University, Chung-Li 32003, Taoyuan, Taiwan
imctpang@saturn.yzu.edu.tw

Abstract. The invention of the Internet makes information sharing and transmission more and more convenient and popular in our daily lives. However, excessive dependence over the Internet turning into addictions often results in serious problems affecting one physically, mentally or interpersonally. We attribute this Internet addiction as an impulse-control disorder, i.e., Internet Obsessive Compulsive Disorder (IOCD). Comparing to teenagers or adults, university students have higher chances to access the Internet for class assignments or projects. Owing to this, our study is to find causes and their relations among various reasons and aspects through the literature to deduce certain rules for Internet Obsessive Compulsive Disorder. We apply Fuzzy Neural Network (FNN) to determine the importance degree of each criterion after collecting data through questionnaires. This study concludes that the threshold as to where the loss of control starts might serve as means for further research to health-care professionals.

Keywords: Internet addiction · Internet obsessive compulsive disorder (IOCD) · Fuzzy neural network (FNN)

1 Introduction

The use of the Internet has grown exponentially during the last two decades, initially in Western countries and more recently worldwide, with nearly three billions of people now having constant and cheap Internet access (Internet World Stats, 2015). In China, as the Internet being ubiquitous, the Internet addiction is increasing exponentially, and treated as one of the most significant public health matters among the widespread

J.-S. Pan et al. (eds.), *Advances in Intelligent Information Hiding and Multimedia Signal Processing*, Smart Innovation, Systems and Technologies 82,
DOI 10.1007/978-3-319-63859-1_29

growth of Internet users. According to a survey conducted by the China Internet Network Information Center (CNNIC) in January 2008 [6], 210 million people had access to the Internet in China, of which 31.8% were youth between 18 and 24 years of age. With the development and spread of the Internet, the Internet use among adolescents pronounces danger in China. The latest National Internet Development Report from China Internet Network Information Center (CNNIC, 2015) [7] showed that the number from 10-year-old to 19-year-old Internet users was about 147.97 million by the end of 2014, which constituted 22.8% of all Internet users in China. At a recent conference, Tao Ran, Ph.D., Director of Addiction Medicine at Beijing Military Region Central Hospital, reported 13.7%, about ten million teenager, meet the Internet addiction diagnostic criteria. Consequently, in 2007, China began restricting computer game use; current laws now discourage more than 3 h of daily game use [16]. The improper frequency carried out in scenarios where one should limit the use is now referring to Internet Obsessive Compulsive Disorder, abbreviated IOCD. Internet Obsessive Compulsive Disorder (IOCD), defined as a psychological dependence on the Internet, is characterized by severe psychosocial problems related to excessive time spent online. The negative health impacts of excessive Internet use have been well documented internationally. The prevalence in university students by Ni et al. [15] showed the IOCD rate about freshmen university students in China was 6.44% in 2009. As one can see in the reports, there is prevalence of Internet addiction among university students in China varying from 4.5% to 15.0% [11, 12].

2 Literature Review

2.1 Internet Addiction

The original concept of addiction is rooted in psychology, mental health and pharmacology research, referred to as "compulsive, uncontrollable dependence on a substance, habit, or practice to such a degree that cessation causes severe emotional, mental, or physiological reactions" [4, 14]. Internet addiction started off as a poorly-hinted joke or satirical hoax made by an American psychologist Goldberg in 1995 [8]. The first research about Internet addiction was presented by Young [20] at the American Psychological Association in 1996. She takes the term "Internet addiction" to describe a portion of population that fascinated on Internet use and suffered from negative consequences and defined Internet addiction as an impulse-control disorder which does not involve an intoxicant by pathological gambling. Beard and Wolf [3] also modified the pathological gambling diagnose criteria to Internet Addiction Diagnostic Criteria which is similar to Young's Diagnostic Questionnaire criteria [20]. Many researches focus on different aspects to explore the Internet addiction [9, 10, 18]. Because excessive dependence over the Internet turning into addictions often results in serious problems affecting one physically, mentally or interpersonally, we attribute this Internet addiction as an impulse-control disorder, i.e., Internet Obsessive Compulsive Disorder (IOCD).

2.2 IOCD Dimension and Criteria

Concluding the literature about Internet addiction, we converted these factors into four dimensions: (1) IOCD Symptom Dimension, (2) IOCD Behavior Dimension, (3) IOCD Problem Dimension, (4) IOCD Psychology Recognition Dimension to discuss the importance degree. Each dimension and related criteria are explained in the later section.

IOCD Symptom Dimension. According to the researches about Internet addiction [1, 5, 8, 10, 13, 20], we developed 4 dimensions, IOCD symptom dimensions included 5 criteria (tolerance, withdraw, planning, lack of control, time-consuming) that reflect Internet addiction symptoms.

IOCD Behavior Dimension. We classify the Internet addiction behavior dimension into five kinds of Internet behavior (cybersexual, cyber-relationship, net compulsions, information overload and computer addiction) proposed by Young [21].

IOCD Problem Dimension. Excessive Internet dependence often jeopardizes interpersonal relationship, vocational opportunities or education pursuits. These problems refer to the individual's mental problem, physical problem, interaction or social relationship with others or the influence to work or school performance [5, 10]. There are five criteria included in this problem dimension mood alteration, conflict, time management, social relationship, health.

IOCD Psychology Recognition Dimension. The Internet evolves into a commodity for many people. What psychological characteristics those addicts possess are the main theme this study focuses. We can draw a preliminary depiction on Internet addicts. They usually have lower self-esteem, own less assurance of self-worth, demonstrate weak interpersonal relationship, adapt poorly to a new environment [2, 19, 22].

3 Research Method

Our approach combines fuzzy numbers and a backpropagation network (BPN) structure to determine the importance degree about IOCD dimensions and criteria. Fuzzy set theory was proposed by Zadeh in 1965 [23], and he applied the concept of the fuzzy sets to emphasize that the degree of things should be described by fuzzy logic for catching up on the difficulty of the direct description in the real life. The Fuzzy-Neural Network (FNN) is a technology combined fuzzy theory and neural network. The performance of a fuzzy system is determined by its membership function, fuzzy logic rules and inference mechanism. This kind of fuzzy neural network expert system can be served as the basis of conclusion, prediction or decision making. There are many researches about hybrid fuzzy sets and neural networks [17, 24]. Compared to a common neural network, connection weights and propagation and activation functions of fuzzy neural networks differ a lot. The architectures correspond to the type of fuzzy inference realized by the fuzzy systems represented by the networks. Neuro-fuzzy systems of this kind are called fuzzy neural networks; their architectures are exactly the same as the connectionist multi-layer architectures of artificial neural networks, but they are fuzzy, as their connection weights as well as the input and the output values are fuzzy numbers. According to the type of inputs and weights, three different types of fuzzy neural networks can be defined as follows: (1) crisp weight and fuzzy inputs; (2) fuzzy weight and crisp inputs; (3) fuzzy weight and fuzzy inputs. According to the property of fuzzy linguistic input variable, our approach deals with the type (1) of fuzzy neural networks. The linguistic input variable direct fuzzification is transforming real inputs and real targets to fuzzy numbers.

4 Research Result Analysis

This study administers 568 questionnaires, in which 56 invalid ones deducted, with the total 512 valid questionnaires taken into consideration in China. Of 512 questionnaires, males take up 47.5% whereas females 52.5%. As in China, all the interviewees are college students, 512 subjects as the total. Among the interviewees in China, 8 times of daily online use takes up the highest proportion of 33.8%, followed by 30.7% of 2 to 3 times of online use, 17.6% of 4 to 5 times of online use, 10.5% of 6 to 7 times of online use, and 7.4% of one time online use. The frequency of the Internet use is similar on both straits, where we can see the proportion of 8 times of daily use plays the dominant part reaching 40% of the both groups. There are twenty IOCD criteria in this questionnaire. We use twenty input variables $x_1, x_2, \ldots, x_{20}$ to represent each ICOD criteria and x_{21} as output variable. The relationship between the input variables and dimension criteria has shown in Table 1.

Table 1. Input variable and criteria

Variable	Criteria	Variable	Criteria
x_1	Tolerance	x_{11}	Mood alteration
x_2	Withdrawal	x_{12}	Conflict
x_3	Planning	x_{13}	Time management
x_4	Lack of control	x_{14}	Social relationship
x_5	Time-consuming	x_{15}	Health
x_6	Cybersexual	x_{16}	Self-Esteem
x_7	Cyber-relationship	x_{17}	Personal Relationships
x_8	Net compulsions	x_{18}	Life Adjustment
x_9	Information overload	x_{19}	Family Function
x_{10}	Computer addiction	x_{20}	Emotions

4.1 Fuzzy Neural Network (FNN)

In the input layer with 20 input variables $x_1, x_2, \ldots, x_{20}$ corresponding with the first 20 question in our questionnaires, the 20 criteria in 4 IOCD dimensions are presented. One output variable x_{21} stands for the addiction degree in the questionnaire as the output layer. For simplicity, there is one layer in the hidden layer of the fuzzy neural network structure. The number of internal neuron nodes is training from the input questionnaire data from China which is determined with the smallest sum of squares due to the error (SSE). Before fuzzifying the input variable value, we normalize the Likert scale value (1 to 5) to the number between 1 to 0. After the preprocessing of the normalization about the value of input variable, we then fuzzify the input variable value to fuzzy numbers according to the rule in Table 2. In order to compute the weight of input variable (criteria), we first convert the crisp value of the input variable to the fuzzy number by the triangular transfer function. The relationship between the input variable weight, normalized input variable weight and the fuzzified of normalized input variable weight is shown in Table 2.

Table 2. Fuzzy weight

Input weight	Neural weight	Fuzzy weight
1	0.00	[0.00, 0.00, 0.25]
2	0.25	[0.00, 0.25, 0.50]
3	0.50	[0.25, 0.50, 0.75]
4	0.75	[0.50, 0.75, 1.00]
5	1.00	[0.75, 1.00, 1.00]

4.2 Data Analysis by the FNN in China

We construct the fuzzy neural network structure from the 512 China college students' questionnaire data based on the smallest sum of the squared error (SSE). There are 20 input variables (x_1, x_2, …, x_{20}) in the input layer, in which one output variable x_{21} stands for the addiction degree in the questionnaire as the output layer. For the smallest sum of squares due to the error (SSE), the internal nodes in the hidden layer are determined as 9. The fuzzy neural network structure is shown in Fig. 1. Instead of

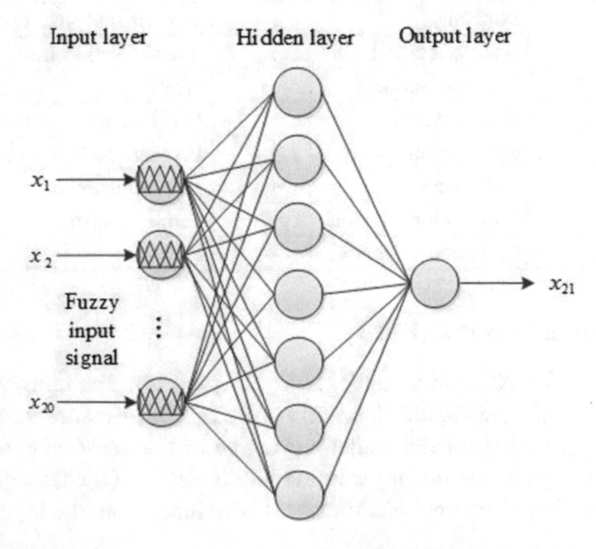

Fig. 1. Fuzzy neural network structure for the data collected from Taiwan students.

k-fold cross validation technique to determine the performance of our fuzzy neural network structure, 10 times replication process is conducted to get the average weight of IOCD criteria. After the data running in the fuzzy neural network, the result is derived and shown in Table 3.

Table 3. The average weight of dimension and criteria by the FNN in China

Variable	Criteria (Factor)	Weight	Ranking	Dimension	Weight
x_1	Tolerance	0.036	20	Symptom	0.263
x_2	Withdrawal	0.075	2		
x_3	Planning	0.045	12		
x_4	Lack of control	0.057	5		
x_5	Time-consuming	0.051	8		
x_6	Cybersexual	0.036	19	Behavior	0.267
x_7	Cyber-relationship	0.040	17		
x_8	Net compulsions	0.043	14		
x_9	Information overload	0.060	3		
x_{10}	Computer addiction	0.087	1		
x_{11}	Mood alteration	0.041	16	Problem	0.231
x_{12}	Conflict	0.045	11		
x_{13}	Time management	0.060	4		
x_{14}	Social relationship	0.048	9		
x_{15}	Health	0.038	18		
x_{16}	Self-Esteem	0.054	6	Psychology cognition	0.239
x_{17}	Personal Relationships	0.044	13		
x_{18}	Life Adjustment	0.053	7		
x_{19}	Family Function	0.045	10		
x_{20}	Emotions	0.043	15		

Under the same IOCD dimension structure, the criteria importance analysis is conducted by the Fuzzy Neural Network (FNN). The results are shown in Table 3. For the importance of the IOCD dimension in China, the weight of behavior dimension is 0.2670, the weight of symptom dimension is 0.2630, the weight of cognition dimension is 0.2387, the weight of problem dimension is 0.2313. It reveals that the most important IOCD dimension is behavior dimension, the behavior caused by the computer addiction is thought as the most important factor for the college students in China. For the IOCD behavior dimension, the most important criteria is "Computer addiction" criteria whose weight is 0.087 (rank 1). The weight of "Information overload" criteria is 0.060 (rank 3), playing an important role in the IOCD behavior dimension. For the IOCD symptom dimension, the weight of "Withdrawal" criteria is 0.075 (rank2), which means that the addiction to the Internet causing too much time spent on the Internet is considered as the second important factor. Two criteria of the IOCD recognition dimension, are more important than the other three criteria, which is "Self-Esteem" criteria whose weight is 0.054 (rank 6) and "Life Adjustment" criteria whose weight is 0.053 (rank 7).

The IOCD problem dimension, only "Time management" criteria is more important than the other four criteria in this dimension, whose weight is 0.060 (rank 4). To determine the most important criterion in the IOCD dimension, we choose the first 8 highest weight in this dimension, whose weights are the summation to 0.497, "Computer addiction", "Withdrawal", "Information overload", "Time management", "Lack of control", "Self-Esteem", "Life Adjustment" and "Time-consuming".

5 Conclusion

In our research, we propose a suggestion to the deficiency about the current Internet addiction research topic. First, the addiction factors and dimensions about the IOCD problem is discussed, since the current researches mainly focus on the addiction scale or diagnose criteria design. The scale score or the matched diagnose criteria is the main method to determine the degree of the Internet addiction. However, there are some common properties about the impulse-control disorder to the Internet addiction which cannot be accurately reflected as the result of IOCD. To determine the degree of importance about the IOCD criteria, MCDM methodology is used to construct the dimension of the Internet addiction that we call the IOCD dimension. We define four dimensions of the IOCD which are symptom, behavior, problem and psychology cognition dimensions.

And then, the FNN is used to determine the importance degree about the IOCD dimensions and criteria. For the FNN approach, three same criteria, "Withdrawal", "Lack of control" and "Time-consuming" criteria are both more important and the same criteria are applicable to China students. Although the fuzzy neural network gets a good result for determining the importance degree of dimension and criteria in our IOCD research, the neural network structure is a black box, easy to get results but explanations. In the meantime, the tuning process about the neural network is challenging when determining a best structure in the neural network. It also takes more time to try different combinations of learning rates, starting weights, hidden layer structures, activation functions, transfer functions and so on. We want to replace the internal neural network structure with genetic algorithm by using fuzzy theory. Another data mining technique may be included in our research to enhance the capability to self-learning in our framework.

References

1. American Psychiatric Association. Diagnostic and statistical manual of mental disorders, 4th edn., text rev. Author, Washington, DC (2000)
2. Armstrong, L., Phillips, J., Saling, L.: Potential determinants of heavier internet usage. Int. J. Hum. Comput. Stud. **53**, 537–550 (2000)
3. Beard, K.W., Wolf, E.M.: Modification in the proposed diagnostic criteria for Internet addiction. Cyber Psychol. Behav. **4**, 377–383 (2001)

4. Byun, S., Ruffini, C., Mills, J.E., Douglas, A.C., Niang, M., Stepchenkova, S., Lee, S.K., Loutfi, J., Lee, J.K., Atallah, M., Blanton, M.: Internet addiction: metasynthesis of 1996–2006 quantitative research. Cyber Psychol. Behav. **12**(2), 203–207 (2009)
5. Chen, S., Weng, L., Su, Y., Wu, H., Yang, P.: Development of a chinese internet addiction scale and its psychometric study. Chin. J. Psychol. **45**, 279–294 (2003)
6. China Internet Network Information Center (CNNIC). Statistical report of Internet development of China (2008). Accessed 17 Jan 2008
7. China Internet Network Information Center (CNNIC), report in (2015)
8. Goldberg, I.: Internet Addictive Disorder (IAD), Diagnostic Criteria (1996). http://www.psycom.net/iadcriteria.html. Accessed 21 Feb 2011
9. Griffiths, M.D.: Internet addiction: An issue for clinical psychology? Clin. Psychol. Forum. **97**, 32–36 (1996)
10. Griffiths, M.D.: Internet addiction: does it really exist?. In: Psychology and the Internet: Intrapersonal, Interpersonal, and Transpersonal Implications. pp. 61–75. Academic Press, New York (1998)
11. Jiang, D., Zhu, S., Ye, M., Lin, C.: Cross-sectional survey of prevalence and personality characteristics of college students with Internet addiction in Wenzhou China. Shanghai Arch. Psychiatry **24**(2), 99–107 (2012)
12. Liu, C.Y., Chen, Y.H.: An investigation of Internet dependence among college students and its psychological analysis. China Youth Study **11**, 64–67 (2009)
13. Meerkerk, G.J., Van Den Eijnden, R.J., Garretsen, H.F.: Predicting compulsive Internet use: it's all about sex! Cyberpsychology Behav. **1**, 95–103 (2006)
14. Mosby's Medical Nursing and Allied Health Dictionary, 5th edn., p. 321. Mosby, St. Louis (1998)
15. Ni, X., Yan, H., Chen, S., Liu, Z.: Factors influencing internet addiction in a sample of freshmen university students in China. Cyber Psychol. Behav. **12**(3), 327–330 (2009)
16. The more they play, the more they lose. People's Daily Online, 10 April 2007. http://en.people.cn/200704/10/eng20070410\underline364977.html
17. Liu, M., Quan, T.F., Luan, S.H.: An attribute recognition system based on rough set theory-fuzzy neural network and fuzzy expert system. In: Fifth World Congress on Intelligent Control and Automation (WCICA), vol. 3, pp. 2355–2359 (2004)
18. Shaw, M., Black, D.W.: Internet addiction: definition, assessment, epidemiology and clinical management. CNS Drugs **22**(5), 353–365 (2008)
19. Yang, C.K.: Sociopsychiatric characteristics of adolescents who use computers to excess. Acta Psychiatr. Scand. **104**, 217–222 (2001)
20. Young, K.S.: Internet addiction: the emergence of a new clinical disorder. Cyber Psychol. Behav. Soc. Netw. **1**(3), 237–344 (1996)
21. Young, K.S., Rogers, R.C.: The relationship between depression and internet addiction. Cyber Psychol. Behav. **1**(1), 25–28 (1998)
22. Young, K.S.: Internet addiction: evaluation and treatment. Student Br. Med. J. **7**, 351–352 (1999)
23. Zadeh, L.A.: Fuzzy sets. Inf. Control **8**(3), 338–353 (1965)
24. Wang, S.Q., Li, Z.H., Xiao, Z.H., Zhang, Z.P.: Application of GA-FNN hybrid control system for hydroelectric generating units. In: International Conference on Machine Learning and Cybernetics, vol. 2, pp. 840–845 (2005)

Optimal Economic Dispatch of Fuel Cost Based on Intelligent Monkey King Evolutionary Algorithm

Jing Tang[1,2], Jeng-Shyang Pan[1,2], Yen-Ming Tseng[1,2(✉)], Pei-Wei Tsai[3], and Zhenyu Meng[4]

[1] School of Information Science and Engineering, Fujian University of Technology, Fuzhou, China
tangjingdndx@163.com, jengshyangpan@gmail.com, swk1200@qq.com

[2] Fujian Provincial Key Laboratory of Big Data Mining and Application, Fujian University of Technology, Fuzhou, China

[3] Department of Computer Science and Software Engineering, Swinburne University of Technology, Melbourne, Australia
peri.tsai@gmail.com

[4] Harbin Institute of Technology Shenzhen Graduate School, Shenzhen, China
mzy1314@gmail.com

Abstract. Monkey King evolutionary algorithm (MKEA) is a new type and innovation of gene method that can be more effective evolution of the algorithm to reach goal or objective function. In this study and research is applied the monkey king evolutionary algorithm is used to apply the evolutionary particle to find the optimal power flow of system and calculate the complex power of each line, bus and to minimize power generation cost of the power plant. In order to study the practicability of the algorithm, it is applied to the standard IEEE 5bus load flow test system, and its convergence characteristic curve is observed and compared with the genetic algorithm (GA). The experimental results show that the MKEA can effectively solve the power system optimal power flow problem and this method is find the global solution not local solution that be confirmed in minimum fuel cost of generator of power plants. The minimum fuel cost obtained by MKE and GA is 5369.55 and 5422.0 US Dollars, respectively, when the number of population particles is 100 and the number of iterations is 300 that compared with GA which is 7.6% lower than GA. The results show that MKE has the obvious superiority to find the global solution.

Keywords: Monkey king evolutionary algorithm · Optimal power flow · Minimum fuel cost · IEEE 5bus · Convergence characteristic

1 Introduction

The optimal scheduling and cost problem of the generator under the optimal power flow (OPF) of the power system is a complex nonlinear programming problem. It is required to adjust the control system that can be used under the condition of satisfying

J.-S. Pan et al. (eds.), *Advances in Intelligent Information Hiding and Multimedia Signal Processing*, Smart Innovation, Systems and Technologies 82,
DOI 10.1007/978-3-319-63859-1_30

the specific power system operation and security [1] constraint to achieve the desired target stability [2] of the system stable operation state. In the existing optimization algorithm research which mainly divided into two categories, one for the traditional mathematical algorithms and another is artificial intelligence methods. The conventional mathematical methods are nonlinear programming method [3], linear programming method [4], mixed planning method [5] and interior point method. The above classical optimization methods are more mature and can be successfully applied to the power system optimization problem but they exist some problem as below: (1). it is difficult to control in real time depends on the exact model seriously. (2). the initial requirements of the more stringent, otherwise prone to suboptimal solution or even infeasible solution. (3). on the objective function and constraints have certain restrictions, such as continuous and different by accrue consideration to be simplified and approximate treatment. n recent years, with the development of computer and artificial intelligence technology which as artificial intelligence algorithms for differential evolution algorithm (DEA) [6, 7], particle swarm algorithm (PSO) [8, 9], bee colony algorithm [10, 11], gene algorithm (GA) [12, 13], Frog leaping algorithm (FLA) [14, 15] and so on, have been applied to power system problems. The artificial intelligence algorithm does not rely on the exact mathematical model and it can deal with continuous and discrete variables at the same time and achieved good results for the global solution to search the best path. The monkey king evolution (MKE) algorithm is a new gene evolutionary algorithm. Compared with other intelligent algorithms, the MKE algorithm uses a kind of affine transformation on the change and spatial expansion of the particle and the parameters to be set in the affine transformation that is only one parameter. MKE algorithm has a significant advantage in enhancing the cooperation between particles and solving large-scale optimization problems by the results.

2 Monkey King Evolutionary Algorithm

Monkey king evolution (MKE) algorithm [16, 17] is a new gene evolutionary algorithm. In the monkey king evolutionary algorithms which apply both monkey macros and ordinary particles by using the same affine transformation to achieve the search and expansion of particles at the same time. That mean in the monkey king evolutionary algorithm particle equivalent and enhance the search ability of all particles. Monkey King Algorithm uses a new affine transformation, the corresponding matrix elements of the multiplication and the specific process of change as shown in Eqs. (1)–(4).

$$\hat{X}_{diff} = \left(\hat{X}_{r1} - \hat{X}_{r2}\right) \tag{1}$$

$$\hat{X}_{gbest,G} = \hat{X}_{gbest,G} + FC * \hat{X}_{diff} \tag{2}$$

$$X_{G+1} = M \otimes \hat{X}_G + Bias \tag{3}$$

$$Bias = \overline{M} \otimes \hat{X}_{gbest,G} \tag{4}$$

Where

$\hat{X}_{diff}$: represents the extension matrix, which is subtracted by two different random matrices; where the matrix obtained by randomly substituting the row vector and each matrix has a setting number of row vectors by programmer.

FC: fluctuation coefficient of the expansion matrix that represent the position coordinate matrix of the *Gth* generation population.

M: transformation matrix which is transformed by the matrix obtained by orthogonal multiplication of the eigenvector matrix P and the diagonal eigenvalue matrix.

$\overline{M}$: The binary inverse matrix of M and that is the corresponding value of the nonzero element in M is zero in the matrix and the corresponding value of the zero element is become 1.

Solve steps and procedure in this paper to obtain the optimal economic dispatch of fuel cost of selected IEEE 5 bus system that including two generator.

Step 1: Define the search space;
Step 2: Generate the initial population particle position coordinates between the minimum and maximum values of the control variable;
Step 3: Call the Newton-Raphson method to calculate the power flow, judge and choose the variables within the limitation interval, otherwise abandon the particle;
Step 4: Calculate the fitness function value of the OPF problem and update it;
Step 5: For each particle based on the current optimal value is compared with the global optimal value then reserve the better value and update.
Step 6: If not meet the conditions then repeat three, four, five, six steps;
Step 7: Meet the conditions after, furthermore, output the results and stop the program.

3 Simulation System Description

The optimal dispatch and cost of power system problem basically in mathematic is a constrained optimization problem which consists mainly of variable set, equality equation, inequality constraint set and objective function. This paper chooses the minimum operating cost of the system as the objective function such as Eq. (5).

$$F = \min. \sum_{i \in S_G} \left(a_{2i}P_{Gi}^2 + a_{1i}P_{Gi} + a_{0i}\right) \tag{5}$$

Where

P_{Gi}: Real power of the *i*-th generator.
a_{2i}, a_{1i}, a_{0i}: Consumption characteristics curve parameters of the *i*-th generator.

3.1 Condition of Constraints

Constraints include equality constraints and inequality constraints. Equality constraints are node power balance equations and inequality constraints with control variables and state variables upper and lower bounds such as equations.

1. Equality constraints – node power balance.

$$P_{Gi} - P_{Di} - V_i \sum_{j=1}^{n} V_j \left(G_{ij} \cos\theta_{ij} + B_{ij} \sin\theta_{ij}\right) = 0, \qquad i \in S_B \tag{6}$$

$$Q_{Gi} - Q_{Di} + V_i \sum_{j=1}^{n} V_j \left(G_{ij} \sin\theta_{ij} - B_{ij} \cos\theta_{ij}\right) = 0 \qquad i \in S_B \tag{7}$$

2. Inequality constraints – state variables upper and lower bounds.

$$\underline{P_{Gi}} \le P_{Gi} \le \overline{P_{Gi}}, \qquad i \in S_G \tag{8}$$

$$\underline{Q_{Ri}} \le Q_{Ri} \le \overline{Q_R}, \qquad i \in S_G \tag{9}$$

$$\underline{V_i} \le V_{Ri} \le \overline{V_i}, \qquad i \in S_G \tag{10}$$

$$|P_l| = \left|P_{ij}\right| = \left|V_i V_j (G_{ij} \cos\theta_{ij} + B_{ij} \sin\left|\theta_{ij}\right) - V_i^2 G_{ij} \le \overline{P_l} \tag{11}$$

Where

S_B : the set of all nodes of the system.
S_G : The set of all generators.
S_R : The set of all the reactive power
S_l : The set of all branches;
P_{Gi}, Q_{Gi}: The active and reactive power of the generator
P_{Di}, Q_{Di}: The active and reactive load of the node
V_i, θ_i: For the node voltage amplitude and phase angle
G_{ij}, B_{ij}: The node admittance matrix *ith* row and *jth* column elements of the real and imaginary part
P_l: The line of real power set the line at both ends of the node

In order to ensure the stable operation of the system, requiring the transmission line at both ends of the voltage phase difference does not exceed 20 electricity degrees such as Eq. (12).

$$\left|\theta_i - \theta_j\right| < \left|\theta_i - \theta_j\right|_{\max} = 20^{\circ} \tag{12}$$

4 Experimental Results

This paper is for the IEEE 5-node power system such as Fig. 1, where all data is expressed in terms of per unit value and the upper and lower limits of the bus voltage are 1.1 and 0.9… The parameters of the generator consumption characteristic curve are shown in Table 1.

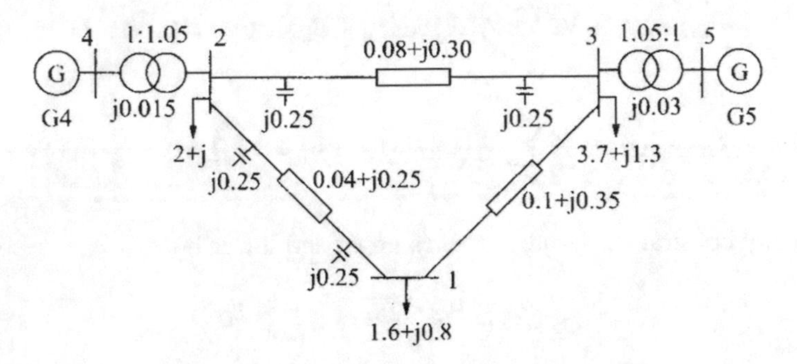

Fig. 1. IEEE 5-node power system diagram

The algorithm uses MATLAB2010b to verify the algorithm which parameters are as follows: MKE algorithm in the FC value is about 0.7, GA algorithm calls MATLAB comes with GA toolbox and which reorganization crossover probability is about 0.7. Under this setting, if the continuous 60 algebraic value is stable and convergence till the program cycle 10 times, furthermore, the final convergence results take average of 10 times of the convergence results.

Table 1 shows the parameters of the generator consumption curve.

Table 1. Generator fuel consumption and cost data

Generator number	BUS No.	Fuel consumption and cost curve parameters		
		Quadratic coefficient	Primary coefficient	Constant
1	4	50.4359	200.4335	1200.6485
2	5	200.55	500.746	1857.201

Table 2 shows the results of the optimization of the two algorithms for MKE and GA.

Figure 2 is the convergence curve of the MKE algorithm, and Fig. 3 is the comparison between of the convergence of MKE and GA.

It can be seen from Table 2 and Fig. 3 that the minimum fuel cost obtained by the MKE algorithm is 5369.55 U.S. Dollars in the case of initial setup under the number of population particles is 40 and the number of iterations is 300 which is 7.6% lower than that of GA. In the case of initial setting under the number of population particles is 100 and the number of iterations is 300 that minimum fuel cost obtained by MKE is 5339.56 U.S. Dollars, which is 1.9% lower than that of GA.

Table 2. Comparison of fuel cost results

Algorithm	Fuel cost (U.S. Dollar)		
	Minimal	Maximum	Average
(A) The MKE algorithm and GA algorithm comparison chart under the number of population particles is 40 and the number of iterations is 300.			
MKE	5304.25	5442.42	5369.55
GA	5334.82	6515.72	5809.41
(B) The MKE algorithm and GA algorithm comparison chart under the number of population particles is 100 and the number of iterations is 300.			
MKE	5304.14	5399.32	5339.56
GA	5346.98	5661.35	5422.03

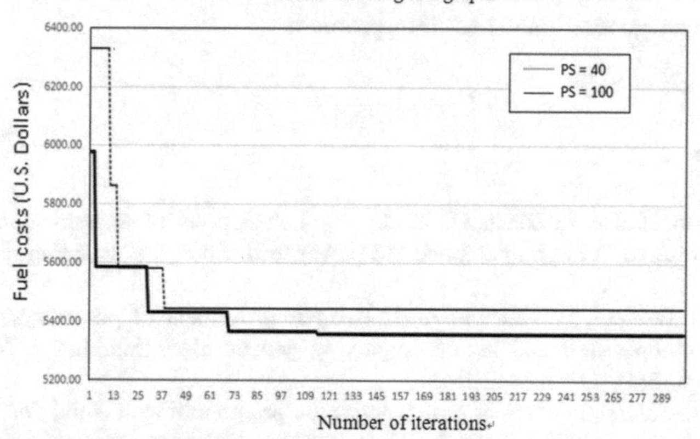

Fig. 2. Convergence characteristic curve of MKE algorithm

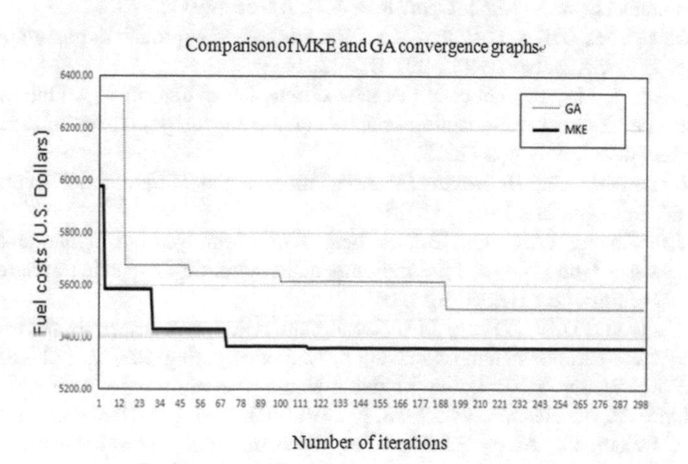

Fig. 3. Comparison of MKE and GA convergence graphs

It is shown that MKE algorithm is better than GA algorithm in solving optimal dispatch and cost of power system problem. It can be seen from Fig. 2, MKE algorithm in the initial setting of the appropriate increase in the number of particles can have a better convergence effect. Therefore, the experimental results of MKE algorithm show that our algorithm has better convergence speed and convergence accuracy.

5 Conclusion

Monkey King Evolution (MKE) algorithm is a new gene evolutionary algorithm with fewer parameters to setting. Monkey King Evolution (MKE) algorithm as a new gene evolutionary algorithm with fewer parameters and by mean of the IEEE-5 node power system is shown to be better robust and convergent. MKE algorithm has the characteristics of stable performance, fast convergence speed, strong global optimization ability, and has certain practical value for MKE and OPF to solve the problem of economic power generation scheduling problem.

References

1. Liu, M., Sun, H., He, J., Zhang, H., Yi, J.: Research on security assessment index system for operating reserve in large interconnected power grid. Energy Power Eng. **5**(4B), 785–791 (2013)
2. Keskes, S., Bahloul, W., Kammoun, M.B.A.: Improvement of power system stability by static var compensator and tuning employing genetic algorithm. Int. J. Mod. Nonlinear Theor. Appl. **3**(3), 113–123 (2014)
3. Li, D.F.: Closeness coefficient based nonlinear programming method for interval-valued intuitionistic fuzzy multiattribute decision making with incomplete preference information. Appl. Soft Comput. **11**(4), 3402–3418 (2011)
4. Hatami-Marbini, A., Tavana, M.: An extension of the linear programming method with fuzzy parameters. Int. J. Math. Oper. Res. **3**(1), 44–55 (2011)
5. Wöhrmann, A.M., Deller, J., Wang, M.: A mixed-method approach to post-retirement career planning. J. Vocat. Behav. **84**(3), 307–317 (2014)
6. Yi, W., Gao, L., Li, X., Zhou, Y.: A new differential evolution algorithm with a hybrid mutation operator and self-adapting control parameters for global optimization problems. Appl. Intell. **42**(4), 642–660 (2015)
7. Brest, J., Zamuda, A., Bošković, B.: Adaptation in the Differential Evolution, vol. 18, pp. 53–68. Springer, Heidelberg (2015)
8. Liang, J.J., Zhang, W.X., Qu, B.Y., Chen, T.J.: Multiobjective Dynamic Multi-Swarm Particle Swarm Optimization for Environmental/Economic Dispatch Problem, vol. 7389, pp. 657–664. Springer, Heidelberg (2012)
9. Zhang, Y., Gong, D.W., Zhang, J.H.: Robot path planning in uncertain environment using multi-objective particle swarm optimization. Neurocomputing **103**(2), 172–185 (2013)
10. Ling, W.X., Wang, Y.X.: Using Modular Neural Network with Artificial Bee Colony Algorithm for Classification, vol. 7928, pp. 396–403. Springer, Heidelberg (2013)
11. Kisi, O., Ozkan, C., Akay, B.: Modeling discharge–sediment relationship using neural networks with artificial bee colony algorithm. J. Hydrol. **428–429**, 94–103 (2012)

12. Chinnasri, W.: Adaptive probability of crossover and mutation in genetic algorithm on university course timetabling problem. In: 2013 IEEE International Conference on Computer Science and Automation Engineering, vol. 24(4), pp. 656–667 (2002)
13. Wang, S., Lu, Z., Wei, L., Ji, G., Yang, J.: Fitness-scaling adaptive genetic algorithm with local search for solving the multiple depot vehicle routing problem. Simulation **92**(7), 601–616 (2016)
14. Sarkheyli, A., Zain, A.M., Sharif, S.: The role of basic, modified and hybrid shuffled frog leaping algorithm on optimization problems: a review. Soft. Comput. **19**(7), 2011–2038 (2015)
15. Roy, P., Chakrabarti, A.: Modified shuffled frog leaping algorithm for solving economic load dispatch problem. Energy Power Eng. **334068**(4), 551–556 (2011)
16. Meng, Z., Pan, J.S.: Monkey king evolution: a new memetic evolutionary algorithm and its application in vehicle fuel consumption optimization. Knowl. Based Syst. **97**, 144–157 (2016)
17. Meng, Z., Pan, J.S.: A Simple and Accurate Global Optimizer for Continuous Spaces Optimization, Genetic and Evolutionary Computing, pp. 121–129. Springer, Heidelberg (2015)

Optimum Design and Control Research of Direct Drive Hub Motor

Zhong-Shu Liu(✉)

The Key Laboratory for Automotive Electronics and Electric Drive of Fujian Province, School of Information Science and Engineering, Fujian University of Technology, Fuzhou, China
lzs@fjut.edu.cn

Abstract. Using the improved genetic algorithm (IGA) and according to the characteristics of the hub motor, a novel direct drive outer rotor hub motors which is known as permanent magnet synchronous motor (PMSM) used in vehicle was developed. Aiming to the design requirements, the overall design scheme of the outer rotor hub motor was given, and the main size and the overall electromagnetic structure were analyzed and calculated. By using ANSYS software, the internal electromagnetic field analysis of the machine was made. Finally, applying the direct torque control method on the hub motor (PMSM) based on SPWM through MATLAB/SIMULINK, and the simulation results show that the performance of the prototype designed reasonably.

Keywords: IGA · Direct drive · Hub motor · PMSM · Simulation

1 Introduction

In the context of global energy conservation and emission reduction, electric cars have been the focus of research in recent years. The driving motor is the core component of electric car, and its performance is critical to the performance of electric car [1]. The technology of hub motor is also known as the hub motor built-in technology, which integrate power, drive and brake together on the wheel hub. It could eliminate the intermediate transmission mechanism and save space, which lead to the high efficiency. As a power equipment, hub motor should meet the requirements of starting frequently and acceleration or deceleration frequently. After considering a number of motors, three-phase permanent magnet synchronous motor was chosen to be hub motor for its high power density, high speed and high efficiency. This paper will optimize a kind of new type direct drive hub motor, the process is to transform the problem of engineering design into the optimal design, it is a problem of complicated, nonlinear, constrained, discrete and multivariate, which is difficult to use traditional optimization algorithm to meet the overall optimal solution of the hub motor. In this paper, basing on the research of the direct drive hub motor, according to the characteristics of the motor, its optimization design mathematical model was established, meanwhile, using the improved genetic algorithm (IGA), the structural dimensions of motor was determined.

J.-S. Pan et al. (eds.), *Advances in Intelligent Information Hiding and Multimedia Signal Processing*, Smart Innovation, Systems and Technologies 82,
DOI 10.1007/978-3-319-63859-1_31

2 The Optimum Design of the Direct Drive Hub Motor

2.1 The Motor Structure

Three-phase permanent magnet synchronous motor (PMSM) with low speed and great torque characteristics was used as direct drive hub motor, the outer rotor of the motor fitted at wheel rim directly was used to drive the vehicle, the overall structure was shown in Fig. 1.

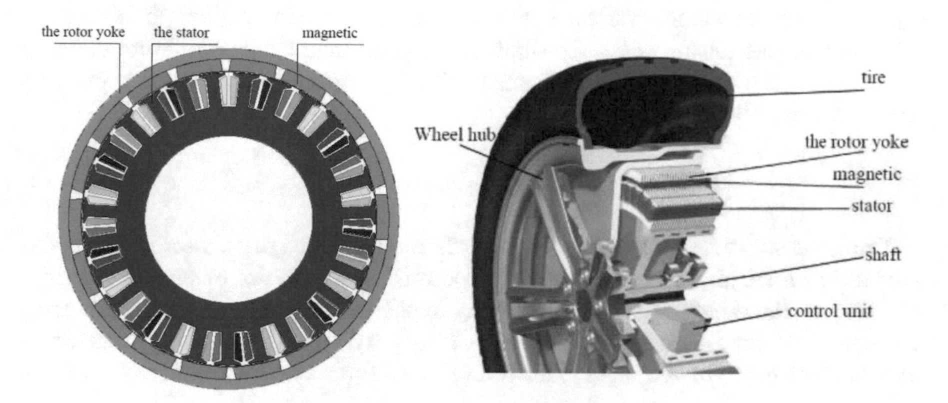

Fig. 1. Structure of three-phase PMSG

2.2 Optimum Design of the Hub Motor and the Determination of the Main Dimensions

The main dimensions of the rotor refers to as inside diameter D_{i1} and effective length of the stator core l_{eff}, which determines the motor quality and material cost [4]. The main dimensions of the motor design choice of reasonable has a decisive role on the technical and economic indexes. In this paper, we use the improved genetic algorithm (IGA) for calculating the main dimensions of the three-phase permanent magnet synchronous motor and the size of permanent magnet [2].

2.3 Mathematical Model for Optimal Design

The mathematical model of electrical motor optimization includes objective function, design variable and constraint conditions. Then, we will discuss these three terms in the following [3].

First of all, motor design optimization results reflected in whether achieving optimization goal finally. In recent years, the problem of energy crisis has become a global focus, therefore, under the premise of the motor performance, how to reduce the cost of motor has become our primary thinking. Due to expensive materials such as rare earth permanent magnet used in motor, we select the corresponding cost of unit efficiency as objective function [3] namely:

$$f(x) = \min \frac{F(x)}{\eta} = \min \frac{(m_{cu}t_{cu} + m_{Fe}t_{Fe} + m_m t_m)U_N I_N}{U_N I_N - \sum p} \tag{1}$$

In above, m_{cu} is the price for the copper materials; m_{Fe} is the price of steel materials; m_m is the price of ND-FE-B materials; t_{cu} is the total weight of copper materials used; t_{Fe} is the total weight of the steel materials used for; t_m is the total weight of ND-FE-B permanent magnet materials.

Second, we select the following parameters as optimization design of permanent magnet synchronous motor variables: the stator inner diameter D_{i1}, length of core l_{eff}, permanent magnet thickness h_m, the width of the permanent magnet b_m, rotor diameter D_2, rotor diameter D_{i2}, stator outer diameter D_1, the length of air l_δ and the coil turns per slot N_s, namely:

$$x = \left[D_{i1}, l_{eff}, h_m, b_m, D_2, D_{i2}, D_1, l_\delta, N_s\right]^T, x \in R^9 \tag{2}$$

Third, the constraint condition of rare earth permanent magnet motor mainly refers to the technical requirements and ensure the performance of motor of some constraints, according to three-phase permanent magnet synchronous motor design engineering experience and the required performance index [4], we select constraint parameters as follows: efficiency, voltage regulation, voltage waveform sine distortion rate and heat load.

2.4 Mathematical Model for Optimal Design

Optimum design of three-phase PMSM by using improved genetic algorithm (IGA).

In this paper we used the improved genetic algorithm (IGA) for motor structure design. By using MATLAB running in batch mode, we could achieve the goal of optimization [5]. The algorithm implementation process is shown in Fig. 2.

2.5 Fitness Function, Cross Operator and Variable Operator

First, the object function should be minimized to reduce the cost of the PMSM. The improved genetic algorithm is the most adaptive solution, so the work is to convert the minimum value problem into a maximum value problem. Hence, the fitness function is:

$$F = \frac{1}{f(x)} \tag{3}$$

Then, in order to get rid of disadvantages of conventional sort standard genetic algorithm selection method, the improved genetic algorithm (IGA) is used in front of the crossing method of selection sort, at the same time, using adaptive crossover and mutation of genetic algorithm for operation, it can make the crossover operator and mutation operator regulate with the increase of genetic algebra and ongoing automatically [5]. As the group has a tendency to into local optimal solution, the corresponding improved crossover probability and mutation probability; when spread group in the

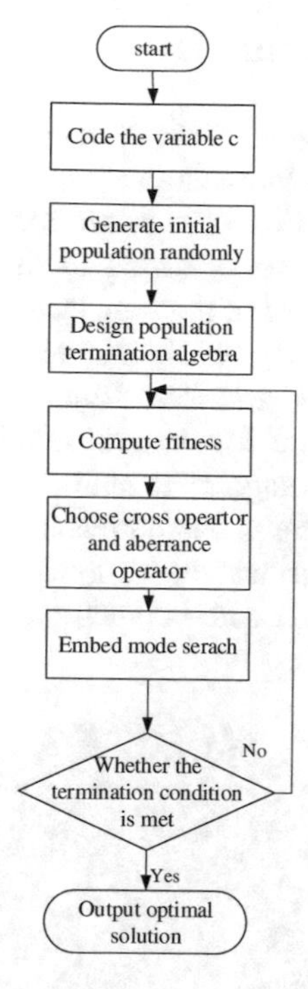

Fig. 2. IGA algorithm flow chart of PMSM

solution space, decrease the crossover probability and mutation probability. This improvement occurs at the same time in order to maintain the diversity of the population which ensure the convergence ability of genetic algorithm which make each generation of the individual is uneasy to be destroyed, moreover, to improve the optimization ability effectively.

2.6 Fitness Function, Cross Operator and Variable Operator

Based on the theory and method above, we can optimize the outer rotor hub motor which has rated voltage 230 v and rated power 8 kW by using the improved genetic algorithm (IGA). The results are shown in Table 1.

Table 1. Structure parameters of the prototype

Parameter	Value	Parameter	Value
Outer diameter of the stator	175/mm	Outer diameter of the rotor	204/mm
Inner diameter of the stator	100/mm	Inner diameter of the rotor	187/mm
The length of the core	75/mm	The air gap length	1/mm
The thickness of the magnet	7.5/mm	Slot number	27
The width of the magnet	27/mm	Pole	18

3 Finite Element Field Analysis

Now it is easy to analyze the magnetic field of the prototype by using two-dimensional finite element method, it has the following two prominent advantage makes it is particularly suitable for the calculation of the electromagnetic field distribution in the interior of the hub motor. One is that owing to the unification of each calculation section, it is easy to achieve the goal of program standardization. The other is that it can deal with internal medium boundary conditions very convenient. Then we first use the Maxwell 2D package to realize the grid dissection, which is the most critical step of the finite element discretization. Figure 3 is the grid subdivision diagram of the hub motor designed in this paper. From the diagram, it can be seen clearly that the grid density of the stator core and outer rotor core is much greater than the air gap [6]. The reason is that comparing with the air the permeability of the iron core is much greater. Therefore, the boundary condition of the iron core is much more complicated.

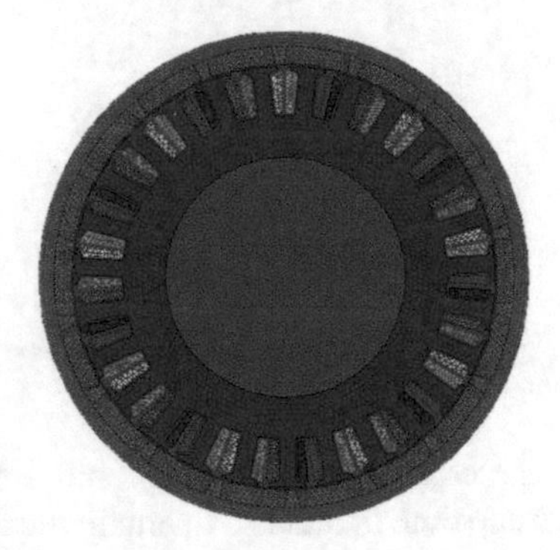

Fig. 3. Grid subdivision diagram of PMSM

Electromagnetic field analysis. Using Ansoft Maxwell we can analyze the flux density distribution of cloud of rotor shown in Fig. 4 when it is in no-load. It can be seen from the diagram, the flux density value of stator yoke of motor is maximum, the iron core of the motor is close to saturation meanwhile. The purpose of this optimize is to increase the utilization rate of ferromagnetic materials so as to improve the power density of the hub motor. It can be seen from the diagram that the distribution of motor is uniform. Only the stator yoke is saturated, while the whole motor magnetic circuit design is reasonable. We can get the flux density distribution of cloud of rotor shown in Fig. 5 when it is in load. Compared with Fig. 4, it can be seen that when the motor in load, the iron core saturation degree increase relative to the case in no-load, motor stator yoke of flux density value increase to a certain value, at the same time [6], all parts of the stator iron core flux density amplitude is increasing, the center line of the air-gap magnetic field of electric skewed compared with no-load, it is because the armature reaction cause gap magnetic field distortion [7].

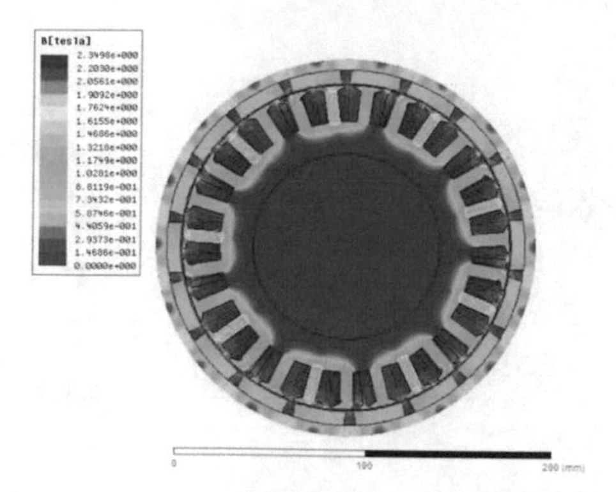

Fig. 4. No-load flux density cloud diagram of PMSM

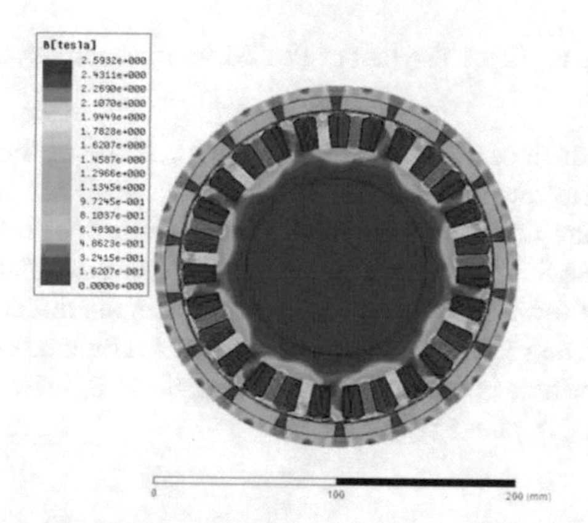

Fig. 5. Load flux density cloud diagram of PMSM

4 Finite Element Field Analysis

There are two kinds of control strategies for hub motor. When the hub motor is running below the basic speed, we can take maximum torque/current control strategy which can not only make the resistance losses minimum, but also reduce the losses of inverter and finally raise the efficiency of system. While the motor is running above the basic speed, the back electric motive force motor is increase, as a result, the armature current would be equal to zero, no electromagnetic torque can generate, and the motor will stop. In order to maintain a certain armature current under the condition of constant voltage above the basic speed, a weak magnetic control method should be adopted.

4.1 Mathematical Model of Hub Motor

The most common method used to analyze hub motor is the dq axis mathematical model which can be used not only to analyze the steady-state of the hub motor, but also

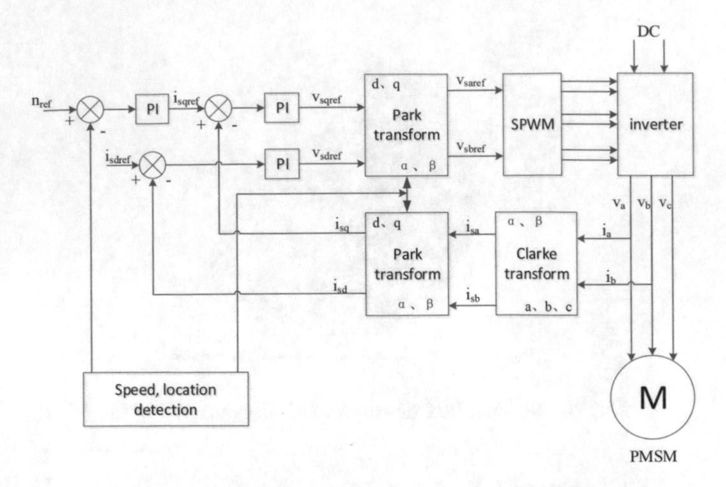

Fig. 6. Block diagram of PMSM vector control system

the transient performance of the hub motor. Figure 6 is the mathematical model of the three-phase permanent magnet synchronous motor (PMSM). In the mathematical model, a_s, b_s and c_s are the axis of the three-phase winding of the stator, assumed that the rotor rotate counterclockwise, take the axis of the fundamental wave magnetic field of permanent magnet for d axis, and q axis leads d axis the direction of rotation 90° in electrical angle [8]. The rotation velocity of the rotor reference coordinate is the rotor velocity, and its space axis is determined by the angle of θ_r between the q axis and the fixed axis (the axis of A phase winding).

4.2 The Voltage Equation and the Flux Linkage Equation of the Hub Motor

Using the above model, the voltage equation represented by the rotor reference coordinate are as follows:

$$\begin{cases} u_d = \dfrac{d\psi_d}{dt} - \omega_r\psi_q + r_1 i_d \\ u_q = \dfrac{d\psi_q}{dt} - \omega_r\psi_d + r_1 i_q \\ u_{2d} = 0 = \dfrac{d\psi_{2d}}{dt} + r_{2d} i_{2d} \\ u_{2q} = 0 = \dfrac{d\psi_{2q}}{dt} + r_{2q} i_{2q} \end{cases} \tag{4}$$

In above, r_1 is the stator winding resistance per phase; r_{2d} is the rotor direct axis damping resistor; r_{2q} is the rotor quadrature axis damping resistor; ω_r is the motor rotor electric angular velocity [9]. The flux linkage equation are as follows:

$$\begin{cases} \psi_d = L_d i_d + L_{ad} i_{2d} + \psi_f \\ \psi_q = L_q i_q + L_{ad} i_{2q} \\ \psi_{2d} = L_{2d} i_{2d} + L_{ad} i_d + \psi_f \\ \psi_{2q} = L_{2q} i_{2q} + L_{ad} i_q \end{cases} \tag{5}$$

In above, L_d and L_q are the stator direct axis and the quadrature axis inductance; L_{2d} and L_{2q} are the rotor direct axis and the quadrature axis inductance, L_{ad} and L_{aq} are the direct axis and quadrature axis mutual inductance between the stator and the rotor. Ψ_f is permanent magnet flux linkage [10]

4.3 Motion Equation of Hub Motor

The mechanical motion equations of the hub motor is as follows in case of iron loss and additional loss taking into account:

$$J\frac{d\Omega_r}{dt} = T_{em} - T_L - R_\Omega \Omega_r - \frac{p_{Fe} + p_s}{\Omega_r} \tag{6}$$

In above, T_L is load torque; $R_\Omega$is viscous friction coefficient; Ω_r is mechanical angular velocity; J is the total rotational inertia of the rotor and the load.

4.4 Coordinate Transformation

The relation between variable s of dq axis system and the variables of three-phase system can be realized by coordinate transformation. The coordinate transformation should be restrained by the condition of power unchanged. The expression for the three phase voltage of stator before the coordinate transformation is as follows [11]:

$$\begin{bmatrix} u_a \\ u_b \\ u_c \end{bmatrix} = \sqrt{2}U_N \begin{bmatrix} \cos\omega_1 t \\ \cos(\omega_1 t - \frac{2}{3}\pi) \\ \cos(\omega_1 t + \frac{2}{3}\pi) \end{bmatrix} \tag{7}$$

After changing to dq system, the expression is as follows:

$$\begin{bmatrix} u_d \\ u_q \\ u_0 \end{bmatrix} = T(\theta) \begin{bmatrix} u_a \\ u_b \\ u_c \end{bmatrix} = \sqrt{3}U_N \begin{bmatrix} \cos(\theta - \omega_1 t) \\ \sin(\theta - \omega_1 t) \\ 0 \end{bmatrix} \tag{8}$$

4.5 The Vector Control Strategy of the Hub Motor Based on SPWM

The basic concept of vector control theory of is to take the rotating rotor flux space vector as the parameter coordinate, the stator current is decomposed into two mutually orthogonal component, one is in the same direction as flux, which represents stator current excitation component, another orthogonal to flux linkage, which represents the stator torque component of the electric circulation, to control them respectively, we could obtain a good dynamic characteristic like DC motor. Because its control structure is simple, the control software implementation is easy to realize, so it has been widely applied to the speed regulating system.

Figure 6 is a block diagram for the vector control system of the permanent magnet synchronous motor [12].

In above rotor magnetic field orientation permanent magnet synchronous motor speed control system, the measured stator currents and rotor permanent magnet flux are independent to each other, the control system is very simple, and the torque constancy is good, which made it very suitable for realizing permanent magnet synchronous motor speed regulation control.

4.6 Control Strategy for Hub Motor Running at High Speed

Field-weakening control is used when the hub motor is running at high speed. When permanent magnet synchronous motor running steadily, the terminal voltage and the armature current are controlled. When the motor running at high speed, the resistance value is far less than reactance value, the voltage drop of resistance could be negligible. When the speed of the motor reaches a certain value, the stator field must be attenuated if we want to achieve a high speed [13].

The magnetic field of permanent magnet synchronous motor is established by a permanent magnet, so it cannot be directly reduced. The only way to achieve the goal of weak magnetic and speed growth is to increase the stator direct axis current by using direct axis armature to realize field weakening effect [14, 15].

4.7 Performance Simulation

Because the adjusting time, maximum speed, current, torque and the output power of the hub motor directly affect the performance of electric vehicles, so we analyze the speed and speed-torque characteristic of the prototype and obtain the simulation curves as follows. The simulation system based on the MATLAB software, it is composed of Simulink modules, the simulation results is shown in Figs. 7, 8 and 9 respectively.

It could be seen from the Fig. 7 that the adjust time is 1.93 s, which reflected that the hub motor had better starting performance, and from Fig. 8 we know that it take about 1.2 s for the current and the torque to reach its steady value. Finally it can be seen from the Fig. 9, the speed-torque curves reflected that the maximum output power of motor is 8 kW, so the prototype meet the design requirements fully.

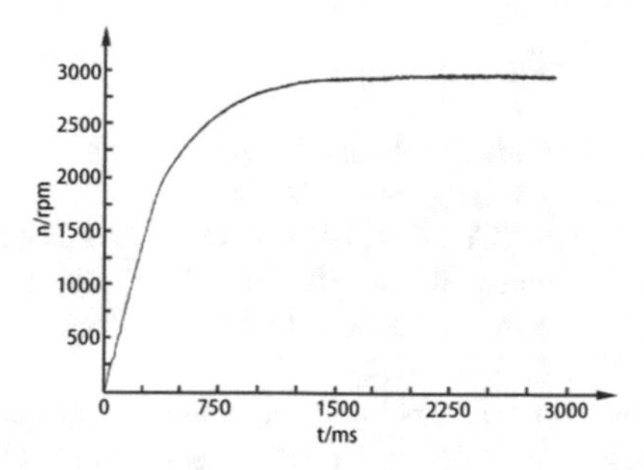

Fig. 7. Speed response simulation curve

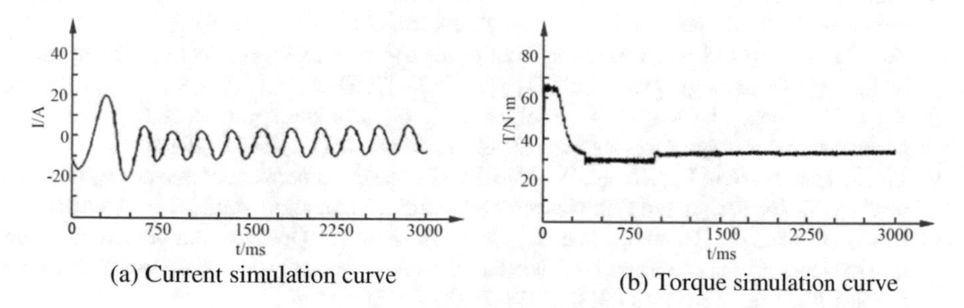

(a) Current simulation curve

(b) Torque simulation curve

Fig. 8. Current and torque simulation curve

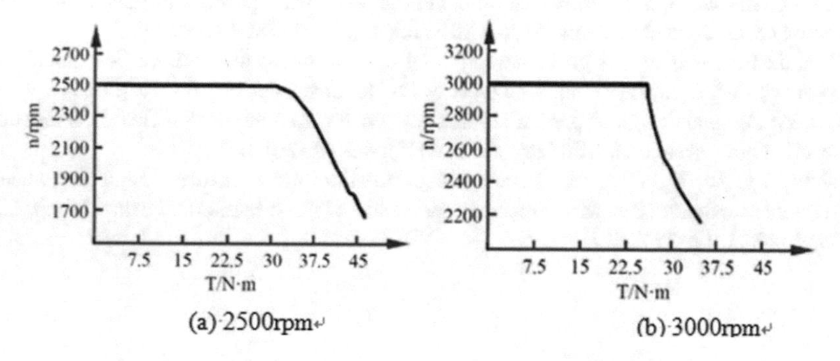

(a) 2500rpm

(b) 3000rpm

Fig. 9. Speed-torque characteristic curve

5 Conclusion

By using improved genetic algorithm(IGA) a novel direct drive hub motor was design, the simulation results show that the prototype had very good starting performance and speed- torque characteristic, so it meet the design requirements fully.

Acknowledgment. This research was supported by the Natural Science Foundation of Fujian Province of China. (Grant No. 2017J01667).

References

1. Xing, X., Shi, C.: Design of a fault-tolerant direct-drive hub motor used by electric vehicle. Small Special Electr. Mach. **44**(10), 22–25 (2016)
2. Jia, Y., Tong, L., Lu, T., Du, C.Y.: Simulation analysis and design of speed control system for direct drive motor. J. Beijing Inf. Sci. Technol. Univ. **28**(6), 5–9 (2013)
3. Lei, L., Hu, Y., Song, Z., Chen, Y., Zhang, H.: Design of outer rotor hub motor for electric vehicle. Micromotors **49**(10), 6–9 (2016)
4. Lu, D., Ouyang, M., Gu, J., Li, J.: Field oriented control of permanent magnet brushless hub motor in electric vehicle. Electr. Mach. Control **16**(11), 76–83 (2016)
5. Wang, X., Gao, P.: Analysis of 3-D temperature of in-wheel motor with inner-oil improved cooling for electric vehicle. Electr. Mach. Control **20**(3), 36–41 (2012)
6. Yane, Z., Jianwu, Z.: Study on electronic differential control system of independent in-wheel motor drive electric vehicle. J. Syst. Simul. **20**(18), 4767–4771 (2008)
7. Sun, M.: Design and flux-weakening control for in-wheel motor of electric vehicle. Harbin Institute of Technology, Harbin, Ph.D., pp. 25–32 (2009)
8. Shen, X., Zhao, Q., Xu, P.: Control research on maximum power of the direct-drive permanent magnet synchronous rotor. J. Electr. Power **29**(1), 28–31 (2014)
9. Liu, J., Hao, S., Sun, M., Zheng, W., Hao, M.: Research on permanent magnet synchronous in-wheel motor design and flux-weakening control. Micromotors **43**(2), 17–21 (2010)
10. Chen, Q., Shu, H., Ren, K., Chen, L., Chen, B., Xin, A.: Optimization design of driving in-wheel motor of micro-electric vehicle based on improvement genetic algorithm. J. Central South Univ. (Sci. Technol.) **43**(8), 2013–3018 (2012)
11. Shen, J., Miao, D.: Variable speed permanent magnet synchronous rotor systems and control strategies. Trans. China Electrotechn. Soc. **28**(3), 2–8 (2013)
12. An, Q., Sun, L., Sun, L.: Research on novel open-end wind permanent magnet synchronous motor vector control systems. Proc. CSEE **35**(22), 5891–5898 (2015)
13. Gao, J.: Research on design technology and application of direct-drive permanent magnet rotor with wind turbine. Hunan University Ph.D., Changsha, pp. 36–55 (2013)
14. Liu, Z.: Optimal design of permanent magnet synchronous motor vehicle characteristics and speed control system. J. Minjiang Univ. **31**(5), 44–48 (2010)
15. Weng, M., Li, Q., Cao, M.: Parameters calculation and structure design for wheel hub permanent magnet brushless direct current motor of drive method. Electric Mach. Control Appl. **42**(7), 12–15 (2015)

Building of a Practical Monitoring System for the Small Wind Turbine

Rong-Ching Wu[1], Yen-Ming Tseng[2(✉)], En-Chih Chang[1], and Chih-Yang Hsiao[3]

[1] Department of Electrical Engineering, I-Shou University, Kaohsiung, Taiwan
{rcwu, enchihchang}@isu.edu.tw

[2] School of Information Science and Engineering, Fujian University of Technology, Fuzhou, China
swk1200@qq.com

[3] Department of Electrical Engineering, Chung Yuan Christian University, Taoyuan, Taiwan
o1003689@gmail.com

Abstract. The small wind power system is a renewable energy system that is more applicable to general residential or other applications. Most users of the small-scale wind power generation system come from the general population, who do not have professional technologies and concepts. A good monitoring system can provide the user with quick understanding of the operational status, efficient troubleshooting, and improvement of the whole system. The paper proposes a wireless remote monitoring system for wind systems. This system includes a host computer and a remote monitoring module. The host computer is used for command control, data collection, and analysis. It can order the remote monitoring module to operate electrical equipment, gather measured data from the remote monitoring module, and analyze these data. The operation screen of the host computer is accomplished by the LabVIEW front panel. Via interactive man-machine interface, the user can input control commands and monitor operating conditions, so the operation is simpler and more straightforward. The remote monitoring module can measure voltage, current, generator speed, and wind speed data; in addition, it can provide an electromagnetic brake for the wind turbine.

The communication of the host computer and remote monitoring module is constituted through a wireless network, so the host computer can fully understand the system status wherever placed, and remotely monitor and control the entire system. The specific functions of the remote monitoring module are performed by an AVR single chip and its surrounding instrumentation.

The monitoring system of this paper is expected to effectively manage the wind turbine system, improve the efficiency of the system, and provide a reference to further improve the system. This paper monitors a practical 400 W small wind turbine as an object, which verifies the practicability of this paper in practical applications.

Keywords: Wind turbine · Monitoring system · Microprocessor

J.-S. Pan et al. (eds.), *Advances in Intelligent Information Hiding and Multimedia Signal Processing*, Smart Innovation, Systems and Technologies 82,
DOI 10.1007/978-3-319-63859-1_32

1 Introduction

Taiwan has many suppliers of wind turbine components that are ranked tenth in the world. At least twenty-two of these are involved in development or production of small and medium-sized wind turbine components. At least nine of these companies participate in the production and manufacture of wind turbines producing wattage below 10 KW. The domestic supply chain and repair maintenance of the wind turbine system components are almost completely in place. In order to support the development of the wind turbine industry, the Ministry of Economic Affairs of R.O.C in the "Draft of Renewable Energy Electricity Purchase Rate and Calculation" hearings suggested the wholesale payment of electric energy for about 1-200 KW is 5.344 NTD/degree and 2.1826 NTD/degree over 200 KW. The development of distributed generation and centralized generated renew energy form will be based on 10-200KW wind turbines that is also oriented to a low-carbon model society, and off-shore island renewal energy. Besides, small wind turbine use on farms and recreation areas will also be one of the future directions of development, for both tourism and energy alternative benefits [1]. Taiwan's leisure farms number over five hundred with about sixty recreation areas; half of them have potential for using small wind turbine devices. Each device can potentially produce about 25-200 KW, which has considerable development potential. As the small wind turbine generation system users are mostly from the general population and lacking professional technique and concept of wind power generator, a functional monitoring system as a supplement will play a very important role. A monitoring system can not only present operating conditions for preventing the generator from exceeding a safe speed limit and measuring battery overcharge, but also provide the setting of mains electricity in parallel under allowable operating conditions, thereby enhancing the overall efficiency of the wind turbine [2]. The purpose of this study is to establish a monitoring system for a wind turbine. By using the intelligent network-monitoring device to establish practical monitoring technology, we can control the operating condition of the system, thereby providing a useful, effective solution for difficulties facing researchers and engineers when encountered in live monitoring.

2 Hardware and Software Systems

A wind power generator is the device that uses air flow through rotor blades to make it rotate, transforming the rotating blade's mechanical power into electrical power. The wind power generation system of this paper focuses on the independent power generator that is originally sold on the market. Such generators transform wind power into electric energy and directly supply the load used [3].

By modifying the generator with grid-tie inverter, it can enable the small wind turbine to generate mains electricity in parallel to form a micro-grid [4]. In the condition without main electricity supply, it can still provide electricity through the stored energy device. Actual foundation is shown in Fig. 1. The wind turbine set is composed of five rotor blades, a permanent synchronous generator, tower, rectifier, regulator circuit, battery, grid-tie inverter and a braking system. Summarized as follows:

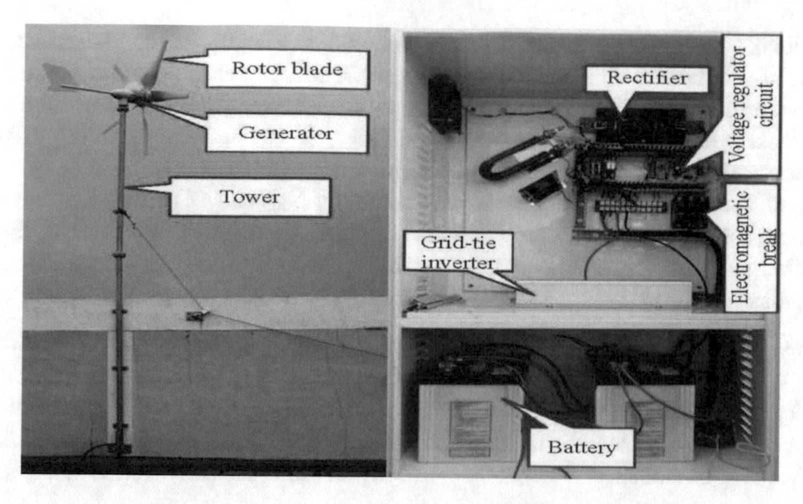

Fig. 1. Actual foundation of wind turbine generator set

(a) Rotor blade: Wind turbine designs have two types: one is based on efficacy of the wind power transformation; the other requires large torque. The first one has high revolution speed characteristics. The bigger the tip speed ratio between speed of blade tip and wind flow velocity, the more effective the wind power transformation. The efficient type will normally design tip speed ratio of 6 to 7 and without too many blades. According to the experimental data, the machine's efficacy increases with the blade numbers, but having over three blades appears to cause other factors to increase, significantly reducing its efficacy. However, at about five blades, the machine will reach maximum efficacy. This paper is applied to a five-bladed wind rotor, which has very low start-up and cut-in point, with extremely high wind-power efficiency.

(b) Generator: The types of generators can be divided into inductor-type generator and permanent-magnet synchronous generators. Among them, the construction of a permanent-magnet synchronous generator is simpler than an inductor-type generator, having advantages of not needing an additional field power supply, possessing high efficiency and stability, etc. With the advances in rectifier and transformer technology, most wind turbines recently have adopted synchronous generators. As the wind turbine rotates, it creates a rotary magnetic field that interacts with a stator circuit, producing relative motion into the result of induced electromotive force. The magnetic field caused by the motion is fixed, so input voltage changes with the rotating speed of the wind turbine, as does the frequency.

(c) Rectifier: three-phase voltage, which is output by the wind turbine, normally transforms into DC voltage through a three-phase bridge-diodes rectifier circuit. The diode rectifier is suitable for high-efficiency conditions, having the advantage of low cost [5].

(d) Battery: The battery is a stored-energy device, making the system not only able to build mains electricity in parallel but also independently supply power to the regional load. Batteries now widely use lead-acid batteries as the stored energy device. Before a larger, more environmentally friendly battery is invented, the lead-acid battery with all its characteristics is still first choice for the stored energy battery. Although the lead-acid battery doesn't have the advantage of high energy density and lacks quick charge modalities, it has refined technology, widely adopted operating temperature, high efficiency discharge, and no memory effect, etc. Therefore, the lead-acid battery so far in high-power energy storage is still paid considerable attention. For small container lead-acid battery control, there are charge technologies like constant-voltage charging, constant-current charging, pulse-reverse charging, and pulse charging methods.

In view of the importance of the monitoring system for wind power generation, this paper uses a host computer, a wind turbine monitoring module and a wind flow velocity monitoring module to establish a monitoring system for wind turbines, applying a wireless network on the monitoring system to create the communication between host computer and remote monitoring module. As shown in Fig. 2 [6].

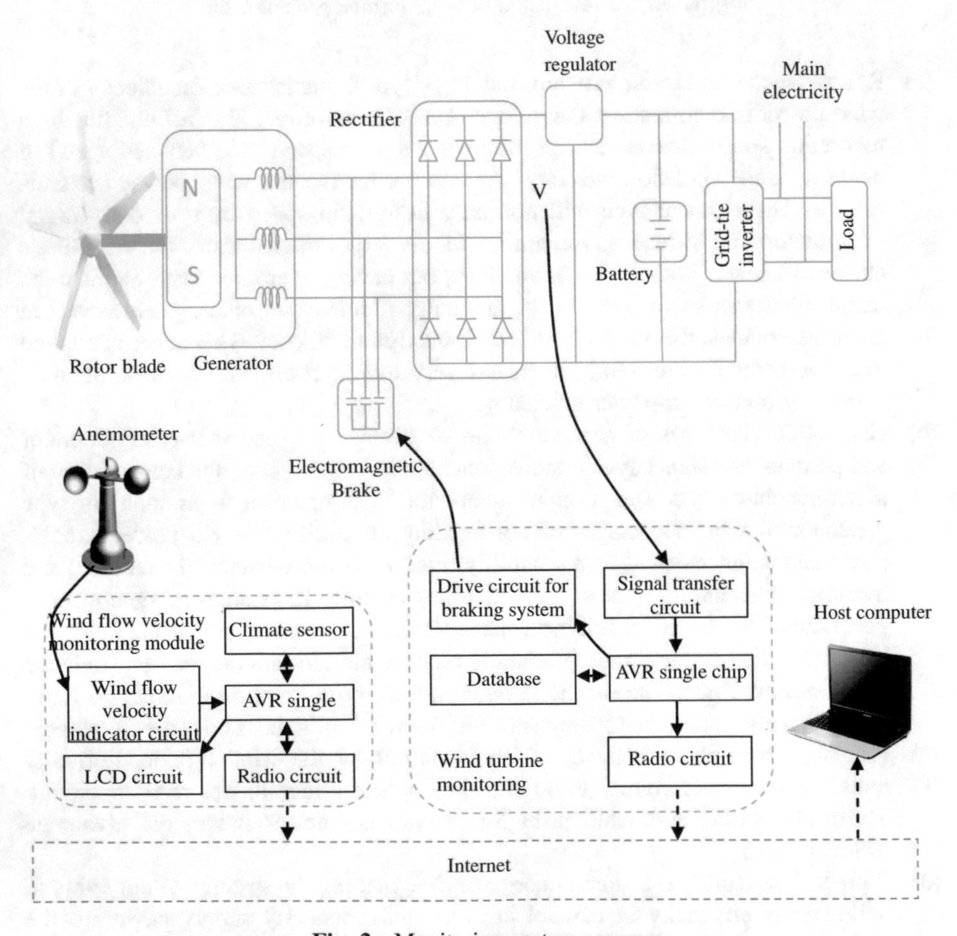

Fig. 2. Monitoring system program

The host computer can be used to gather the measured data from the remote monitoring module and analyze such data. The operation interface of the host computer is accomplished by the LabVIEW [7]. The user can input control commands and monitor operation conditions simply through a man-machine interface. The remote monitoring module can measure voltage; in addition, it can apply electromagnetic brakes for the wind turbine. In order to be applied on long-term monitoring conditions, the voltage data adopts effective value and records it once every two seconds. The wind turbine-monitoring module is constituted through a signal transfer circuit, AVR single chip, memory card and wireless network [8].

The wind flow velocity, which can detect pulse wave from anemometer and transfer this into corresponding wind speed, can output the result to a liquid crystal display circuit and link the wireless internet with the computer online. The LCD circuit can also show all the measured data and manual operation conditions.

The communication of the host computer and remote monitoring module is constituted through a wireless network, so the host computer can fully understand the system status wherever placed with the measured data being transported through the wireless network, allowing the maintenance staff to monitor all the time.

3 Result and Discussion

This study uses a wind turbine, host computer, wind turbine monitoring module and a wind flow velocity-monitoring module to establish a remote monitoring system for wind turbines. The completed function is described as follows:

Wind turbine system

Wind turbine and electricity structure as shown in Fig. 1. This paper applies to small vertical-axis wind turbines. The specifications are as follows:

(a) Rotor blade: The rotor blades are made of carbon fiber and fiberglass-reinforced composite materials; blade diameter is 1.4 m; consists of a five-bladed wind rotor.
(b) Generator: the generator adopted a permanent-magnet synchronous machine; 3-phase, 18 poles; rated power at 400 W; maximum output power 450 W; and rated voltage 12 V.
(c) Battery: it requires two 100-150Ah/12 V lead-acid batteries.
(d) Grid-tie inverter: equips 110 V, 60 Hz inverter; it can be in parallel with mains electricity.
(e) Operating wind flow velocity: the start-up speed of the wind turbine is 2.3 m/s; cut-in speed is 3 m/s. It reaches rated output capacity when the wind speed is 12 m/s. The wind turbine needs to be stopped at the speed of about 20-25 m/s to prevent generator damage from exorbitant wind speed [9].

A. Wind turbine monitoring module. The wind turbine monitoring module as shown in Fig. 3, is equipped with a Wi-Fi module that records machine rotating data once every two seconds and uploads data to the internet. The user can get the operating state of the wind turbine as long as they enter the specific website. Long-term measurement is also available through the online database, gaining more referential wind turbine operation result.

B. Wind velocity-monitoring module. The wind flow velocity-monitoring module is shown in Fig. 4. It measures the pulse signal of an anemometer. By breaking off the program to calculate actual rotating speed, it can not only transfer instantaneous value of the rotating speed but also show the average rotating speed [10]. In addition, this module adds a climate detection circuit by using a BMP180 climate sensor as an operating core. This can observe atmospheric pressure and temperature at once, providing the user with information concerning weather variation. The measured data could be transferred through Wi-Fi to the host computer for further processing.

On the other side, the server receives data from the remote monitoring module by UART and transfers it through Wi-Fi. It can also let the Wi-Fi module to connect the router to expand signal range; only the IP and port for receiving the message from the website needs to be inputted, as shown in Fig. 5 [11].

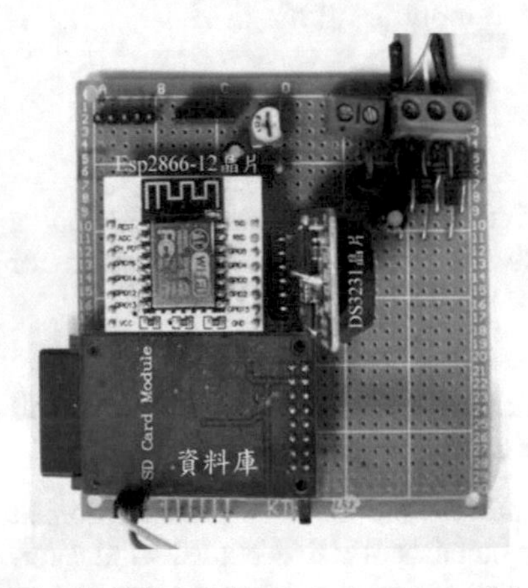

Fig. 3. Wind turbine monitoring module

Fig. 4. Wind flow velocity monitoring module

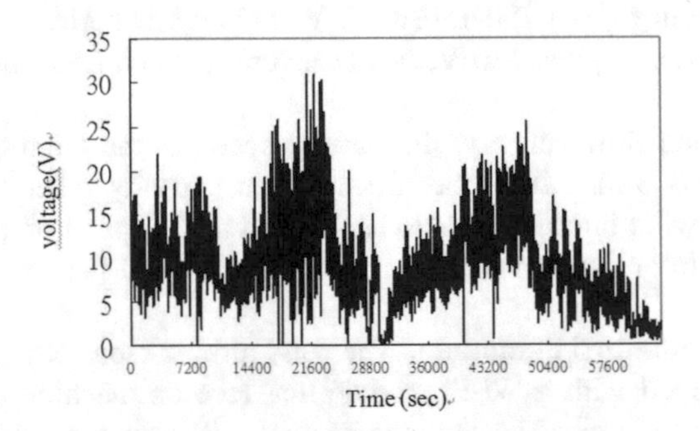

Fig. 5. Time-voltage curve in 24 h

C. Host computer. The user can use any computer to connect to the website to observe the operating state of the wind turbine. Long-term measurement can also be obtained through the online database, obtaining referential wind turbine operation results. The website display state is shown in Fig. 6.

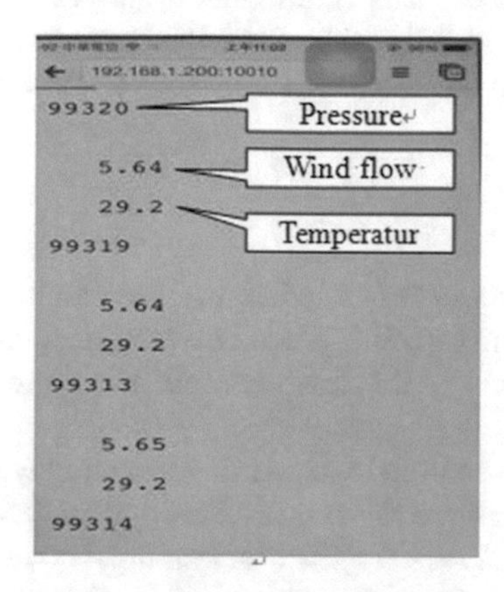

Fig. 6. Received data on website

Besides monitoring wind turbine operating voltage, the host computer can retrieve wind speed, temperature and atmospheric pressure immediately for data collection through Wi-Fi and analysis these data, as shown in Fig. 7.

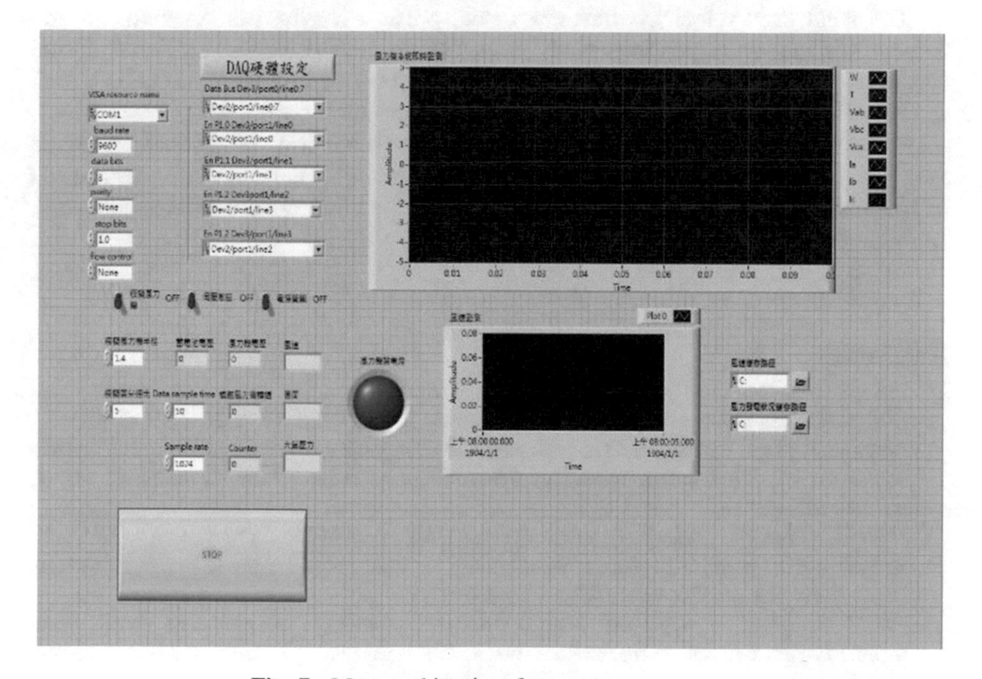

Fig. 7. Man-machine interface on host computer

By establishing a database, the user can look up historical records without letting the computer work constantly. It is merely necessary to connect the host computer to the remote monitoring module when the user needs to remotely monitor or analyze historical records.

With receiving wind speed, temperature and atmospheric pressure through wireless transformation, the ability to fully understand the system situation in whatever locale, monitoring and controlling the whole system from a distance is achieved.

4 Conclusion

The paper uses a host computer, a wind turbine-monitoring module and a wind speed-monitoring module to establish a wind power generation monitoring system that uses the wireless network to the host computer and remote monitoring module communication.

The host computer can be used to gather the measured waveform from them remote monitoring module and analyze these data. The operation interface of the host computer is accomplished by LabVIEW. The user can input control commands and monitor operation conditions simply through a man-machine interface.

The remote monitoring module can measure voltage; in addition, it can apply an electromagnetic brake for the wind turbine. In order to be applied in long-term monitoring conditions, the voltage data adopts effective value and records it once every two seconds. The wind turbine-monitoring module is constituted through a signal transfer circuit, AVR single chip, memory card and wireless network.

The wind flow velocity, which can detect pulse wave from an anemometer and transfer this into corresponding wind speed, can output the result to a liquid crystal display circuit and link the wireless internet with a computer online. The LCD circuit can also show all the measured data and manual operation conditions.

The communication of host computer and remote monitoring module is constituted through a wireless network, so the host computer can fully understand the system status in whichever locale with the measured data transported through the wireless network, allowing the maintenance staff to monitor constantly.

References

1. Bull, S.R.: Renewable energy today and tomorrow. Proc. IEEE **89**, 1216–1226 (2002)
2. Singh, B.N., Jain, P., Joos, G.: Three-phase AC-DC regulated power supplies: a comparative evaluation of different topologies. In: Proceedings of IEEE APEC Conference, vol. 1, pp. 518-523 (2005)
3. Johnson, G.L.: Wind Energy Systems. Prentice Hall, Englewood Cliffs (1985)
4. Knight, A.M., Peters, G.E.: Simple wind energy controller for an expanded operation range. IEEE Trans. Energy Convers. **20**(2), 459–466 (2005)
5. Tan, F.K., Islam, S.: Optimum control strategies in energy conversion of PMSG wind turbine system without mechanical sensors. IEEE Trans. Energy Convers. **19**(2), 392–399 (2004)

6. Luo, C., Banakar, H., Baike, S., Ooi, B.T.: Strategies to Smooth Wind Power Fluctuations of Wind Turbine Generator. IEEE Trans. Energy Convers. **22**(2), 341–349 (2007)
7. Wu, R.-C., Tsai, J.-I., Huang, S.-H., Tseng, W.-C.: Establishing monitoring system of advanced metering infrastructure by power line carrier. In: Proceedings of the First International Conference on Robot, Vision and Signal Processing, Kaohsiung City, Taiwan, R.O.C., November 2011
8. Wu, R.-C., Zhu, K., Chang, E.-C., Lee, J.-C.: Intelligent control via power-line carrier for illumination and air condition in buildings, IE & EM 2012, Changsha, China, pp. 1209–1213 (2012)
9. Koutroulis, E.: Design of a maximum power tracking system for wind-energy-conversion applications. IEEE Trans. Industr. Electron. **53**, 486–494 (2006)
10. Zinger, D.S., Muljadi, E., Miller, A.: A simple control scheme for variable speed wind turbines. In: IEEE Industry Applications Conference, vol. 102(8), pp. 2793–2798 (1983)
11. Ali, M., Milanovic, J.V., Ilie, I.-S., Chicco, G.: Wind farm model aggregation using probabilistic clustering. IEEE Trans. Power Syst. **28**(1), 309–316 (2013)

Recent Advances in Security and Privacy for Multimodal Network Environments

A Survey of Secret Sharing Schemes Based on Latin Squares

Raylin Tso[1](✉) and Ying Miao[2]

[1] Department of Computer Science, National Chengchi University, Taipei, Taiwan
raylin@cs.nccu.edu.tw

[2] Graduate School of Systems and Information Engineering, University of Tsukuba, Tsukuba, Japan
miao@sk.tsukuba.ac.jp

Abstract. Secret sharing schemes are wildly used in many applications where the secret must be recovered by joint work of certain amount of participants. There are many techniques to construct a secret sharing scheme, one of them is the construction using critical sets of Latin squares. In this paper, we will investigate the features of back circulant Latin squares, their corresponding critical sets and show how a secret sharing scheme can be constructed using such kind of critical sets. Finally, we will point out the constraints and future research on such kind of secret sharing schemes.

Keywords: Back circulant latin square · Critical sets · Multilevel scheme · Multi-department scheme · Secret sharing schemes

1 Introduction

A *secret sharing scheme* is a method whereby n pieces of information called *shares* or *shadows* are assigned to a secret key K. The shares have the property that certain authorized groups of shares can be used to reconstruct the secret key. The secret cannot be reconstructed from an unauthorized group of shares. The recipients of the shares are called the *participants* in the scheme. A (t, w)-*threshold access structure* has a basis consisting of all t-subsets of w participants. A (t, w)-*threshold scheme* with *threshold* t is a secret sharing scheme for a (t, w)-threshold access structure. Threshold schemes were the first type of secret sharing schemes that were constructed — by Shamir [6] using polynomial interpolation, and by Blakley [1] using finite geometry — in 1979. A (t, w)-threshold scheme is *perfect* if knowledge of fewer than t shares provides no information about the secret K.

There are many ways to construct a secret sharing scheme, and there are many extensions of secret sharing schemes. In this chapter, we will first introduce the way Cooper, Donovan and Seberry [3] adapted to construct secret sharing schemes from Latin squares, describe the schemes they constructed, present new

J.-S. Pan et al. (eds.), *Advances in Intelligent Information Hiding and Multimedia Signal Processing*, Smart Innovation, Systems and Technologies 82,
DOI 10.1007/978-3-319-63859-1_33

multilevel and multi-department schemes by means of their techniques, then point out the disadvantages of the secret sharing schemes constructed from Latin squares, and try to overcome these problems to some extent.

2 Preliminaries

This section gives some definitions and structures required for our construction.

2.1 Back Circulant Latin Squares and Their Critical Sets

A *Latin square* L of *order* n is an $n \times n$ array with entries chosen from a set, N, of size n such that each entry occurs precisely once in each row and column. For convenience, we sometimes talk of the Latin square L as a set of ordered triples $(i, j; k)$ and take this to mean that the element k occurs in the cell (i, j) of the Latin square L. If we index the rows and column of the array by the set $N = \{0, 1, \ldots, n-1\}$, with $n > 1$, then the array with integer $i+j \bmod n$ in cell (i, j) is said to be a *back circulant* Latin square. Table 1 shows a back circulant Latin square of order 7.

Table 1.

0	1	2	3	4	5	6
1	2	3	4	5	6	0
2	3	4	5	6	0	1
3	4	5	6	0	1	2
4	5	6	0	1	2	3
5	6	0	1	2	3	4
6	0	1	2	3	4	5

A *critical set* in a Latin square L of order n is a set $C = \{(i, j; k) \ : \ i, j, k \in \{0, 1, \ldots, n-1\}\}$ such that

1. L is the only Latin square of order n which has element k in cell (i, j) for each $(i, j; k) \in C$, and
2. no proper subset of C satisfies the above property 1.

A *minimal critical set* in a Latin square L is a critical set of minimum cardinality. For example, the Latin square in Table 1 has the minimal critical set $\{(0, 0; 0), (0, 1; 1), (1, 0; 1), (0, 2; 2), (1, 1; 2), (2, 0; 2), (4, 6; 3), (5, 5; 3), (6, 4; 3), (5, 6; 4), (6, 5; 4), (6, 6; 5)\}$, shows in Table 2.

There is not a lot known about critical sets for Latin squares in general. Results on critical set for Latin squares have appeared in papers by, for example, Cooper, Donovan and Seberry [2], Donovan, Cooper, Nott and Seberry [5], Smetaniuk [7], Stinson and van Rees [8], and Street [9]. However, a class of critical sets are known for back circulant Latin squares. Many results was proved by Donovan and Cooper [4].

Table 2.

0	1	2	3	4	5	6
1	2	3	4	5	6	0
2	3	4	5	6	0	1
3	4	5	6	0	1	2
4	5	6	0	1	2	3
5	6	0	1	2	3	4
6	0	1	2	3	4	5

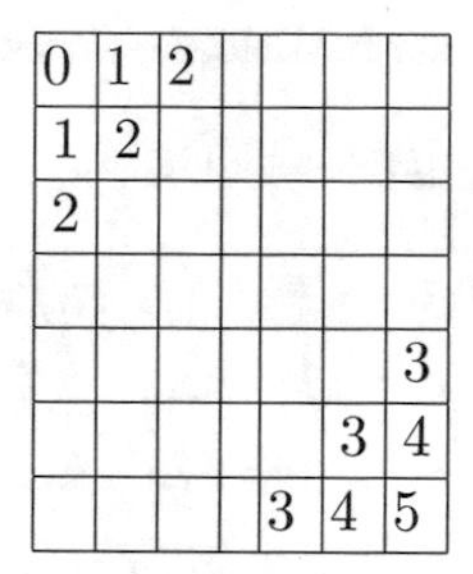

0	1	2				
1	2					
2						
						3
					3	4
				3	4	5

3 A Basic Secret Sharing Scheme

A secret sharing scheme can be constructed from Latin square L of order n in which the secret key is the Latin square and the shares in the scheme are based on a partial Latin square $\mathcal{S} = \{\cup C_i \ : \ C_i$ is a critical set in $L\}$. The number of critical sets used will be dependent on the order of the Latin square and the number of participants in the secret sharing scheme. The access structure will be the set $\Gamma = \{B \ : \ B \subseteq \mathcal{S}$ and $B \supseteq C$ where C is some critical set of L in $\mathcal{S}\}$.

The protocol for a secret sharing scheme involving l participants and based on a Latin square is as follows.

Protocol:

- A Latin square L of order n is chosen. The number n is made public, but the Latin square L is kept secret and taken to be the key.
- A set $\mathcal{S}$ which is the union of a number of critical sets in L is defined.
- For each $(i, j; k) \in \mathcal{S}$, the share $(i, j; k)$ is distributed privately to a participant.
- When a group of participants whose shares constitute a critical set come together, they can reconstruct the Latin square L and hence the secret key.

We can easily see that only the authorized groups with these critical sets can reconstruct the unique Latin square L. If pooled shares can not constitute a critical set, then they can not uniquely recover the Latin square L. This is the basic idea behind the secret sharing schemes constructed from Latin squares by Cooper, Donovan and Seberry [3].

4 Multilevel Schemes

In many situations we require a secret sharing scheme to be hierarchical. It means that the shares in any authorized group are divided into several levels, and a participant with a share at level i can be replaced by two or more participants at lower levels. In general, we can construct a Latin square based multilevel scheme according to the following protocol.

- Construct a back circulant Latin square of order n by $L_n = \{(i,j;i+j) : 0 \leq i,j \leq n-1\}$ with addition reduced modulo n, and define it to be the secret key.
- Find the minimal critical set(s) of L_n. When n is even, the two minimal critical sets of L_n are

$$\begin{aligned} A_0 = \{(i,j;i+j) \ &: \ i=0,\ldots,n/2-1, \text{ and} \\ &\quad j=0,\ldots,n/2-1-i\} \cup \\ \{(i,j;i+j) \ &: \ i=n/2+1,\ldots,n-1, \text{ and} \\ &\quad j=3n/2-i,\ldots,n-1\}, \end{aligned}$$

$$\begin{aligned} B_0 = \{(i,j;i+j) \ &: \ i=0,\ldots,n/2-2, \text{ and} \\ &\quad j=0,\ldots,n/2-2-i\} \cup \\ \{(i,j;i+j) \ &: \ i=n/2,\ldots,n-1, \text{ and} \\ &\quad j=3n/2-1-i,\ldots,n-1\}. \end{aligned}$$

When n is odd, the unique minimal critical set of L_n is

$$\begin{aligned} A_0 = B_0 = \{(i,j;i+j) \ &: \ i=0,\ldots,(n-3)/2, \text{ and} \\ &\quad j=0,\ldots,(n-3)/2-i\} \cup \\ \{(i,j;i+j) \ &: \ i=(n+1)/2,\ldots,n-1, \text{ and} \\ &\quad j=(3n-1)/2-i,\ldots,n-1\}. \end{aligned}$$

- Find the other critical sets of L_n. When n is even, these critical sets are

$$\begin{aligned} A_k = \{(i,j;i+j) \ &: \ i=0,\ldots,n/2-1+k, \text{ and} \\ &\quad j=0,\ldots,n/2-1+k-i\} \cup \\ \{(i,j;i+j) \ &: \ i=n/2+1+k,\ldots,n-1, \text{ and} \\ &\quad j=3n/2+k-i,\ldots,n-1\}, \\ B_k = \{(i,j;i+j) \ &: \ i=0,\ldots,n/2-2-k, \text{ and} \\ &\quad j=0,\ldots,n/2-2-k-i\} \cup \\ \{(i,j;i+j) \ &: \ i=n/2-k,\ldots,n-1, \text{ and} \\ &\quad j=3n/2-k-1-i,\ldots,n-1\}, \end{aligned}$$

for $0 \leq k \leq n/2-1$.

When n is odd, these critical sets are

$$\begin{aligned} A_k = \{(i,j;i+j) \ &: \ i=0,\ldots,(n-3)/2+k, \text{ and} \\ &\quad j=0,\ldots,(n-3)/2+k-i\} \cup \\ \{(i,j;i+j) \ &: \ i=(n+1)/2+k,\ldots,n-1, \text{ and} \\ &\quad j=(3n-1)/2+k-i,\ldots,n-1\}, \end{aligned}$$

$$\begin{aligned} B_k = \{(i, j; i+j) \ &: \ i = 0, \ldots, (n-3)/2 - k, \text{ and} \\ & \qquad j = 0, \ldots, (n-3)/2 - k - i\} \cup \\ \{(i, j; i+j) \ &: \ i = (n+1)/2 - k, \ldots, n-1, \text{ and} \\ & \qquad j = (3n-1)/2 - k - i, \ldots, n-1\}, \text{ for} \\ 0 \le k \le (n-1)/2. & \end{aligned}$$

- Define elements of A_0 and B_0 to be shares of rank 1.
- Define elements of $A_k \backslash A_0$, $B_k \backslash B_0$ to be shares of rank $k+1$ for $1 \le k \le (n-1)/2$.
- When participants pool their shares and if their shares can constitute one of A_k or B_k, $1 \le k \le (n-1)/2$, then they can reconstruct the secret key L_n.

5 Multi-department Schemes

In addition to multilevel schemes, sometimes we also need those schemes in which one or some of the shares are also shares of other schemes. That is, a number of secret sharing schemes contain one or more common participants, or, in other words, one or more secret shares are common to a number of different secret sharing schemes. The protocol for the multi-department scheme is as follows.

- Construct r back circulant Latin squares $L_{n_l}, 1 \le l \le r$, of order $n_1, n_2, \ldots, n_r$ by $L_{n_l} = \{(i, j; i+j) \ : \ 0 \le i, j \le n_l - 1\}$ with addition reduced modulo n_l, respectively, and define L_{n_l} to be the secret key for department l, $1 \le l \le r$.
- Choose one critical set A_{k_l} or B_{k_l} (for their definitions, see Sect. 4.3) for $1 \le l \le r$, where when B_{k_l} is chosen, $k_l \ne n_l/2 - 1$ if n_l is even, and $k_l \ne (n_l - 1)/2$ if n_l is odd (so that there is at least one element in the upper triangle of the Latin square).
- Distribute the elements in the upper triangles common to all such critical sets to key participants as primary shares.
- Distribute the remaining elements in the critical set A_{k_l} or B_{k_l} to minor participants as minor shares for each l, $1 \le l \le r$.

Then, without the key participants, the minor participants in each department cannot reconstruct their secret key. The secret key can be recovered only when the key participants and minor participants in a department work together. The key participants are necessary for each department to reconstruct its secret key.

6 Conclusion and Future Research

We have investigated secret sharing schemes from Latin squares. By means of critical sets in back circulant Latin squares, we have constructed multilevel and multi-department secret sharing schemes.

Since we have applied Latin squares in Experimental Designs to construct some secret sharing schemes, many questions are raised from both points of view. We list some interesting open problems for future research in the following.

- All secret sharing schemes constructed so far from Latin squares are not perfect, since we used Latin squares as secret keys, and the elements of their critical sets as shares. Is it possible to obtain perfect secret sharing schemes from Latin squares? For example, instead of using a Latin square itself as the secret key, we use some of its entries, which is not an element of the critical set used as shares, as the secret key, then the knowledge on shares might not leak out any information on this secret key. But is there any Latin square satisfying such requirement?
- Because of the particular form of the back circulant Latin square, it may be easy to be decrypted when using it as the secret key. Though we can conceal its particular form by interchanging their rows and columns, we are not sure that there is no trapdoors for finding the original back circulant Latin square. Therefore, a suitable way to construct a secret sharing scheme from a Latin square is to use a Latin square of general form, not a back circulant Latin square. However, some difficulties still remain. We do not know about critical sets of Latin squares in general. Is there any property of critical sets of Latin squares in general form which can be used to construct secret sharing schemes so that we could make the schemes more practical?

References

1. Blakley, G.R.: Safeguarding cryptographic keys. AFIPS Conf. Proc. **48**, 313–317 (1979)
2. Cooper, J., Donovan, D., Seberry, J.: Latin squares and critical sets of minimal size. Australas. J. Combin. **4**, 113–120 (1991)
3. Cooper, J., Donovan, D., Seberry, J.: Secret sharing schemes arising from Latin square. Bull. Inst. Combin. Appl. **12**, 33–43 (1994)
4. Donovan, D., Cooper, J.: Critical sets in back circulant Latin squares. Aequationes Math. **52**, 157–179 (1996)
5. Donovan, D., Cooper, J., Nott, D.J., Seberry, J.: Latin squares: critical sets and their lower bounds. Ars Combin. **39**, 33–48 (1995)
6. Shamir, A.: How to share a secret. Comm. ACM **22**, 612–613 (1979)
7. Smetaniuk, B.: On the minimal critical set of a latin square. Util. Math. **16**, 97–100 (1979)
8. Stinson, D.R., van Rees, G.H.J.: Some large critical sets. Congr. Numer. **34**, 441–456 (1982)
9. Street, A.P.: Defining sets for *t*-designs and critical sets for Latin squares. New Zealand J. Math. **21**, 133–144 (1992)

A NFC-Based Authentication Scheme for Personalized IPTV Services

Kuo-Hui Yeh[1(✉)], Nai-Wei Lo[2], and Chun-Kai Wang[2]

[1] Department of Information Management, National Dong Hwa University, Hualien 97401, Taiwan
khyeh@gms.ndhu.edu.tw

[2] Department of Information Management, National Taiwan University of Science and Technology, Taipei 10607, Taiwan
nwlo@cs.ntust.edu.tw, m10209122@mail.ntust.edu.tw

Abstract. Internet Protocol Television (IPTV) has promptly changed the way providing information and entertainment. In this study, we present a novel authentication scheme for IPTV services in which a robust hash function is adopted as the major crypto-module for securing IPTV systems. The security analysis shows that the proposed mechanism can meet the critical security requirements for personalized IPTV services. In brief, the proposed mechanism is suitable for personalized IPTV services and is able to be easily deployed within current IPTV systems.

Keywords: IPTV · Personalized services · Authentication · NFC · HCE

1 Introduction

Thanks to the rapid development and successful advancement of Internet applications, the computational performance has been similarly enhanced in consumer electronic devices. A rich and pervasive computing application environment, consisting of sensing equipment and modern communication technologies, is able to provide instant customized online services, such as information retrieval or data analysis. According to a recent report released by Groupe Speciale Mobile Association [1], in modern countries, mobile devices have reached a high penetration rate, and people may use handheld mobile devices at any time, in any place they have the ability to access resources. Mobile IPTV is a combination of modern computing, networking and storage technology that can be accessed through an Internet Protocol. It provides high-quality audio and video content and a variety of services, the intention being to

This work was supported in part by the Academia Sinica, in part by the Taiwan Information Security Center (TWISC) and in part by the Ministry of Science and Technology, Taiwan, under Grant MOST 105- 2221-E-259-014-MY3, Grant MOST 105-2221-E-011-070-MY3, Grant MOST 105-2221-E-011-080-MY3, Grant MOST 105-2923-E-182-001-MY3, Grant MOST 104-2923-E-011-005-MY3, Grant MOST 104- 2218-E-001-002, Grant MOST 105-2218-E-011-015, Grant MOST 105-2218-E-001-001 and Grant MOST 106-3114-E-011-003.

J.-S. Pan et al. (eds.), *Advances in Intelligent Information Hiding and Multimedia Signal Processing*, Smart Innovation, Systems and Technologies 82,
DOI 10.1007/978-3-319-63859-1_34

digitize all multimedia content so that it can be provided to a large number of end-users through a wide range of IP network technologies which support its quality of service (QoS), quality of experience (QoE), security, mobility and interactivity. Today, mobile IPTV has become a platform that has changed the way people receive information and entertainment [2]. Two approaches can be used to construct a mobile IPTV platform. One is to add mobility to traditional fixed IPTV to make it become a mobile IPTV. The other is to combine mobile TV with IP technology in the form of a mobile IPTV [3]. In brief, mobile IPTV is an extension of IPTV services in which end-users can utilize it to retrieve an IPTV service anywhere and anyplace. This capability enhances the service level and value [4]. A set-top box (STB) is a device on the customer side that connects an ordinary TV to the external communication network and converts the received signal to a display on the TV screen [5]. It has a unique hardware identifier, which is registered by the service provider to offer a basic subscriber identity capability that is used to authenticate a family [6]. This indicates that subscriber authentication based on STB-level identification is inconsistent with IPTV's main intention: to provide personalized services. For instance, if an IPTV service provider wants to offer personalized recommendation services to a viewer, it must be able to identify who is actually watching a certain program to build historical data for analysis [7]. However, existing IPTV authentication schemes are based on STB-level, leading to the situation wherein all the members of a family get the same service and the same level of access to IPTV services.

To overcome this problem, various user authentication schemes for IPTV personalized services have been proposed. These methods are password-based [8], biometrics-based [9], RFID-based [10], USIM-based [11] and Bluetooth-based systems [12]. First, a password-based system can be simply implemented by software alone, but it is not secure enough and is highly vulnerable to attack. Second, a biometrics-based system requires additional hardware support, which leads to a higher cost of STB and increases the complexity during system implementation. In RFID-based approaches, authentication can be performed easily through STB, a RF reader and unique RF tags owned by the specific user. However, there exist potential security risks, such as the possibility that a tag may be stolen and cloned. Recently, mobile phones have become indispensable accessories in people's daily lives, so solutions based on USIM and Bluetooth technology and operated on intelligent mobile handheld devices have been proposed. However, these solutions may still be insecure against various malicious attacks, as the adopted cryptography modules are not currently powerful enough. In this paper, we propose a novel user authentication scheme based on Near Field Communication (NFC) technology with host card emulation (HCE) for personalized IPTV services. Based on the proposed scheme, IPTV service providers can identify the viewer who is actually sitting in front of the TV, and can thus provide more personalized services, such as a personalized electronic program guide (EPG), video on demand (VOD) recommendation services, resource access, parental controls and integrated payment services.

2 Notations

The notations used in this study are described in Table 1.

Table 1. Notations used in this study.

Notation	Description
SP	The IPTV service provider
U_i	The user *i*, who is an IPTV subscriber
PC_i	The personal computer of the user *i*
NP_i	The NFC-enabled mobile phone owned by the user *i*
STB_j	The set-top box *j*, which is equipped with a NFC reader
AS_{SP}	The Application Server, *SP*
APP_{SP}	The HCE-enabled mobile app, which is developed by *SP*
$Email_i$	The email address of the user *i*
PW_i	The password chosen by the user *i*
$Info_i$	The personal information of the user *i*
$Cell_i$	The cell phone number of the user *i*
ID_A	The unique identifier of the entity *A*
$K_{A,B}$	The secret key, only known to the entities *A* and *B*
PK_A	The public key of the entity *A*
SK_A	The security (private) key of the entity *A*
R_A	A random number that is generated by the entity *A*
N_A^n	The *n* th nonce value that is generated by the entity *A*
T_A^n	The *n* th timestamp that is generated by the entity *A*
$TDuring$	The maximum allowed time interval for transmission delay
$Valid(M)$	To check if the message *M* is valid
$HMAC_K(M)$	A keyed-hash message authentication code of message *M* using security key *K*
$A \rightarrow B : M$	The entity *A* sends the message *M* to the entity *B*
$\|\|$	A concatenation operator

3 The Proposed Authentication Scheme

In this section, we introduce the proposed authentication mechanism for IPTV systems, where a keyed-hash message authentication code (HMAC) is generated via a cryptographic hash function combined with a secret key. It can be used to guarantee the integrity and authentication of a message.

3.1 Registration Phase

Step 1 ($U_i \rightarrow PC_i$): First, U_i operates PC_i by using the web browser to visit SP's website (AS_{SP}). Once accessing the site, an HTTP session is established between PC_i and AS_{SP}.

Step 2 ($PC_i \rightarrow AS_{SP}$: $request, Email_i, PW_i, Cell_i, Info_i$)**:** In order to register, U_i has to fill out the registration form with required information (i.e. $Email_i$, PW_i, $Cell_i$, $Info_i$) on the website. Then, U_i sends the registration request to AS_{SP}.
Step 3 ($AS_{SP} \rightarrow PC_i$: $Info_{SP}$)**:** Once AS_{SP} receives the registration request, it stores personal information of U_i and generates a verification code R_{SP} and a timestamp T^1_{SP}. The verification code R_{SP} is sent to $Cell_i$ via SMS, and the response message is displayed on the web page to notify U_i receiving SMS.
Step 4 ($U_i \rightarrow NP_i$)**:** U_i starts APP_{SP} on NP_i after he/she obtains the verification code R_{SP} from the SMS. Once APP_{SP} is opened, an SSL/TLS connection is established between NP_i and AS_{SP}.
Step 5 ($NP_i \rightarrow AS_{SP}$: $Email_i, PW_i, R_{SP}$)**:** U_i enters his/her $Email_i$, PW_i and the verification code R_{SP} on APP_{SP}. Then, he/she sends them to AS_{SP}.
Step 6 ($AS_{SP} \rightarrow NP_i$: $ID_i, K_{i,SP}$)**:** As soon as AS_{SP} receives the message form NP_i, it generates another timestamp T^2_{SP}, and then checks Valid ($Email_i$, PW_i, R_{SP}) and $T^2_{SP} - T^1_{SP} \leq$ TDuring. If both conditions are met, AS_{SP} generates ID_i and $K_{i,SP}$ to send them to NP_i. Otherwise, AS_{SP} responses an error message to NP_i.
Step 7 ($NP_i \rightarrow AS_{SP}$: ID_i, M)**:** If NP_i receives the message from AS_{SP} without error, it stores ID_i and $K_{i,SP}$ in the secure storage of NP_i. NP_i uses $K_{i,SP}$ to compute $M = HMAC_{K_{i,SP}}(ID_i)$, and then sends ID_i and M to AS_{SP}. This is to make AS_{SP} confirm that NP_i has received ID_i and $K_{i,SP}$.
Step 8 ($AS_{SP} \rightarrow NP_i$: $result$)**:** AS_{SP} obtains M and ID_i, from the message sent by NP_i, and retrieves the corresponding secret key $K_{i,SP}$ from its database according to ID_i. After that, AS_{SP} uses $K_{i,SP}$ to compute $M' = HMAC_{K_{i,SP}}(ID_i)$. If M = M', which means that NP_i has received ID_i and $K_{i,SP}$, AS_{SP} responses result back to NP_i shows the registration process is completed for U_i. Otherwise, AS_{SP} responses an error message to NP_i.

3.2 Authentication Phase (Fig. 1)

Step 1 ($U_i \rightarrow NP_i$)**:** First, U_i taps NP_i to the NFC reader attached to STB_j for login request. Note that U_i should execute APP_{SP} and enter the PIN code before NP_i is able to communicate with the NFC reader. If the entered PIN code is correct, NP_i enables HCE mode and it can react to APDU commands from the NFC reader.
Step 2 ($STB_j \rightarrow NP_i$: $request$)**:** If STB_j is ready, the NFC reader begins scanning. As soon as the NFC reader detects NP_i, it sends a "SELECT AID" APDU command to NP_i. The Application ID (AID) is chosen by *SP*, let NP_i know which HCE service the NFC reader actually wants to talk to. Additionally, it can be confirmed whether APP_{SP} has been installed on NP_i.
Step 3 ($NP_i \rightarrow STB_j$: $response$)**:** Once NP_i receives the APDU command, if APP_{SP} is running on NP_i, it sends a APDU response with status word "9000", which stand for "command executed without error", back to the NFC reader. Otherwise, stop the request.

Step 4 ($STB_j \rightarrow NP_i : Query$): If the STB_j receives the APDU response without error, the NFC reader sends a APDU command to NP_i again. What this command does is that it asks for ID_i and a nonce.
Step 5 ($NP_i \rightarrow STB_j : ID_i, N_i^1$): Once NP_i receives the *Query* request from STB_j, it generates N_i^1, and then sends ID_i and N_i^1 back to STB_j.
Step 6 ($STB_j \rightarrow AS_{SP} : Request, ID_i, N_i^1, ID_j, N_j^1$): STB_j generates N_j^1 after it receives the response with ID_i and N_i^1 from NP_i. Then, STB_j sends a login request along with ID_i, N_i^1, ID_j and N_j^1 to AS_{SP}.
Step 7 ($AS_{SP} \rightarrow STB_j : M_1, M_2, N_{SP}^1$): As soon as AS_{SP} receives the login request, it generates N_{SP}^1 to compute $M_1 = HMAC_{K_{i,SP}}(N_i^1 \parallel N_j^1 \parallel N_{SP}^1)$ and $M_2 = HMAC_{K_{j,SP}}(N_j^1 \parallel N_i^1 \parallel N_{SP}^1)$. After that, AS_{SP} sends M_1, M_2 and N_{SP}^1 back to STB_j.
Step 8 ($STB_j \rightarrow NP_i : M_1, N_j^1, N_{SP}^1, N_j^2$): STB_j obtains M_2 and N_{SP}^1 from the message sent by AS_{SP}. Then, STB_j uses $K_{j,SP}$ to compute $M_2' = HMAC_{K_{j,SP}}(N_j^1 \parallel N_i^1 \parallel N_{SP}^1)$. If $M_2 = M_2$', STB_j generates N_j^2 and sends a APDU command along with M_1, N_j^1, N_{SP}^1 and N_j^2 to NP_i. Otherwise, stop the request.
Step 9 ($NP_i \rightarrow STB_j : M_3, N_i^2$): Once NP_i receives the APDU command to get M_1 and N_{SP}^1, and then it uses $K_{i,SP}$ to compute $M_1' = HMAC_{K_{i,SP}}(N_i^1 \parallel N_j^1 \parallel N_{SP}^1)$. If $M_1 = M_1$', NP_i generates N_i^2 to compute $M_3 = HMAC_{K_{i,SP}}(N_i^1 \parallel N_j^1 \parallel N_{SP}^1 \parallel N_j^2 \parallel N_i^2)$ and sends a APDU response along with M_3 and N_i^2 back to STB_j. Otherwise, stop the request.
Step 10 ($STB_j \rightarrow AS_{SP} : M_3, M_4, N_i^2, N_j^2$) : STB_j uses $K_{j,SP}$ to compute $M_4 = HMAC_{K_{j,AS}}(N_j^1 \parallel N_i^1 \parallel N_{SP}^1 \parallel N_i^2 \parallel N_j^2)$ after it receives the APDU response to get N_i^2. Then, STB_j sends M_3, M_4, N_i^2 and N_j^2 to AS_{SP}.
Step 11 ($AS_{SP} \rightarrow STB_j : Accept/Reject$) : AS_{SP} uses $K_{i,SP}$ and $K_{j,SP}$ to compute $M_3' = HMAC_{K_{i,AS}}(N_i^1 \parallel N_j^1 \parallel N_{SP}^1 \parallel N_j^2 \parallel N_i^2)$ and $M_4' = HMAC_{K_{j,AS}}(N_j^1 \parallel N_i^1 \parallel N_{SP}^1 \parallel N_i^2 \parallel N_j^2)$, respectively. If $M_3 = M_3'$ and $M_4 = M_4'$, U_i is authenticated successfully. Otherwise, reject the request.

3.3 Key Update Phase

Step 1 ($U_i \rightarrow PC_i$) : First, U_i operates PC_i by using the web browser to visit *SP*'s website (AS_{SP}). Once accessing the site, an HTTP session is established between PC_i and AS_{SP}.
Step 2 ($PC_i \rightarrow AS_{SP} : request, Cell_i$): In order to apply for key update, U_i must log into his/her account on the website. After that, U_i fills in the key update form with $Cell_i$ that wants to receive the new key. Then, U_i sends the key update request to AS_{SP}.
Step 3 ($AS_{SP} \rightarrow PC_i : Info_{SP}$): Once AS_{SP} receives the key update request, it generates a verification code R_{SP} and a timestamp T_{SP}^1. The verification code R_{SP} is

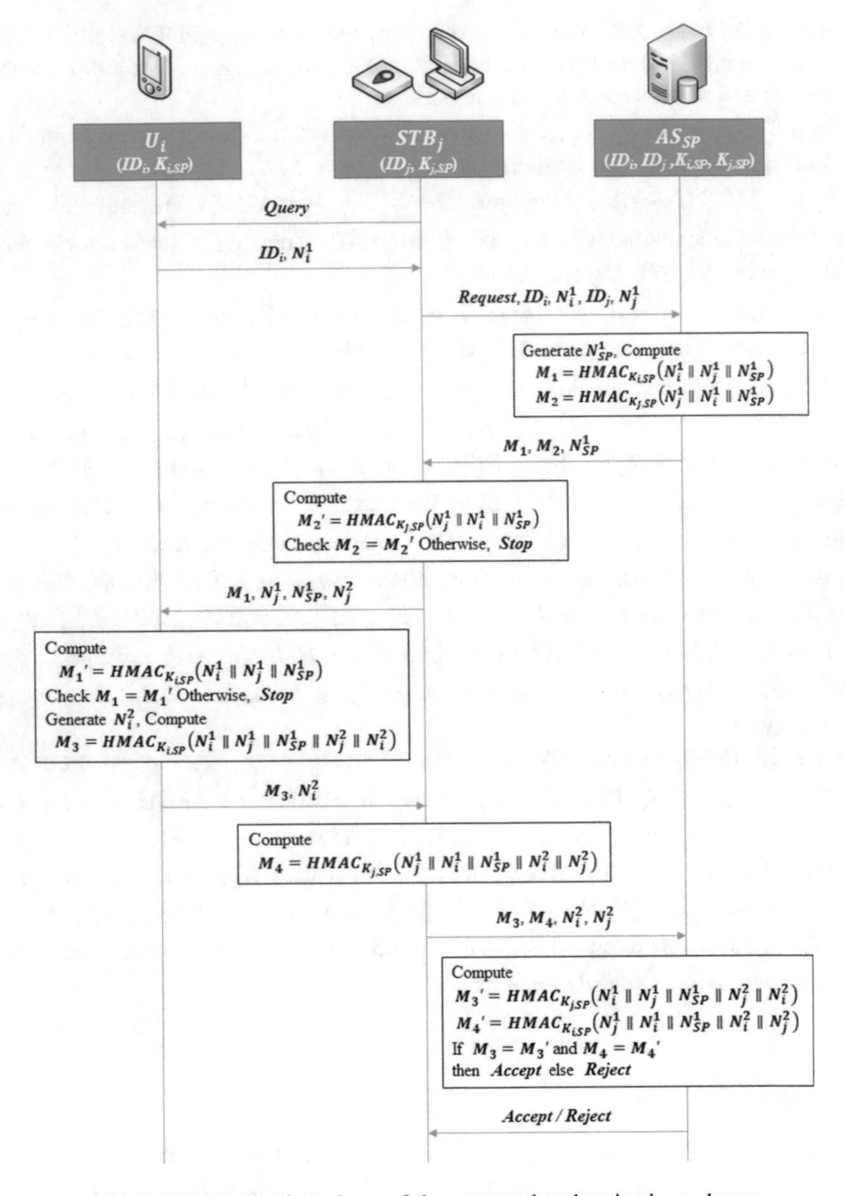

Fig. 1. Authentication phase of the proposed authentication scheme

sent to $Cell_i$ via SMS, and the response message is displayed on the web page to notify U_i receiving SMS.

Step 4 ($U_i \rightarrow NP_i$) : U_i starts APP_{SP} on NP_i after he/she obtains the verification code R_{SP} from the SMS. Once APP_{SP} is opened, an SSL/TLS connection is established between NP_i and AS_{SP}.

Step 5 ($NP_i \rightarrow AS_{SP} : Email_i, PW_i, R_{SP}$): U_i enters his/her $Email_i$, PW_i and the verification code R_{SP} on APP_{SP}. Then he/she sends them to AS_{SP}.

Step 6 ($AS_{SP} \rightarrow NP_i : ID_i, K_{i,SP}$): As soon as AS_{SP} receives the message form NP_i, it generates another timestamp T^2_{SP}, and then checks *Valid* ($Email_i$, PW_i, R_{SP}) and $T^2_{SP} - T^1_{SP} \leq TDuring$. If both conditions are met, AS_{SP} generates ID_i and a new secret key $K_{i,SP}$ to send them to NP_i. Otherwise, AS_{SP} responses an error message to NP_i.

Step 7 ($NP_i \rightarrow AS_{SP} : ID_i, M$): If NP_i receives the message from AS_{SP} without error, it stores ID_i and $K_{i,SP}$ in the secure storage of NP_i. NP_i uses $K_{i,SP}$ to compute $M = HMAC_{K_{i,SP}}(ID_i)$, and then sends ID_i and M to AS_{SP}. This is to make AS_{SP} confirm that NP_i has received ID_i and $K_{i,SP}$.

Step 8 ($AS_{SP} \rightarrow NP_i$: *result*): AS_{SP} obtains M and ID_i, from the message sent by NP_i, and retrieves the corresponding secret key $K_{i,SP}$ from its database according to ID_i. After that, AS_{SP} uses $K_{i,SP}$ to compute $M' = HMAC_{K_{i,SP}}(ID_i)$. If $M = M'$, which means that NP_i has received ID_i and the new secret key $K_{i,SP}$, AS_{SP} responses result back to NP_i shows the key update process is completed for U_i. Otherwise, AS_{SP} responses an error message to NP_i.

4 Security Analysis

An authentication system is practical only if it is proven to be secure enough. The proposed mechanism can resist most attacks and provide mutual authentication.

- Mutual Authentication & Impersonation Attack

Mutual authentication means that the two communicating entities are authenticated by each other. In the proposed authentication mechanism, AS_{SP} gets authenticated by U_i if it can correctly compute HMAC value by using secret key $K_{i,SP}$, and U_i gets authenticated by AS_{SP} if it can correctly compute HMAC value by using secret key $K_{i,SP}$. Consequently, the proposed protocol provides mutual authentication in which U_i and AS_{SP} can authenticate each other. Impersonation attack refers to an attacker that can masquerade as an authenticated user and deceive the server in order to obtain unauthorized services. Due to STB_j existing in an open environment, an invader may attack it to capture authentication sequences and obtain user identifiers. In the proposed authentication scheme, the secret key $K_{i,SP}$ is known only to U_i and ***SP***. An attacker cannot calculate the correct HMAC value without $K_{i,SP}$. Thus, an attacker is unable to masquerade as a genuine user.

- Replay Attack, Man-in-the-Middle Attack & Server Spoofing Attack

Replay attack is a malicious approach whereby an attacker gets a copy of the messages sent by an authenticated user and later tries to replay it. The proposed authentication scheme is based on a challenge-response mechanism and works by using pseudo-random values (i.e. nonce values) to prevent replay attack. In the proposed authentication protocol each message (i.e. M_1, M_2, M_3, M_4) contains nonce values. The nonce is a random value that is used only once and not repeated. If the nonce has been used previously, AS_{SP} can detect the message is a replay attack, and then reject the request. Therefore, malicious attackers attempting to use replay attack won't succeed.

Man-in-the-middle attack is an active attack whereby the attacker intercepts and selectively modifies communicated data for the purpose of masquerading as one or more of the entities involved in a communication. The connection between STB_j and AS_{SP} is based on SSL/TLS, so the data sent through this connection cannot be modified or snooped. On the other hand, the NFC channel is a hard link, and in very close proximity to the sender and receiver. By virtue of the fact that the active device is constantly sending a signal to power the passive device, it is essentially very difficult for an attacker to remove sent messages, insert new messages, or modify legitimate messages without detection. For these reasons, it is impossible to apply man-in-the-middle attack in the proposed protocol. As was mentioned above, the proposed authentication protocol can provide mutual authentication making it possible for U_i to authenticate AS_{SP}. Hence, server spoofing attack on the proposed mechanisms is infeasible.

5 Conclusion

In this paper, a user authentication mechanism for personalized IPTV services based on NFC technology is proposed. Relying on hash-based crypto-techniques, the proposed mechanism is particularly suitable for IPTV services. Security analysis also demonstrate the robustness of the proposed scheme. Future work may focus on actual deployment scenarios, and address further security issues, such as content protection, service protection, user authentication, and confidentiality, in the IPTV area.

References

1. GSMA Intelligence: GSMA Mobile Economy 2015. Mobile Economy 2015
2. Liu, Z., Wei, B., Yu, H.: IPTV, towards seamless infotainment. In: 6th IEEE Consumer Communications and Networking Conference, pp. 1–5 (2009)
3. Park, S., Jeong, S.H.: Mobile IPTV: approaches, challenges, standards, and QoS support. IEEE Int. Comput. **13**(3), 23–31 (2009)
4. Park, S., Jeong, S.H, Hwang, C.: Mobile IPTV expanding the value of IPTV. In: 7th International Conference on Networking, pp. 296–301 (2008)
5. Zeadally, S., Moustafa, H., Siddiqui, F.: Internet protocol television (IPTV): architecture, trends, and challenges. IEEE Syst. J. **5**(4), 518–527 (2011)
6. Jana, R., Chen, Y.F., Gibbon, D.C., Huang, Y., Jora, S., Murray, J., Wei, B.: Clicker - an IPTV remote control in your cell phone. In: 2007 IEEE International Conference on Multimedia and Expo, pp. 1055–1058 (2007)
7. Bambini, R., Cremonesi, P., Turrin, R.: A recommender system for an IPTV service provider: a real large-scale production environment. In: Recommender Systems Handbook. Springer US, pp. 299–331 (2011)
8. Choi, J.H., Jeok, J., Lim, S.Y., Kim, H.C., Lee, H.K., Hong, J.W.: Personalized data broadcasting service based on TV-anytime metadata. In: IEEE International Symposium on Consumer Electronics, pp. 1–6 (2007)
9. Wang, H.L., Wang, J.G., Yau, W.Y.: Automated age regression for personalized IPTV services. In: IEEE International Conference on Multimedia and Expo, pp. 1333–1336 (2010)

10. van Brandenburg, R., van den Berg, H., van Deventer, M.O., Schenk, I.M.: Towards multi-user personalized TV services, introducing combined RFID digest authentication. Graduate Thesis, University of Twente (2009)
11. Park, Y.-K., Lim, S.-H., Yi, O., Lee, S., Kim, S.H.: User authentication mechanism using java card for personalized IPTV services. In: International Conference on Convergence and Hybrid Information Technology, pp. 618–626 (2008)
12. Foina, A.G., Ramirez-Fernandez, J., Badia, R.M.: Cell BE and Bluetooth applied to digital TV. In: IEEE Network Operations and Management Symposium, pp. 825–828 (2010)

On Design and Implementation a Smart Contract-Based Investigation Report Management Framework for Smartphone Applications

Shi-Cho Cha(✉), Wei-Ching Peng, Zi-Jia Huang, Tzu-Yang Hsu, Jyun-Fu Chen, and Tsung-Ying Tsai

Department of Information Management, National Taiwan University of Science and Technology, Taipei, Taiwan
csc@cs.ntust.edu.tw

Abstract. To prevent users from downloading and installing malicious smartphone applications, several countries and organizations have developed security requirements for smartphone applications and associated vetting systems. Certified third parties can inspect whether an application satisfies applicable security requirements and issue inspection reports to notify users of potential risks. However, currently there is no standard method for users to obtain inspection results. Furthermore, as the advances of hacking techniques, a inspecter may discover that an application is vulnerable to a new type of attack and wish to notify application users immediately. To address the issue, this study proposes a Smart Contract-based Investigation Report Management framework for smartphone applications security (SCIRM) to enable smartphone application users to obtain security inspection reports of interested applications with smart contracts. Benefiting from blockchain technology, users can obtain historical inspection reports of an application and verify the integrity of the reports. In addition, this study utilizes smart contract technology to implement the interfaces so that smart contracts will enforce the related actions automatically. This study can hopefully contribute to enabling users to adopt appropriate countermeasures to potential application security risks as users can obtain up-to-dated security information about applications timely.

1 Introduction

To help application developers develop secure smartphone applications, several countries and organizations such as ENISA (the European Union Agency For Network And Information Security) [2] and the Taiwan IDB (Industrial Development Bureau) [6], have developed guidelines for smartphone application security. However, when users download smartphone applications from marketplaces, users usually cannot know whether the developers of the applications follow the

J.-S. Pan et al. (eds.), *Advances in Intelligent Information Hiding and Multimedia Signal Processing*, Smart Innovation, Systems and Technologies 82,
DOI 10.1007/978-3-319-63859-1_35

guidelines to develop the applications. Therefore, organizations such as US NIST (National Institute of Standards and Technology) [4] and OWASP (The Open Web Application Security Project) [3] have developed smartphone application verification guidelines as well as security requirements for smartphone applications. Hence, smartphone application developers can delegate third parties to follow the verification guidelines to check whether their applications satisfy security requirements of the verification guidelines.

Furthermore, organization or government agencies may develop certification programs for smartphone applications. For example, the Taiwan IDB has announced the self regulatory mobile application security certification program [7]. Smartphone application developers can appoint accredited inspection laboratories to inspect whether their applications satisfy applicable security requirements [5]. Consequently, users can decide to only install inspected applications to reduce security risks of using smartphone applications. However, to the best of our knowledge, there is no standard means for users to obtain investigation results of smartphone applications.

In light of this, this study proposes a Smart Contract-based Investigation Report Management framework for smartphone applications security (SCIRM). The framework extends our previous work by utilizing smart contract technologies. Our previous work provides ontologies to store verification reports about smartphone applications in a bitcoin-like blockchain [1], this study further provides standard interfaces based on smart contracts for application verifiers to upload inspection reports to the blockchain and notify users that the inspection results of a smartphone application has changed immediately. As a result, the proposed framework can contribute to enabling users to obtain the latest security information of an application timely. Moreover, users can also obtain historical inspection results on applications by versions, developers, and other application profiles. The proposed framework can also contribute to providing users more information to evaluate application security risks.

The rest of the paper is organized as follows: Sect. 2 overviews the proposed framework. Section 3 describes how to implement the proposed framework with smart contracts. Next, this study illustrates the experiment performed for performance evaluation in Sect. 4. Conclusions are finally drawn in Sect. 5.

2 Overview of the Proposed Framework

Figure 1 overviews the proposed framework. As depicted in Fig. 1, the framework contains two views: the conceptual view and the implementation view. The conceptual view is composed of four major standards:

- The *Specification of Inspection Report* defines the components of an inspection report as well as the format of each component.
- The *Specification of Report Storage* clarifies how to store an inspection report in the blockchain.
- The *Trusted Verifier Ontology Specification* provides properties required to describe a trusted verifier.

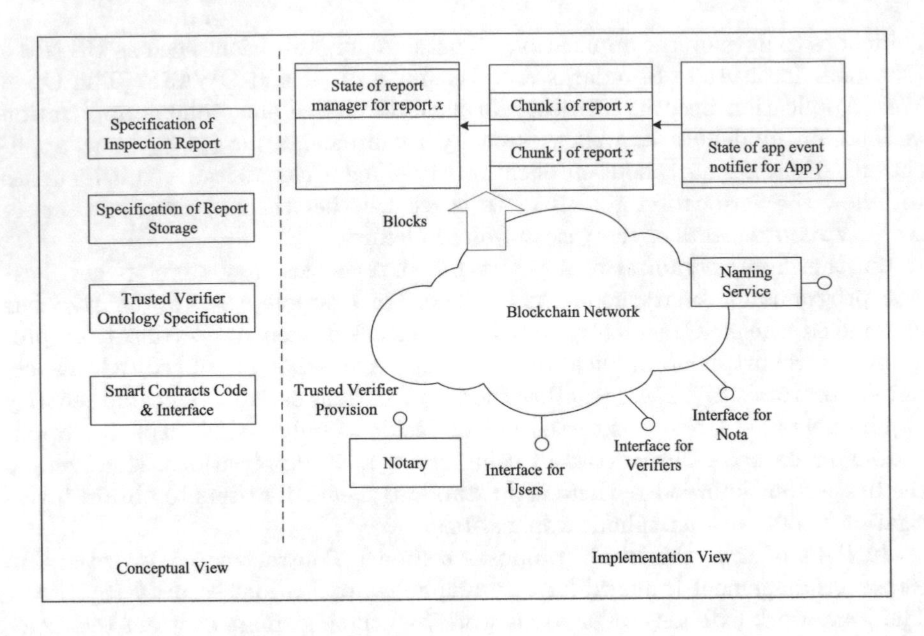

Fig. 1. Overview of the proposed framework

– The standard API and the associated smart contract for report management.

The implementation view illustrates the physical components of the framework. This study deployes the framework on an Ethereum blockchain system. In the Ethereum blockchain system, people can enclose data in transactions. The blockchain system encapsulates the transactions in blocks. Due to the size limits of a transaction, this study requests a verifier to split an inspection report into several chunks and enclose the chunks into different transactions.

This study defines the report manager smart contract to manage the inspection reports. When a verifier submits an inspection report to the blockchain, the verifier obtains the compiled bytecode of the report manager and creates a new instance of the report manager smart contract. The verifier can then bind the contract to the transactions of the inspection report. As depicted in Fig. 1, the Ethereum blockchain system stores the state of each contract in blocks. Therefore, after a verifier has created a report manager smart contract for an inspection report, people can obtain the state of the contract from a block in the blockchain system and use the state information to find associated transactions storing the chunks of the report.

This study also defines the application event notifier smart contract. A notary can create an application event notifier smart contract for an application. Verifiers can then link the report manager smart contract of the application to the application event notifier smart contract. Therefore, users only need to listen to the application event notifier smart contract to obtain events that an inspection report of the application has been created or obsoleted.

In addition to the constructor, a report manager smart contract provides interfaces for users and verifiers to access information of the associated report. Simply speaking, the report manager smart contract of an inspection report enables users to obtain the contents and status (e.g. active, obsolete, or revoked) of the report. Moreover, report manager smart contracts can trigger events to notify users the status changes of reports. On the other hand, a verifier can update the status of a report and append other reports to the report through the related smart contract. Details of the report manager smart contract are provided in Sect. 3.

Although users can obtain all inspection reports and related information of smartphone applications by traversing through every block in the blockchain, finding transactions and contract states of a specific smartphone application could be very time-consuming. Therefore, this study proposes to deploy some supernodes for building indexes on transaction attributes and contract states. The supernodes can provide naming services for users to lookup inspection reports on a specific smartphone application or reports on applications developed by a specific developer.

Finally, in current Ethereum blockchain systems, people only know a transaction or a smart contract is submitted by an identity. Consequently, users cannot determine whether they should trust inspection reports issued by an identity. To address the issue, this study assumes that notaries can provide a list of trusted verifiers as well as information of the verifiers. The notaries may also process complaints of application developers and mark an inspection report as revoked. However, enabling notaries to revoke inspection reports may induce disputes between notaries and verifiers. This study leaves the report revocation mechanism as our future work.

3 Report Management

Before introducing the report management smart contract, this study will first introduce the application event notifier smart contract. Figure 2 illustrates the abstract interface of an application event notifier smart contract. An application event notifier smart contract contains profiles of a smartphone application. In addition, it provides the reportUpdated interface for a report manger smart contract to inform the application event notifier smart contract that the creator of the report manger has just updated the state of the related inspection report. This study provides the following types of update events for an inspection report:

- Creation of the inspection report.
- Obsoleting of the inspection report.
- Being extended by an another report.

Note that because contents of transactions are immutable in blockchains, update operations to an inspection report are not provided. The abstract interface of a report manager smart contract is depicted in Fig. 3. In addition to the profiles of the associated application, a report manager smart contract contains the

```
contract AppEventNotifier {
    address public creator;
    string public appID;
    string public version;
    string public platform;
    string public developerInfo;
    uint public status;
    event statusChangeNotification(string appID, string version, string platform, uint type,
        address rmAddr);
    function AppEventNotifier(string _appID, string _version, string _platform,
            string _developerInfo) { ...}
    function reportStatusChanged(uint type) { ...}
}
```

Fig. 2. Abstract of the application event notifier smart contract.

```
contract ReportManager {
    address public verifier;
    string public appID;
    string public version;
    string public platform;
    string public developer;
    uint public status;
    uint public totalChunks;
    AppEventNotifier appNotifer;
    struct crunkIndex { uint seq; string txAddr; }
    mapping(uint=>crunkIndex) content;
    mapping(uint=>address) extendedBy;
    function ReportManager(string _appID, string _version, string _platform,
            string _developer, unit _totalChunks, AppEventNotifier _notifer) { ...}
    modifier onlyCreator { if (msg.sender == verifier) _ }
    function setChunkInfo(uint _seq, address _txAddress) onlyCreator { ...}
    function extended(address _reportManager) onlyCreator { ...}
    function obsoleteReport() onlyCreator { ... }
    function queryChunkAddress(uint _seqNumber)
            returns (address chunkAddress) { ...}
}
```

Fig. 3. Abstract of the report manager smart contract.

status of an inspection report, total chunks of the report, and the addresses of the chunks. When a verifier wishes to upload an inspection report for an application, the verifier will first split the report into several chunks and submit transactions containing the chunks to the blockchain network. The verifier then generates a report manager smart contract with application profiles, total chunk number of the report, and the address of the application's event notifier smart contract.

Because the current Ethereum platform only allows using fixed size array as function parameters, the verifier needs to use the setChunkInfo function to provide associated transaction addresses one by one. The report manger smart contract will store the addresses in a mapping object. After receiving addresses to access every chunk of the associated report, the report manger smart contract will invoke the reportStatusChanged function of related application event notifier. The notifier can then trigger an event to notify interested users of the creation of the report.

The verifier can also use the obsoleteReport function to obsolete a report, or the extended function to represent the situation that the report is extended by another report. The report manger smart contract will also request the associated application event notifier smart contract to notify users when these two functions are called.

On the other hand, when a user downloads a smartphone application from a marketplace, the user can first query the naming service to obtain the address of the AppEventNotifier smart contract of the application as well as the addresses of associated ReportManager smart contracts. Then, for each ReportManager smart contract, the user can obtain the status of the associated inspection report and the verifier to determine whether or not to download the report. If the user decides to download the report, the user can reassemble the report by collecting all the chunks stored in the transactions. Finally, the user can listen to the AppEventNotifier smart contract of the application for status changing events of the application inspection reports.

4 Experiments

To proof the concept of the proposed framework, this study has implemented a prototype system and performed simulation experiment. As shown in Fig. 4, this study uses Node.js to implement a *Report Upload Server* to handle inspection report uploads. In addition, this study launches a private Ethereum network with a single service node. The report manager communicates with the node via the Ethereum JavaScript API. After receiving an inspection report, the report upload server will split received reports into chucks (Step 1), store the chunks in transactions (Step 2), and send the transaction to the node (Step 3). Next, the report upload server creates a report manager smart contract and send the addresses to the contract (Step 4).

In the experiment, the report manager and the Ethereum service node are both executed on a desktop with Intel Core i7-4790 CPU 3.60 GHz CPU and 12 GB RAM running Ubuntu 16.04 LTS. To simplify the experimental environment, this study fixes the mining difficulty to 0×4000. An inspection report is about 22 Kbytes (based on the inspection items of the Taiwan self regulatory mobile application security certification program). This study calculates the average time for the report update server to perform the tasks (from Step 1 to Step 4) in the above paragraph with different maximum chunk size. For each maximum chunk size, this study performs the experiment 20 times. Table 1

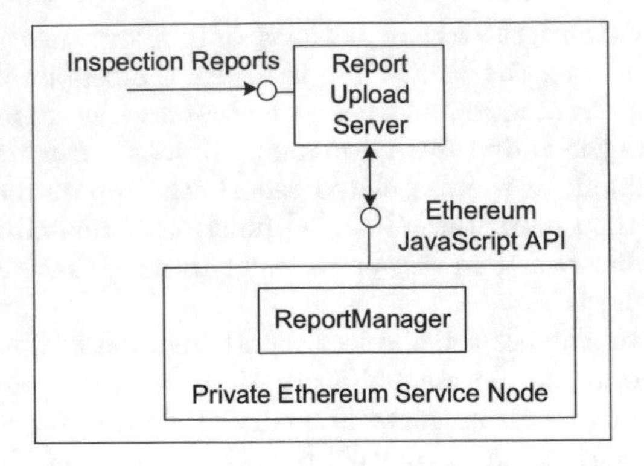

Fig. 4. The experimental scenario

Table 1. Summary of experimental results

Maximum chunk size (bytes)	1 K	2 K	4 K	8 K	16 K	32 K
Average time (seconds)	79.6	42.05	24.94	14.06	10.99	7.41

summarizes the experimental results. As listed in the table, the report update server can usually finish the above tasks in reasonable amount of time. This study leaves executing scalable analysis in our future work.

5 Conclusion

This study has proposed a Smart Contract-based Investigation Report Management framework for smartphone applications security (SCIRM) to enable smartphone application users to obtain security inspection reports of interested applications using smart contracts. Benefiting from the blockchain technology, users can get historical inspection reports of an application and ensure the integrity of the reports. In addition, application users and other volunteers can collaborate to provide resources needed to host the framework. Therefore, as the framework does not rely on a single party, the framework does not need to consider the business interests of a business company.

The proposed framework also provides standard interfaces for a verifier to upload an inspection report of an application and update the status of the report. A user can also follow the interfaces provided to listen for events of report uploading and report status updating. This study utilizes smart contract technology to implement the interfaces so that the smart contracts will enforce related actions automatically. As users can obtain up-to-date security information about applications, the study can hopefully contribute to enabling users to adopt appropriate countermeasures to potential application security risks timely.

Acknowledgement. This work was supported in part by the Taiwan Ministry of Science and Technology under grants MOST 104-2923-E-011-005-MY3 and MOST 105-2218-E-001-001.

References

1. Cha, S.-C., Hung, S.-C., Chen, J.-F., Syu, S.-C., Tsai, T.-Y.: On the design of a blockchain-based reputation service for android applications. In: Preceedings of the 2016 International Conference on Cyber-Society and Smart Computing Communication (The CyberSoc 2016), Yogyakarta, Indonesia (2016)
2. European Union Agency For Network And Information Security (ENISA). Smartphone secure development guidelines (2016). https://www.enisa.europa.eu/publications/smartphonesecuredevelopmentguidelines2016
3. Mueller, B.: Mobile application security verification standard (MASVS) 0.9.2. OWASP Standard (2017)
4. Quirolgico, S., Voas, J., Karygiannis, T., Michael, C., Scarfone, K.: Vetting the security of mobile applications. US National Institute of Standards and Technology (NIST) SP 800-163 (2015)
5. Taiwan Industrial Development Bureau (IDB). Mobile app funtational security requirement v1.1 (2017). http://www.mas.org.tw/news_detail.php?id=38
6. Taiwan Industrial Development Bureau (IDB). Mobile app secure development guidelines v1.0 (2017). http://www.mas.org.tw/news_detail.php?id=38
7. Taiwan Industrial Development Bureau (IDB). Self regulatory mobile app funtational security certification v3.0 (2017). http://www.mas.org.tw/news_detail.php?id=38

Accelerated vBNN–IBS Authentication Scheme for WSN

Danyang Qin(✉), Yan Zhang, Jingya Ma, Songxiang Yang, and Erfu Wang

Key Lab of Electronic and Communication Engineering, Heilongjiang University, Harbin, People's Republic of China
qindanyang@hlju.edu.cn

Abstract. Broadcast authentication is the base of network security in WSN (Wireless Sensor Network) due to versatility task and limited energy. The symmetric encryption has been widely used in WSN because of its high energy efficiency, but the network is vulnerable to DoS (Denial of Service) attacks on account of time delay during the authentication. An accelerated signature mechanism AvBNN-IBS (Accelerated vBNN-IBS), focus on the fix-deployed WSN, through the way of mutual cooperation between nodes is put forward in this paper to solve the problems of low calculation speed and complex algorithm of public key encryption mechanisms. Compared with the traditional authentication mechanisms, the scheme proposed could effectively lower the energy consumption and authentication time.

Keywords: Digital signature · Accelerated authentication · Energy consumption · Wireless Sensor Network

1 Introduction

Wireless sensor network is a large-scale distributed network, and the topology structure of sensor nodes is changing constantly [1]. Compared with traditional network, WSN is more vulnerable to different attacks due to its limited computing power and restricted energy resources [2]. So an effective authentication mechanism should be put forward to ensure the secure communication.

μTESLA is a high-efficiency and tolerate of packet dropout symmetric key encryption protocol based on TESLA, which more suitable for resource-constrained sensor network [3]. It solves the problems of shared key, unidirectional key generation algorithm, announced key packet loss and distribution delay, but it's vulnerable to DoS attacks [4]. Public Key Cryptography (PKC) has more advantages on broadcast authentication compared with symmetric encryption scheme. PKC doesn't need the common key shared between nodes, even though the public key is filched during transfer or release, the invaders cannot get any useful information because of no private key to suit for the public key [5]. However, low encryption and decryption speed leads to higher energy cost and the time delay in the broadcast authentication. The accelerated ECDSA (AECDSA) proposed a fast authentication mechanism by transfer the compute results randomly to its neighbor nodes, in which the neighbor nodes can

J.-S. Pan et al. (eds.), *Advances in Intelligent Information Hiding and Multimedia Signal Processing*, Smart Innovation, Systems and Technologies 82,
DOI 10.1007/978-3-319-63859-1_36

complete the certification quickly by using the results directly. But for WSN, this digital signature mechanism is still high consumption on computing [6]. So this paper proposes a signature authentication scheme AvBNN-IBS (accelerated variant of Bellare Namprempre Neven Identity-based Signature) by mutual cooperation among nodes based on existing vBNN-IBS to reduce the energy and the time cost.

2 vBNN-IBS Authentication Model and Realization

2.1 ECC Algorithm

In the process of elliptic curve cryptography (ECC), the elliptic curve E over a finite field F_q is defined, which use E/F_q to describe, and q is a pretty big prime number. Point of (x, y) satisfies the elliptic curve E: $y^2 = x^3 + ax + b$, where a and b in the field F_q and satisfy $4a^3 + 27b^3 \neq 0$. P is a point of order p on curve E, and P generate a group G [7]. Addition '+' is an operation defined as follows: taking points P and Q on the elliptic curve, l is the straight line containing point P and point Q and intersect with curve E at point R. Point R' is the x axis symmetry point of R, it also mains the result of P'+'Q [8]. Point multiplication can be computed as $nP = P + P + \cdots + P$, where the number of P is n.

2.2 Realization Process of vBNN-IBS

vBNN-ISB is an ID-based encryption scheme. The signature mechanism will reduce the signature size for WSN effectively. The private key of the user will be generated by a credible third party PKG (Private Key Generator), and the public identity information is the public key of the user. The process of the vBNN-IBS can be divided into the follow stages:

Establishment phase: Give the security parameters k, PKG will follow three steps. Firstly, an elliptic curve E is defined over field F_q and a point P of order p. Secondly, a system private key x is selected randomly and the public key $P_0 = xP$ is calculated. Thirdly, two Hash functions $\mathrm{H}_1 : \{0,1\} \times G_1^* \rightarrow Z_P$ and $\mathrm{H}_2 : \{0,1\}^* \rightarrow Z_P$ are selected, where Z_P is a field composed by prime number 0 to $p-1$, and then the system parameter $\left(E/F_q,\ P,\ p,\ P_0,\ \mathrm{H}_1,\ \mathrm{H}_2\right)$ is released so as to make x safe.

User keys extraction phase: $ID_a \in \{0,1\}^*$ refers to user A's unique identity, and P_{ria} is A's private key generated by PKG. Then in this phase, the steps taken are as follows: Select a random number r and calculate $R = rP$, then calculate $s = r + cx$, where $c = \mathrm{H}_1(ID_a||R)$. After that, user A's private key Pair $P_{ria} = (R, S)$ is generated by PKG and be sent to user A.

Generate the signature phase: When user A with the identification ID_a carries the message m, signature can be generated. First, select a random number y and calculate $Y = yP$. Second, calculate $h = \mathrm{H}_2(ID_a, m, R, Y)$ and $z = y + hs$, and then (R, h, z) is user A's signature when carrying message m.

Signature verification phase: Signature verification can be implemented according to the given signature (R, h, z), ID_a and message m. Calculate $c = \mathrm{H}_1(ID_a||R)$, and then verify whether $h = \mathrm{H}_2(ID_a, m, R, zP - h(R + cP_0))$ is valid to complete signature verification or reject it.

3 Accelerated vBNN-IBS Scheme

In traditional vBNN-IBS scheme, every node is required to calculate the value of zP, hR and hcP_0 in the process of signature verification, which means that it needs three times of multiplication operations and twice addition operations for one node completing once authentication. During the broadcast authentication, all sensor nodes need to do the same operation, which not only waste the nodes energy, but also prolong the authentication time. So an improved vBNN-IBS scheme, named AvBNN-IBS, is put forward to speed up the verification by one node transmitting the intermediate results to its neighbor nodes. In other words, when there are some nodes willing to consume a little energy to transmit the intermediate calculation results to its neighbor nodes, the neighbor nodes can save more energy by using the result directly to reduce the authentication time and get higher efficiency.

User needs to send data packet $\{m, ID_u, ct, Sig_m\}$ in broadcasting stage, where m is the message; ID_u is the user's identification; ct means the current time; and Sig_m is the signature (R, h, z) of the user who carries the message m. When the user broadcast data packet, nodes check whether ct is the real time first, if not, the packet is invalid. In Fig. 1, nodes A, B and C will receive data packet from the user and verify if it is effective. They will transmit the intermediate results to their neighbor nodes.

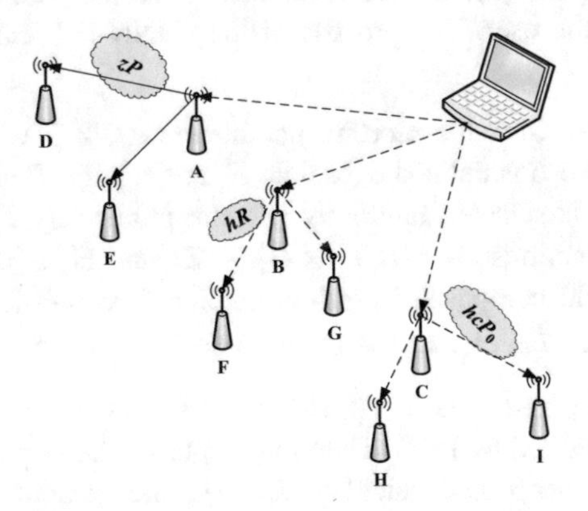

Fig. 1. AvBNN-IBS authentication by cooperation among nodes

For example, node D needs to calculate $zP - hR - hcP_0$ before complete the validation, the value of zP is from node A directly, node D can complete the certification quickly after calculating the value of hR, hcP_0 and twice addition operations. Due to node E and D are both node A's neighbor, so the authentication of E is similar to node D. Similarly, node B transfers the intermediate result hR to its neighbor node F and G, node F and G complete the certification after calculate zP, hcP_0 and twice addition operations. In the same way, node H and I complete accelerated authentication by using intermediate result hcP_0 from node C. In traditional vBNN-IBS authentication scheme, sensor nodes need three times multiplication and twice addition to complete the certification, but only twice multiplication and twice addition can complete in the proposed AvBNN-IBS, so that save the certification time and energy.

Further, if the nodes transmit the sum of two multiplication-operating results to its neighbor nodes, one neighbor node only needs once multiplication and once add to complete the authentication. For example, if node H successfully received $hR + hcP_0$ from node C, the verification will complete with 1Add+1Mul, rather than 2Add+3Mul. This case greatly higher the authentication efficiency and reduce the time and energy consumption. However, nodes can't transfer all the three intermediate results to its neighbor nodes, because of it vulnerable to attacks. Table 1 shows the operations of the attacker and the victim respectively.

Table 1. Operations of the attacker and the victim

Operation of attacker	Operation of victim
1. Generate false message m'	**1.** Calculate $c' = \mathrm{H}_1(ID_a \| R')$
2. Choose randomly R', z', h', hc'	**2.** Count $\mathrm{H}_2(ID_a, m', R', z'P - h'R' - hc'P_0)$
3. Calculate $z'P - h'R' - hc'P_0$	**3.** Compare the result from step 2 with h'
4. Calculate $h' = \mathrm{H}_2(ID_a, m', R', z'P - h'R' - hc'P_0)$	**4.** If consistent, m' will be regarded as effective information
5. Transmit (R', h', z') and $z'P$, $h'R'$, $hc'P_0$	

Assume that the sensor nodes transfer the operation result $hR + hcP_0$ or even no result to its neighbor nodes, so as to ensure the safety of the whole wireless sensor network and guarantee the authentication efficiency. Taking node C as an example, node C will transmit a packet of $\{m, ID, ct, Sig_m\}$ or $\{m, ID, ct, Sig_m, hR + hcP_0\}$ to its neighbor nodes. If node C sends the packet of $\{m, ID, ct, Sig_m\}$, node H will first calculate zP, and wait for a short period time of α_t to judge if it can receive intermediate results $hR + hcP_0$ and caches τ data packets from node C. In AvBNN-IBS authentication scheme, assume that node A has η neighbor nodes, in which $\eta/2$ nodes can receive packets from A. For the $\eta/2$ nodes, there are ϑ nodes may be captured and produce bogus data packets $\vartheta \cdot \lambda$ in all. The number of cache data packets τ should satisfy the following formula:

$$\eta/2 \geq \tau \geq \vartheta \cdot \lambda + 1 \tag{1}$$

Assume that the neighbor nodes of A have τ packets received from node A, then the waiting time α_t is decided by the values of transmission rate, packet size, initial backoff and congestion backoff. α_t satisfies the following formula:

$$\alpha_t \geq (B_{iMAX} + B_{cMAX} + PS_{MAX}/R_{MAX}) \cdot \tau \tag{2}$$

where B_{iMAX}, B_{cMAX}, PS_{MAX} and R_{MAX} are maximum initial backoff, maximum congestion backoff, maximum data packet size and maximum transmission rate respectively.

4 Simulation Results and Performance Analysis

Assuming there is no attack in theoretical analysis and simulation experiments, and the topology structure as shown in Fig. 2. The simulation adopts the manner of work in 8 MHz MICAz motes transmit the intermediate result $hR + hcP_0$ on TinyOS.

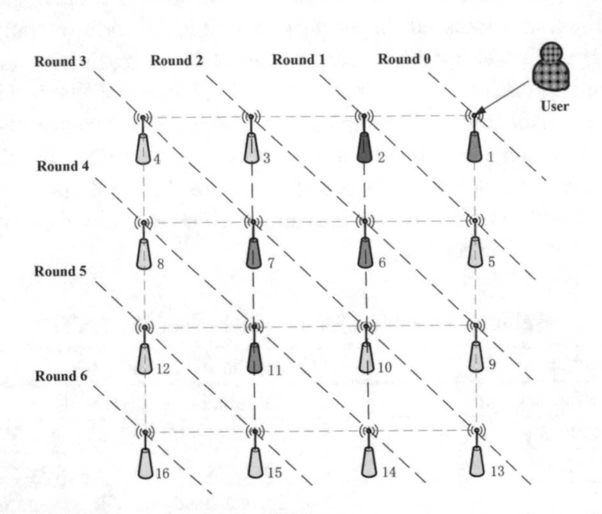

Fig. 2. AvBNN-IBS authentication nodes topology structure in simulation

4.1 Theoretical Analysis

Under the standard of IEEE802.15.4, the analysis of detecting the energy consumption and time cost on AvBNN-IBS is performed. The nodes topology structure is shown in Fig. 2. An 8-bit processor ATmega128L will cost 0.81 s to finish once multiplication with the working voltage as 3 V, transmitting the data at the rate of 250 kbps, which will consume 17.4 mA and 19.7 mA when sending and receiving the information respectively. The known smallest reachable sizes of p and q are 166 bits and 168 bits to achieve the same security strength with 1024 bit RSA, so the packet of $\{m, ID, ct, Sig_m, hR + hcP_0\}$ is 139 bytes, where message m being expected to transfer is 10 bytes, ID and ct are both 2 bytes and the intermediate result $hR + hcP_0$ is 42 bytes, the signature is 83 bytes. The packet must be divided when transfer because MICAz motes at most transmit 128 bytes in the physical layer [9].

The energy consumption of each node can be obtained theoretically based on the above parameters. $E_{s1} = 3.0 \times 17.4 \times 8/250 = 1.67\,\mu\text{J}$, $E_{r1} = 3.0 \times 19.7 \times 8/250 = 1.89\,\mu\text{J}$, where E_{s1} and E_{r1} are the sending and receiving energy consume of each node, respectively. E_{mul} and E_{ver} are the energy cost by each node for once multiplication and a completed verification. There will be $E_{mul} = 3.0 \times 8.0 \times 0.81 = 19.44\,\text{mJ}$ and $E_{ver} = 58.32\,\text{mJ}$. The intermediate result $hR + hcP_0$ will cost the energy of E_s and E_r in sending and receiving respectively, where $E_s = 3.0 \times 17.4 \times 128 \times 8/250 = 0.214\,\text{mJ}$ and $E_r = 3.0 \times 19.7 \times 128 \times 8/250 = 0.466\,\text{mJ}$.

Not all the nodes will transmit the intermediate results to their neighbor nodes to fasten the authentication in the distribution structure in Fig. 2. For example, the user transfers the data packet to node 1 who is the nearest neighbor, and node 2 and 5 will complete the accelerated authentication by using the intermediate result at Round 1 and go into sleeping after that. In this way, the authentication of 16 nodes can be accomplished after 6 rounds. Node 7 will only release the intermediate results to node 8 and 11, while node 3 and 6 can perform the authentication through receiving the calculating results from node 2. So there will be 6 nodes releasing the intermediate results, and that will consume $E_{\cos t(s)} = 6 \times 0.124 = 1.284\,\text{mJ}$. The nodes who receive the results may consume $E_{\cos t(r)} = 11 \times 0.466 = 5.126\,\text{mJ}$, so $E_{sav} = 11 \times 2 \times 19.44 = 427.68\,\text{mJ}$ will be saved by using the intermediate results. It turns out theoretically that the traditional vBNN-IBS will consume the energy of about $16 \times 58.32 = 933.12\,\text{mJ}$ to make $E_{theo} = 427.68 - 1.284 - 5.126 = 421.27\,\text{mJ}$, which shows that there is about 45.15% energy reduction obtained by AvBNN-IBS at least.

4.2 Simulation Results Analysis

The performance of AvBNN-IBS will be simulated from two aspects. One is from the function of nodes, which is aiming at the energy consumption during sending and receiving data packet. The other is from nodes' state, in other words, how much energy will be consumed when a node is in active, idle and sleeping status respectively. The WSN grid scale of the experiment is 4×4 as shown in Fig. 2.

4.2.1 Simulation Analysis from the First Aspect

To detect how much energy consumption and time cost that AvBNN-IBS can reduce, NesC application is executed to send and receive IEEE802.15.4 packets. The 139 bytes data packet including the intermediate result will be divided into two parts during transferring process, because MICAz motes can at most transmit 128 bytes in the physical layer. The first packet is *SIG*, which includes the user's information and signature. The packet *Inter* contains the intermediate results. Packet *SIG* will consume $E_{sig(s)} = 491.4\,\mu\text{J}$ and $E_{sig(r)} = 598\,\mu\text{J}$ when being sent and received respectively. On the other hand, it will cost $E_{\text{inter}(s)} = 387.1\,\mu\text{J}$ and $E_{\text{inter}(r)} = 467\,\mu\text{J}$ when the packet *INTER* is transmitted and received, respectively. The simulation adopts the result that one multiplication will consume 51.795 mJ in Projective Coordinate System (PCS). Through series of simulations, it can be found that vBNN-IBS, AECDSA, ECDSA will consume 2503 mJ, 2783.3 mJ, 3343.5 mJ, respectively. However, AvBNN-IBS will only consume 1371 mJ. In terms of the time overhead, it will cost AvBNN-IBS 2228 ms, but 6699 ms, 6863 ms and 4797 ms for other three schemes, respectively.

Comparing experiments are designed between AvBNN-IBS and other three schemes in time cost and energy consumption, with the results are shown in Fig. 3. The histogram in Fig. 3 shows that AvBNN-IBS scheme proposed in this paper performed better both in energy consumption and time cost, while the traditional ECDSA is inferior compared with others. The proposed AvBNN-IBS authentication mechanism consumes only 54.77% energy and 33.26% time overhead of the traditional vBNN-IBS, which is even better than ECDSA and AECDSA.

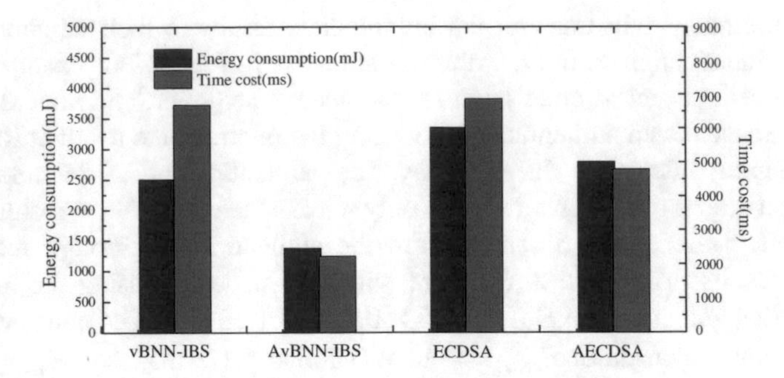

Fig. 3. Energy consumption and time cost of different schemes from the first aspect.

4.2.2 Simulation Analysis from the Second Aspect

The energy consumption is different when sensor nodes are in different states as shown in Fig. 4, which describes the different status changes in the run-time. In active state, the node will send and/or receive data packet. Idle nodes keep radio transceiver on, responsible for listening and certificate packet. In sleeping state node can't able to send and get any data packet. Initially, the node is idle. The node will be in active state when receiving the certification packet, and it will fall in sleeping state to save the energy after being authenticated until all nodes complete the certification.

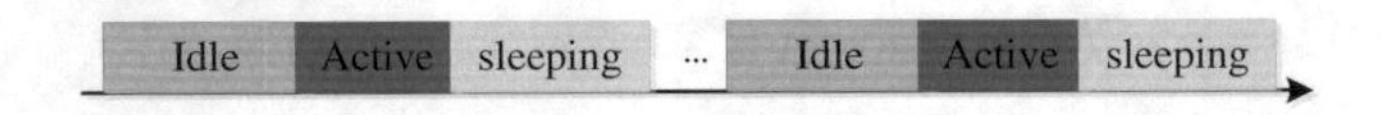

Fig. 4. States of sensor node during authentication

The energy consumption is negligible when the node is in sleeping state, since the node cannot send and receive any data [10]. Although the nodes are idle, it still cost a large part of the whole energy because of the open transceiver. Simulation is implemented from the perspective of node's different states with the results shown in Fig. 5.

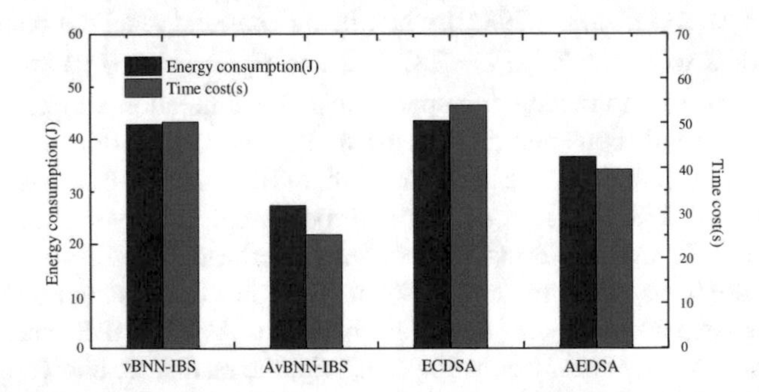

Fig. 5. Energy consumption and time cost of different schemes from the second aspect

The total energy cost in the experiment can be calculated according to each node's energy consumption. AvBNN-IBS, vBNN-IBS, ECDSA and AECDSA's energy consumption is 27.26 J, 42.79 J, 43.46 J and 36.49 J, respectively. Four schemes will cost 25.17 s, 50.30 s, 53.87 s and 39.60 s in sequence.

5 Conclusion

Aiming at the problem of security in fix-deployed WSN, a signature authentication scheme AvBNN-IBS is proposed by mutual cooperation among sensor nodes to accelerate the authentication process based on the existing scheme vBNN-IBS. The main idea of the proposed scheme lies in the fact that all sensor nodes transmit the intermediate results to their neighbor nodes, so that the neighbor nodes could faster the authentication using the results directly. Simulation experiments from nodes' function and nodes' different states in 4×4 grid verify the correctness and effectiveness of AvBNN-IBS, since it can save more energy and reduce time cost so that to speed up the authentication compared with the traditional schemes.

Acknowledgement. This work was supported by the National Natural Science Foundation of China under Grant No. 61302074, 61571181, Natural Science Foundation of Heilongjiang Province under Grant No. QC2013C061, Modern Sensor Technology Research and Innovation Team Foundation of Heilongjiang Province under Grant No. 2012TD007, and Postdoctoral Research Foundation of Heilongjiang Province under Grant No. LBH-Q15121.

References

1. Snasel, V., Kong, L.: Sink node placement strategies based on cat swarm optimization algorithm. J. Netw. Intell. **1**(2), 52–60 (2016)
2. Li, J.-P., Dong, Z.-Q.: Uneven clustering and data transmission strategy for energy hole problem in wireless sensor networks. J. Inf. Hiding Multimedia Signal Process. **8**(2), 500–509 (2017)
3. Perrg, A., Szewczyk, R., Tygar, J.D., Wen, V., Culler, D.E.: SPINS: security protocols for sensor networks. ACM Wireless Netw. **8**, 521–534 (2002)
4. Mansouri, D., Mokdad, L., Ben-othman, J., Ioualalen, M.: Detecting DoS attacks in WSN based on clustering technique. In: 2013 IEEE Wireless Communication and Networking Conference (WCNC), pp. 2214–2219 (2013)
5. Kim, D., An, S.: PKC-based DoS attacks-resistant scheme in wireless sensor networks. IEEE Sens. J. **16**, 2217–2218 (2016)
6. Kodali, R.K: Implementation of ECDSA in WSN. In: 2013 International Conference on Control Communication and Computing (ICCC), pp. 310–314 (2013)
7. Saqib, N., Iqbal, U.: Security in wireless sensor networks using ECC. In: 2016 IEEE International Conference on Advances in Computer Applications (ICACA), pp. 270–274 (2016)
8. Singh, L.D., Singh, K.M.: Implementation of text encryption using elliptic curve cryptography. Procedia Comput. Sci. **54**, 73–82 (2015)
9. Yusof, Y.M., Muzahidul Islam, A.K.M., Baharum, S.: An experimental study of WSN transmission power optimization using MICAz motes. In: 2015 International Conference on Advances in Electrical Engineering (ICAEE), pp. 182–185 (2015)
10. Lu, X., Cen, J., Zhang, X.: Node state optimization based coverage control algorithm for wireless sensor networks. In: 2014 IEEE 7th Joint International Information Technology and Artificial Intelligence Conference, pp. 163–166 (2014)

A Pre-assigned Key Management Scheme for Heterogeneous Wireless Sensor Networks

Danyang Qin(✉), Jingya Ma, Yan Zhang, Songxiang Yang, and Zhifang Wang

Key Lab of Electronic and Communication Engineering, Heilongjiang University, Harbin, People's Republic of China
qindanyang@hlju.edu.cn

Abstract. The studies on key management mostly are focusing on Heterogeneous Wireless Sensor Networks (HWSN), since the security problem of HWSN is complex due to different functions of nodes. In order to solve this problem, a pre-assigned key management scheme combining Schnorr authentication is proposed for HWSN in this paper based on the asymmetric method. Simulation results show that, compared with traditional key management scheme, the proposed scheme can not only enhance the network security but reduce the storage requirements.

Keywords: HWSN · Pre-assigned key management · Authentication · Network security · Storage space

1 Introduction

Security issues of wireless sensor networks (WSNs) cause increasingly attendance, because of its widely application [1]. And the key management occupies a fundamental position in the security mechanisms for WSN [2]. Eschenauer and Gligor proposed the E-G key management scheme with low computational complexity and storage requirements that is easy to achieve, however, its security is not high [3]. A q-composite scheme which is the modification of E-G is proposed by Chan, but the scheme becomes worse with the increasing of compromised nodes [4]. Bloom proposed a random key management scheme, which can make any nodes setup the pair-wise key directly, however, the safely connected of the network topology leads to a large resource overhead, etc [5]. The schemes mentioned above are all for homogeneous wireless sensor network, but this kind of network has the bottleneck in the application of complex security mechanisms [6]. Therefore, it is of great importance to design an effective key management scheme for HWSN using the heterogeneity features [7]. To solve this problem, an asymmetric key pre-assignment scheme is proposed, and to take precautions against the leakage of shared keys, Schnorr authentication [8] is introduced. Finally, the storage space and security performance is analyzed compared with traditional wireless sensor networks. Simulation results show that SAP scheme provides better security with low storage requirements.

J.-S. Pan et al. (eds.), *Advances in Intelligent Information Hiding and Multimedia Signal Processing*, Smart Innovation, Systems and Technologies 82,
DOI 10.1007/978-3-319-63859-1_37

2 Construction of Clustering Model for HWSN

In this section, the construction of clustering model for HWSN is given. Throughout this paper we shall suppose that ordinary nodes and cluster heads (CHs), where CHs with tamper-resistant hardware are energy-gathered nodes presumed secure enough to access the network, are randomly and uniformly deployed. During the process of network initialization, each CH broadcasts Msg_{hel} (Msg_{hel} is a Hello message composed of its ID and location information) to its neighbors with maximum power and random delay that is to avoid Msg_{hel} collision between two adjacent CHs. Hence the sufficient quantity and wide transmission range of CHs, a great majority of ordinary nodes can receive Msg_{hel} from one or more than one CHs, and the sender with optimal signal intensity will be selected as CHs. The network construction is shown in Fig. 1. After the selection of cluster heads, a large amount of keys is assigned to CHs and a small amount of keys is assigned to ordinary nodes, because CHs have larger storage space than ordinary nodes and are relatively secure.

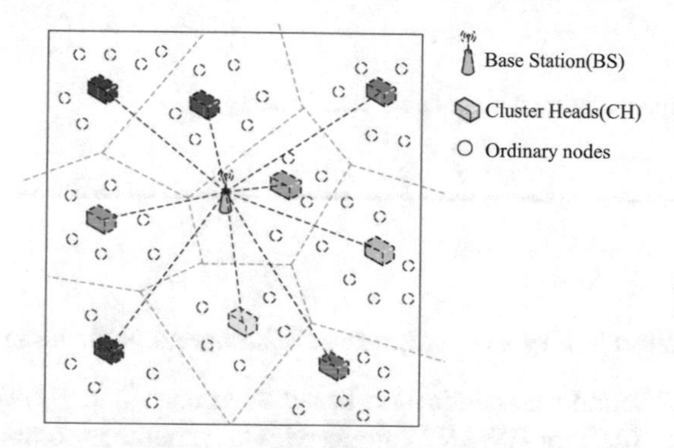

Fig. 1. Construction of clustering model for HWSN

3 A Pre-assigned Key Management Scheme

3.1 Acquisition of Network Certificates

After setting up the clustering model, the signature to the ID of each cluster member will be carried out and reserved by its CH, while each member obtains and reserves the certificate issued by CH. For example, node A is at liberty to select a random number k and calculates $y = g^k \bmod p$ (the parameter k, g and p are shown in Table 1.); node A sends the value y to its CH; and the ID of node A and value y will be signed by CH, that is, $s = sig_A(ID_A, y)$, thus, the certificate composed of ID_A, y and s is obtained, represented as $C_A = (ID_A, y, s)$.

Table 1. Authentication process

(1) **Begin**
(2) **for** i=1; $i < N^*$; i++ % N^* is the nodes' number in the network except CHs
(3) calculate $\alpha_i = g^{k_i} \bmod p$ % p is a large prime number, $p \geq 2^{512}$; k_i is selected by node i randomly, $0 \leq k_i \leq q-1$; q is a large prime number, $q \geq 2^{160}$; g is a q-order element $g \in Z_p^*, g^q \equiv 1(\bmod p)$
(4) calculate $s = sig_i(ID_i, \alpha_i)$, $C_i = (ID_i, \alpha_i, s)$
(5) send C_i and α_i to CH
(6) **if** $ver_i = (ID_i, \alpha_i, s)$ is true
(7) calculate $v_i = k_i + xb_i \bmod q$
(8) send v_i to CH
(9) **if** $\alpha_i \equiv g^{v_i} y^{b_i} \bmod p$
(10) node i is allowed to proceed the third stage
(11) **else**
(12) communication with node i is terminated
(13) **end if**
(14) **else**
(15) communication with node i is terminated }
(16) **end if**
(17) **end for**

3.2 An Asymmetric Key Pre-assignment Management Scheme

In this section, a Schnorr authentication-based Asymmetric key Pre-assignment management scheme (SAP) for HWSN is proposed. Our scheme is composed of four parts: key preprocess stage, authentication stage, shared key acquisition stage and cluster-head based pair-wise key establish stage.

Key Preprocess Stage. The crucial parts related to key preprocess are summarized as follows, which are able to clarify the core idea. Firstly, a large key pool S is generated by network deployment server, in which each key corresponds to a unique ID. Secondly, l and M number of keys are chosen from S to be pre-stored in ordinary nodes and CHs respectively ($M \gg l$), what's more, a special key K_H, which can only be obtained by base station (BS), is assigned to each CH.

Authentication Stage. To enhance the network security, each node should be authenticated in the cluster; the specific process is shown in Table 1.

Arbitrary node 1 in the network selects a random number $k_1 (0 \leq k_1 \leq q-1)$ and calculates $\alpha_1 = g^{k_1} \bmod p$, after that node 1 sends its certificate C_1 and α_1 to corresponding CH; As soon as CH receives the message sent from node 1, it will verify the certificate C_1 held by node 1, that is, whether $ver_1 = (ID_1, \alpha_1, s)$ is true or not; once the verification fails, CH will terminate the communication with node 1; otherwise, a random number $b_1 (0 \leq b_1 \leq 2^t)$ is selected and sent to node 1 by CH; according to the received b_1, node 1 calculates $v_1 = k_1 + xb_1 \bmod q$ and sends v_1 to its CH; CH will verify if $\alpha_1 \equiv g^{v_1} y^{b_1} \bmod p$ in accordance with the received α_1 and v_1; if it is, node 1 is allowed to access the network, and proceeds to the third stage; otherwise, breakup.

Shared Key Acquisition Stage. The legitimate nodes enter into shared key acquisition stage after authenticating. During this phase, each node (say m) will send an unencrypted key list to its CH first, including ID_m and location information of node m; CH will find the shared key for neighbors after receiving the list information. Whether two nodes (e.g. m and n) are neighbors can be confirmed by CH through location information [9]. Shared key information will be sent to its cluster members using Msg_{ks} (shared key message) by CH as shared keys of each neighbors are found.

Cluster-head based Pair-wise Key Establish Stage. Even with the shared key acquisition stage, partial nodes may not establish a shared key with their neighbors. In this case, for any nodes m and n that do not have shared keys, CH will obtain the shared key with node m and n respectively, denoted as K_m, K_n, thus the pair-wise key $K_{m,n}$ can be generated and transmitted to node m, n by CH. The specific process is as follows. CH firstly checks if node m or n has a shared key with it. If CH doesn't share any keys with node m or n, the following measure will be taken.

Since each node sends a list of key message to corresponding CH, the pre-stored keys of each node are known to its CH. In condition that there is no shared key between CH and node m, CH will check if there is a 1st-hop neighbor shares a key with node m. Provided that there is one (or more than one) shared key $K_{m,z}$ between node u and 1st-hop neighbor z, then, shared key $K_{m,z}$, encrypted by K_z, is required to transmit to CH; where K_z represents the shared key between CH and node z, that is, $z \to \mathrm{CH} : \{K_{m,z}\}K_z$.

If there is no shared key between 1st-hop neighbors and node m, CH is going to find it 2nd, 3rd,…, till j th-hop neighbors, where j is a system parameter. Supposed that there is no shared key between node m and $1 \sim j$ th-hop neighbors, CH will send a 'request message' including ID_m to BS; as BS receives the message, a corresponding key encrypted by K_H will be sent to CH. After obtaining the key from BS, to each pair of neighbors (e.g. m and n), a pair-wise key $K_{m,n}$ will be generated and unicasted to m and n respectively, expressed as $\mathrm{CH} \to m : \{K_{m,n}\}K_m, \mathrm{CH} \to n : \{K_{m,n}\}K_n$, so that node m and n have the shared key $K_{m,n}$ and can communicate securely.

4 Performance Analysis and Evaluation

4.1 Analysis on Storage Space

In this section, the analysis on storage space for SAP is given. The effect of compromise nodes on the communication ability of other nodes decreases with the increasing of key pool size P, however, as the increasing of P, the Probability of any Ordinary node Shares at least one key with its CHs (POSC) decreases [10]. In case that there are l and M keys pre-stored in ordinary nodes and CHs respectively, the probability of any ordinary nodes share i keys with its CH be deduced as

$$p(i) = \frac{C_P^i \cdot C_{P-i}^{M+l-2i} \cdot C_{M+l-2i}^{l-i}}{C_P^l \cdot C_P^M} \tag{1}$$

Thus, POSC in SAP scheme can be expressed as

$$p_S = 1 - p(0) = 1 - \frac{C_P^{M+l} \cdot C_{M+l}^l}{C_P^l \cdot C_P^M} = 1 - \frac{(P-l)!(P-M)!}{P!(P-M-l)!} \tag{2}$$

For comparison's sake the POSC in E-G scheme [11] is given

$$p_E = 1 - \frac{C_P^m \cdot C_{P-m}^m}{(C_P^m)^2} = 1 - \frac{[(P-m)!]^2}{P!(P-2m)!} \tag{3}$$

where m is the number of keys pre-stored in each node.

In Fig. 2, different values of M, l and m of p_S and p_E curves are plotted under the condition that key pool size P changes from 1000 to 10000, the increment of which is 500; the parameters $[M, l, m]$ are chosen as [125, 5, 25], [250, 10, 50], [375, 15, 75], [500, 20, 100] from bottom to top, where $[M, l, m]$ values satisfy $M \times l = m^2$. It can be seen that the POSC increases as pre-stored keys in the node become larger, while, the POSC decreases with P increasing under the same $[M, l, m]$. Furthermore, it can be observed that p_S takes on a tendency similar to p_E with m keys pre-stored. From the above values, we can calculate that the storage space of E-G scheme is more than 4 times that of SAP scheme. It can be seen from the above analysis that the HWSN key management scheme can remarkably reduce the burden of storing compared to that of homogeneous wireless sensor network in achieving the same POSC.

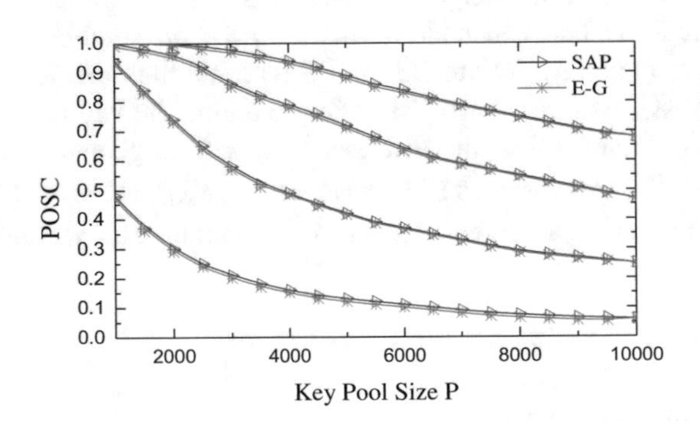

Fig. 2. POSC with different P

4.2 Analysis on Security Performance

Since the CHs have tamper-resistant hardware, CHs are relatively secure; the secure of ordinary nodes in the network is of principal concern. Since there are l keys pre-stored in each ordinary, the probability of any key K in these l keys is l/P. Provide that there are r compromised nodes in the network, the probability of the secure link being exposed can be derived as

$$p_r = 1 - \left(1 - \frac{l}{P}\right)^r \tag{4}$$

Hence, the Probability of any secure Link between two legitimate nodes is Captured when r nodes have been destroyed (PLC) is

$$R(l) = \sum_{i=q}^{l} \left[1 - \left(\frac{l}{P}\right)^r\right]^i \frac{p(i)}{p_{link}} \tag{5}$$

According to reference [12], the PLC in q-composite scheme is

$$R(m) = \sum_{i=q}^{m} \left\{\left[1 - \left(1 - \frac{m}{P}\right)^r\right]^i \times \frac{p'(i)}{p_{link}}\right\} \tag{6}$$

where $p_{link} = p(q) + p(q+1) + \cdots + p(l)$ is the probability of a secure link being set up.

From Eq. (4), we can find that p_r is proportional to l when P and r is fixed. That is why we pre-store a small amount of keys in ordinary nodes. Comparing Eqs. (5) and (6), we observe that $R(l)$ and $R(m)$ increase with the increasing of l and m, l is much less than m, thus, $R(l) \ll R(m)$. It indicates that the proposed SAP scheme has better resistance to compromised attack than that of the traditional way.

In Fig. 3, the PLC curves are plotted. To setup pair-wise directly, the number of required sharing keys is $q = 1$, P is 1000 and the number of compromised nodes r varies from 10 to 200 with an increment of 10; for E-G scheme, we select three different values $m = 20, 30$ and 50; for SAP scheme, let $M = 100$, $l = 10$. The PLC when $q = 3$ is given in Fig. 4, the other parameters are the same as that in Fig. 3.

From Fig. 3 we can observe that for any different r, PLC in SAP scheme is lower than that of E-G scheme. Comparing Figs. 3 and 4, we can get the same result that the PLC in SAP scheme is lower than that of E-G scheme, moreover, the largest PLC of

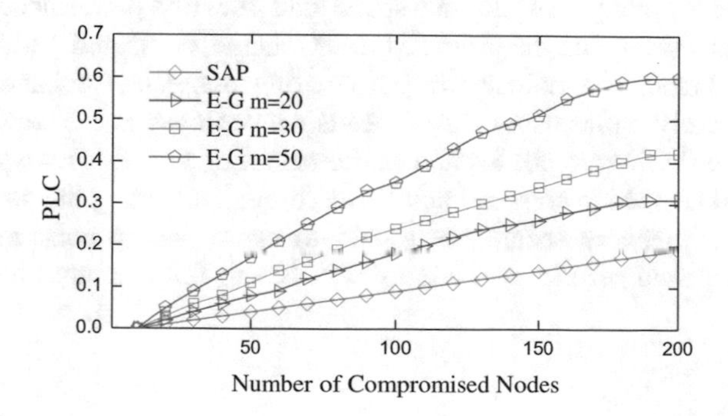

Fig. 3. PLC with $q = 1$

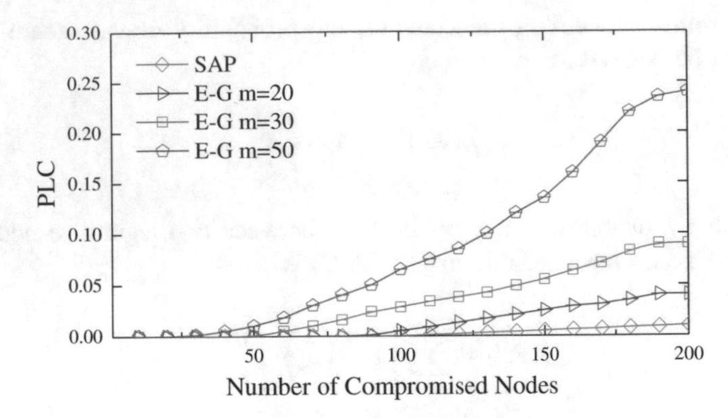

Fig. 4. PLC with $q = 3$

SAP curve in Fig. 3 is 0.2, while it is 0.01 in Fig. 4, thus, we can conclude that PLC decreases with the increasing of q. Above all, it is indicated that the proposed scheme can significantly improve the resistance ability to compromise attack.

5 Conclusion

HWSN constituted by different kinds of sensors with multiple functions is one of the most practical forms in actual application. Relative technologies are focused on during the past few years to prolong the lifetime of network, improve the reliability during transmission and reduce data transmission delay. The security problems, however, become more complex on account of heterogeneity of sensor nodes taking the view of privacy information flooding nowadays. In order to enhance the network security and reduce the burden of network storage as well, this paper proposed an SAP scheme for HWSN. The Schnorr authentication is introduced into key management process, which reduces the probability of illegal nodes access the network, and pseudo code of authentication process is given. The proposed scheme assumes that CHs with tamper-resistant hardware is relatively secure, to reduce the nodes being compromised, a small amount of keys is assigned to ordinary nodes and a large amount of keys is assigned to CHs. Finally, the storage space and security performance is analyzed. Simulation results show that the proposed SAP scheme, compared with traditional key management scheme, can reduce the burden of storage and enhance the network security significantly. Since many cluster heads are deployed in the network, it will not affect the results if CHs are not located at the center of the whole coverage area. Our scheme can also be used in conventional WSN through adjusting the parameters, while the enhancing of network security is bound to cause energy consumption, and the contradiction between security and energy will be our future study direction.

Acknowledgement. This work was supported by the National Natural Science Foundation of China under Grant No. 61302074, 61571181, Natural Science Foundation of Heilongjiang Province under Grant No. QC2013C061, Modern Sensor Technology Research and Innovation Team Foundation of Heilongjiang Province under Grant No. 2012TD007, and Postdoctoral Research Foundation of Heilongjiang Province under Grant No. LBH-Q15121.

References

1. Ye, Q.-Z., Zhang, M.-L., Lin, N.-Y.: Research on secure aggregation scheme based on stateful public key cryptology in wireless sensor networks. J. Inf. Hiding Multimed. Sig. Process. **7**(5), 2073–4212 (2016)
2. Metan, J., Narasimha Murthy, K.N.: Robust and secure key management in WSN using arbitrary key deployment. In: International Conference on Emerging Research in Electronics, Computer Science and Technology, pp. 246–250 (2015)
3. Lu, K., Qian, Y., Hu, J.: A framework for distributed key management schemes in heterogeneous wireless sensor networks. In: IEEE International Performance Computing and Communications Conference, pp. 513–519 (2006)
4. Rahman, M., Samplli, S.: A robust pair-wise and group key management protocol for wireless sensor network. In: IEEE Globecom Workshops, pp. 1528–1532 (2010)
5. Yang, C.J., Zhou, J.M., Zhang, W.S.: Pairwise key establishment for large-scale sensor networks: from identifier-based to location-based. In: Proceedings of the First International Conference on Scalable Information Systems, pp. 55–64 (2006)
6. Mhatre, V., Rosenberg, C.P., Kofman, D., et al.: A minimum cost heterogeneous sensor network with a lifetime constraint. IEEE Trans. Mob. Comput. **4**(1), 4–15 (2005)
7. Sharma, V., Hussasin, M.: Node authentication in WSN using key distribution mechanism. In: International Conference on ICT in Business Industry & Government, pp. 1–7 (2016)
8. Eltoweissy, M., Heydari, H., Morales, L., et al.: Combinatorial optimization of key management in group communications. J. Netw. Syst. Manage. **12**(1), 33–50 (2004)
9. Hussain, S., Kausar, F., Masood, A.: An efficient key distribution scheme for heterogeneous wireless sensor networks. In: ACM International Wireless Communication and Mobile Computing Conference, pp. 388–392 (2007)
10. Chan, H., Perrig, A., Song, D.: Random key pre-distribution schemes for sensor networks. In: Proceedings of the 2003 IEEE Symposium on Security and Privacy, Washington, pp. 197–213 (2003)
11. Eschenauer, L., Gligor, V.: A key management scheme for distributed sensor networks. In: Proceedings of the 9th ACM Conference on Computer and Communication Security, Washingtion, DC, pp. 41–47 (2002)
12. Kuchipudi, R., Qyser, A.A.M.: A dynamic key distribution in wireless sensor networks with reduced communication overhead. In: International Conference on Electrical, Electronic and Optimization Techniques, pp. 3651–3654 (2016)

Multimedia Signal Processing and Machine Learning

A Protocol Vulnerability Analysis Method Based on Logical Attack Graph

Chunrui Zhang[1,2], Shen Wang[1(✉)], and Dechen Zhan[1]

[1] Department of Computer Science and Technology, Harbin Institute of Technology, Harbin 150001, China
shen.wang@hit.edu.cn
[2] Institute of Computer Application, China Academy of Engineering Physics, Mianyang 621900, China

Abstract. The method of analyze the complex protocol vulnerability information from a large number of simple protocol vulnerability information is a tough problem. In this paper, we use attack graph method and construct the protocol vulnerability correlation graph. We also combine the attack target with other information to build the protocol logic attack graph, which is transformed into adjacency matrix. Through the adjacency matrix, we can find and calculate the path of complex attacks and the probability of success and hazard index. The experimental results show that this method can find the correlation among protocol vulnerabilities and can calculate the optimal attack path for protocol vulnerability.

Keywords: Protocol vulnerability analysis · Vulnerability correlation graph · Logic attack graph

1 Introduction and Related Work

In recent years, people pay attention to network attacks, most of which focus on vulnerabilities in computer systems, but less on communication protocols especially on wireless communication protocol security analysis. The main method of analyzing the vulnerability of a protocol is formal analysis [1]. Now attacks against protocols have obtained the characteristics of multi-step and multi-way, which utilizes different stages of communication protocol, and the existing methods lack the ability of correlation analysis of different vulnerabilities. Researchers have used attack graphs [2, 3] to analyze attacking behavior. Common Vulnerability Scoring System (CVSS) [4, 5] can analyze the dependencies of vulnerability, show all the attack path, and finally assessment Comprehensively evaluate system security trends. Holm et al. [6] focused on impact of the basic data provided by CVSS in the assessment of vulnerability. Combined with the shortcomings of CVSS quantification, a specific quantization method was given. But the quantization method was too complicated. Chen et al. [7] proposed an attack graph model for the probability of internal attack intention judgment. Based on the model, he proposed an algorithm to infer the intent of internal attack and maximum probabilistic attack path for that attack target. Liu et al. [8] proposed a game model to obtain optimized attack and defense decisions, which add

J.-S. Pan et al. (eds.), *Advances in Intelligent Information Hiding and Multimedia Signal Processing*, Smart Innovation, Systems and Technologies 82,
DOI 10.1007/978-3-319-63859-1_38

confidence probability to extend vulnerability attributes. Li et al. [9] studied the correlation of vulnerability utilization. They proposed a kind of horizontal and vertical correlation which is suitable for different components of computer system. For the communication protocol vulnerability analysis, horizontal correlation simplification model will lose some information. In this paper, we proposed a method to generate a protocol attack logic diagram from the vulnerability target by using vulnerability of protocol. It can show every route to reach the target of attack, thus showing the possible complex attack, then search and calculate the optimal attack path at the same time.

2 Basic Concepts and Definitions

Attack target and constraints. Attacker takes some of the necessary pre-conditions to achieve the target, only when these preconditions are met, the attacker is possible to achieve the target with one means of attack. These necessary preconditions are the constraint for the attack target.

Protocol Vulnerability. A ternary $\{V_{ID}, C_S, R_S\}$ is used to represent an atomic vulnerability of communication protocol, where V_{ID} is the vulnerability of a protocol against a certain attack, C_S is the set of constraints required for attack and R_S is the set of posterior results caused by attack. It can be expressed as:

$$\xrightarrow{C_s} (V_{ID}) \xrightarrow{R_s}$$

Protocol attack mode. The attack mode for wireless communication protocol usually includes packet discarding, packet hijacking and forwarding, and so on. Protocol attacks include information such as privilege information, vulnerability information, conditional requirements, attack methods, and attack consequences.

3 Protocol Vulnerabilities Correlation Analysis

Because of certain correlations among preconditions and posterior results, atomic vulnerability set is combined into a protocol vulnerability correlation graph according to the correlation among them. The graph supports the use of multiple atomic vulnerabilities to achieve more advanced and complex attack targets. The graph construction process is described in detail.

3.1 Protocol Vulnerability Correlation Graph Construction

Protocol vulnerability correlation graph construction algorithm is divided into the following two steps, as shown in Fig. 1.

1) Traversing the vulnerability of the protocol, generating a sub-tree containing only the current vulnerability and its child nodes by matching the preconditions and posterior results of attack;

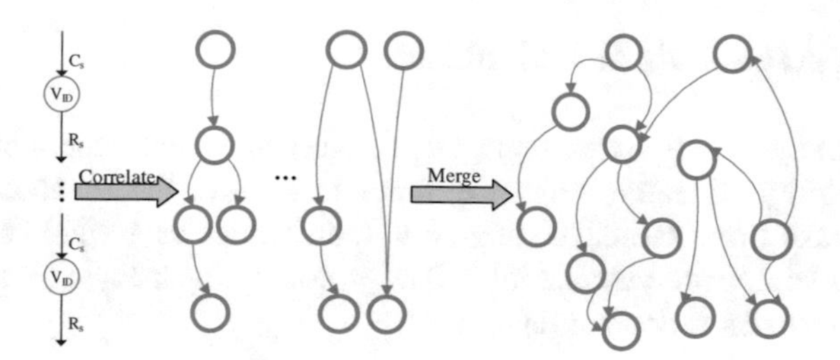

Fig. 1. Protocol vulnerability correlation graph

2) Using a recursive algorithm, which starts with an empty vulnerability correlation graph by accessing sub-tree of a node and traversing all its leaf nodes. If the leaf node has merged into the vulnerability correlation graph, the node is deleted and the connection to the leaf node points to the corresponding node in the vulnerability correlation graph. If the leaf node does not merge into vulnerability correlation graph, the sub-tree of that node is accessed. This algorithm is executed recursively until all nodes are visited again.

3.2 Protocol Vulnerability Correlation Graph Simplification

The original vulnerability graph is relatively large and needs to be simplified. We design the vulnerability reduction method by loop correlation.

Loop Correlation: If attacker has used different types of vulnerability in same constraints, and the formation of loops between the vulnerability to obtain a certain attack results, the vulnerability can be linked, as shown in Fig. 2. This process is called vulnerability loop correlation.

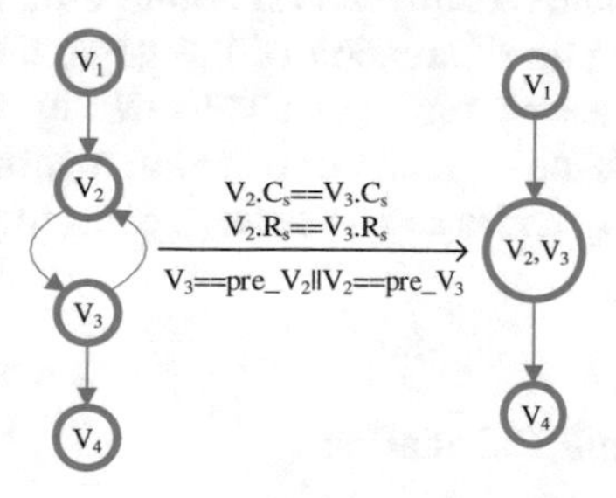

Fig. 2. Vulnerabilities correlation graph simplification using loop-link

4 Protocol Attack Path Calculation

According to the protocol attack logic graph, we can analyze the attack process (or atomic vulnerability utilization sequence) that can achieve the advanced attack target. We can also make these associated atomic vulnerabilities as a whole and constitute a more advanced composite vulnerability. The protocol attack logic graph construction includes three steps as following description.

4.1 Protocol Attack Logic Graph Construction

The input required to construct the protocol attack logic consists of four aspects: target attack result, protocol vulnerability correlation, protocol attack mode, vulnerabilities set. These input information after being calculated will output protocol attack logic graph which can show the influence of vulnerability and constraints, as shown in Fig. 3.

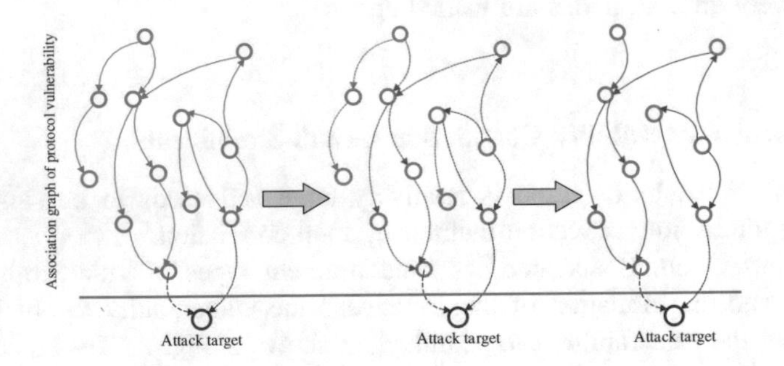

Fig. 3. Protocol attack logic graph generation

Protocol attack logic graph construction algorithm is divided into the following two steps: (1) the corresponding transformation of the target and the vulnerability of protocol. According to the target we can search vulnerability which is used by attacking method. (2) Protocol attack logic graph construction algorithm. It needs to merge the same redundant vulnerability nodes and associated paths to generate attack logic graph the correlation tree.

4.2 Logical Attack Graph Calculation

After obtaining the protocol logical attack graph, it is analyzed and calculated to obtain optimal attack path. Since it reaches an attack target, it needs to calculate the logical graph associated with the target, and end nodes in the logical graph of the target is many, which is similar to finding multiple optimization paths in a directed graph. The main steps include the following aspects.

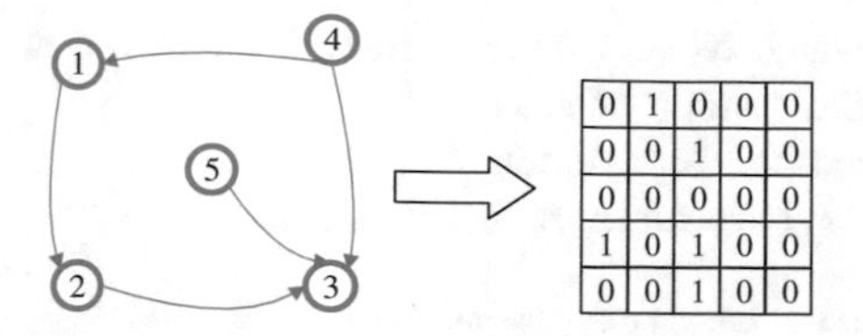

Fig. 4. Logical attack graph and adjacency matrix

1. First convert the directed graph to an adjacency matrix, as shown in Fig. 4. The value of each element in the adjacency matrix is calculated by Eq. (1).

$$a_{ij} = \begin{cases} 0 & i = j \\ 0 & \text{no directed edge from } i \text{ to } j \\ 1 & \text{directed edge from } i \text{ to } j \end{cases} \quad (1)$$

2. Calculate the path cost of the logical attack graph by the adjacency matrix. Each attack path on the composite attack logic graph is composed of a series of atomic attacks. This paper uses CVSS to quantify the attributes of an atomic vulnerability. The probability p_i used by vulnerability i is:

$$p_i = c_i \cdot d_i \cdot ic_i \quad (2)$$

In (2) c_i is the confidence level of node i, d_i is the degree of vulnerability of node i, and ic_i is the influence coefficient of the constraint that node i is used.
Atomic Attack Successful Hazard Index h_i

$$h_i = p_i \cdot k_i \quad (3)$$

In (3) k_i is attacking impact on the confidentiality, integrity, and availability.

The protocol attack logic graph is calculated by adjacency matrix, and the calculation process is shown in the Fig. 5. The steps are shown in Algorithm 1.

a) Find the node with 0 in-degrees; calculate the probability and hazard index of the vulnerability utility from the beginning node to the next node;

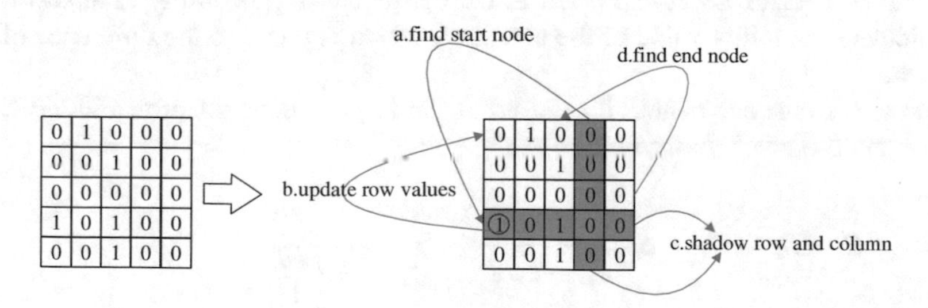

Fig. 5. Logical attack graph calculation

b) Update adjacent value of start node to next node in the adjacency matrix, while shadowing row and column of start node;
c) With the shadowed matrix as input, jump to a;
d) Find end node to get the utility value of attack.

```
Algorithm 1 best path calculate and choose
  Required atomic vulnerability quantify information
  Function Find_start_node(matrix N*N)
    m=0
    for i from 0 to columns of matrix
        if sum of ith column is not equal to 0
          i++
        else
          Save( Vi); m++  //number of start nodes
    Return Vi, m          //return start nodes and its num-
    ber
  Function graph-calculate (adjacency_matrix MN*N)
    While (N!=1)
       Vi, m = Find_start_node(MN*N)
       For n from 1 to m
         For j from 0 to N-1
           If (Vij>0 && Vij <=1)
              Vij.P = Vij.P *pij
              Vij.H = Vij.H *hij
              For k from 0 to N-1
                 Vjk.P= Vjk.P *pij
                 Vjk.H= Vjk.H* hij
              END For
           END If
       END For
       Shadow row and column of start node
       N=N-1
    END While
  return MN*N
```

In Algorithm 1 V_{ij}.P is probability saved, p_{ij} is probability of attack, V_{ij}.H is hazard index saved, h_{ij} is hazard index of attack.

4.3 Optimal Attack Path Calculation

By calculating and continually updating the value of the adjacency matrix, we can calculate the benefit achieved by the attack target. The benefits of an atomic attack B_F are generally expressed as a hazard to the communication protocol by the attack. The cost of atomic attack C_T is expressed as the reciprocal of probability of attack. So we can calculate the utility value of the attack U, which is equal to the difference of gain and cost.

The attack cost and benefit for the attack path L_i can have the probability of attack success and the attack hazard index of the path:

$$U = B_F - C_T = a_{ij}.h_{ij} - \frac{1}{a_{ij} \cdot p_{ij}} = \sum_{j=1}^{pathlen} h_{ij} - \frac{1}{\prod_{j=1}^{pathlen} p_{ij}} \quad (p_{ij} \neq 0) \tag{4}$$

In (4) a_{ij} is the non-zero element and it belongs to the node whose out-degree is 0 (that is, the terminating node), $a_{ij}.h_{ij}$ is the hazard index, $a_{ij}.p_{ij}$ is the probability of the success of the attack on the node.

5 Experiment Analysis

In order to verify the above algorithm, we use C ++ language programming to implement it. Taking the CCSDS protocol as an example, the list of atomic vulnerability inputted in the experiment is shown in Table 1.

Table 1. Atomic attack success probability and hazard index

No.	Vulnerability	Attack Probability	Hazard index
V0	Address information easy to leak	0.6	4.93434
V1	Not dense or weak	0.9	7.40151
V2	Header type field complete check defect	0.64	5.2633
V3	Header length complete check defect	0.456	3.7501
V4	Header field value is not defined	0.655	5.38665
V5	The message is easily modified	0.29	2.38493
V6	Sync head interference	0.85	6.99031
V7	Weak identity authentication	0.627	5.15639
V8	No time validation	0.4	3.28956

We can verify protocol vulnerability correlation graph generation algorithm. After executing the protocol vulnerability correlation graph generation function, the foreground will display the protocol vulnerability and the relationship in graphical form, as shown in Fig. 6.

According to the analysis of the relationship among vulnerability and constraints of the protocol, the experiment first tests the correctness of the logical attack graph

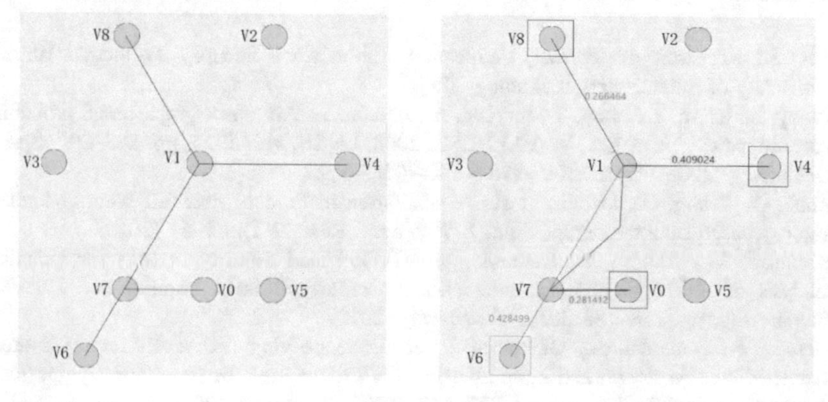

Fig. 6. Protocol vulnerability correlation graph and logic attack graph calculation

generation algorithm. The calculation of protocol vulnerability based on logical attack graphs includes atomic attack, attack path probability and hazard index evaluation, and path optimal decision algorithm. Then in the experiment, we calculate the path attack success rate and hazard index evaluation value. Finally, the optimal decision is made. The final selection of the path with the greatest attack utility is as shown in Table 2:

Path (V6, V7, V1) represents the attack path with the largest attack utility as shown in Fig. 6.

Table 2. Attack success probability and hazard index calculation

Attack path	Attack probability	Hazard index	Attack path utility value
(V0,V7,V1)	0.22572	15.8475	0.281412
(V4,V1)	0.4995	11.9658	0.409024
(V6,V7,V1)	0.479655	19.5482	0.428499
(V8,V1)	0.36	10.6911	0.266464

6 Conclusion

The method is relatively brief by attacking logic graph and adjacency matrix calculation. By expressing complex link graph structure, and removing the repeated complex calculation, this method is made to be extensible. However, the experimental data of this paper is less in number. When the amount of data is large, the time overhead is weak when constructing the vulnerability association graph and the attacking logic graph. Thus the algorithm needs to be optimized to fit the calculation of large-scale data.

Acknowledgement. This work is supported by China Academy of Engineering Physics Project 2014A0403020 and 2015A0403002.

References

1. Shi, S.: Research on Formal Verification Methods of Security Protocols. Huazhong University of Science and Technology (2009)
2. Wang, L., Islam, T., Long, T., Singhal, A., Jajodia, S.: An attack graph-based probabilistic security metric. In: Atluri, V. (ed.) DBSec 2008. LNCS, vol. 5094, pp. 283–296. Springer, Heidelberg (2008). doi:10.1007/978-3-540-70567-3_22
3. Zhao, C., Wang, H., Lin, J., et al.: A generation method of network security hardening strategy based on attack graphs. Int. J. Web Serv. Res. **12**(1), 45–61 (2015)
4. Keramati, M., Akbari, A., Keramati, M.: CVSS-based security metrics for quantitative analysis of attack graphs. In: International Conference on Computer and Knowledge Engineering, pp. 178–183. IEEE, Piscataway (2013)
5. Harada, T., Kanaoka, A., Okamoto, E., et al.: Identifying potentially-impacted area by vulnerabilities in networked systems using CVSS. In: 10th International Symposium on Applications and the Internet, pp. 367–370. IEEE, Piscataway (2010)

6. Holm, H., Ekstedt, M., Andersson, D.: Empirical analysis of system-level vulnerability metrics through actual attacks. IEEE Trans. Dependable Secure Comput. **9**(6), 825–837 (2012)
7. Chen, X., Fang, B., Tan, Q., et al.: Inferring attack intent of malicious insider based on probabilistic attack graph model. Chin. J. Comput. **37**(1), 62–72 (2014)
8. Liu, G., Zhang, H., Li, Q.: Network security optimal attack and defense decision-making method based on game model. J. Nanjing Univ. Sci. Technol. **38**(1), 12–21 (2014)
9. Li, Q., Zhang, L., Zhang, C., Yang, T.: Optimization method for attack graph based on vulnerability exploit correlation. Comput. Eng. **38**(21), 129–132 (2012)

A Preamble Mining Algorithm Oriented to Binary Protocol Using Random Probes

Tingyue Yu, Shen Wang(✉), and Xiangzhan Yu

Department of Computer Science and Technology,
Harbin Institute of Technology, Harbin 150001, China
shen.wang@hit.edu.cn

Abstract. At present, most of the researches on the protocol reverse are on the basis of segmented frames and lack of effective methods to analyze the raw data stream. Several existing frame segmentation algorithms based on AC have the problem of large space overhead and low time efficiency. In this paper, we study on frames segmentation algorithms based on preamble mining and propose a preamble mining algorithm based on random probes oriented to binary protocol. We extract the correct preamble by randomly inserting some probes into the data stream, from which to find continuous short mode strings, after which extracting the most frequently repeated strings as the candidate units, and then filtering them with the help of structural characteristics of the preamble. Experiment shows that the algorithm has higher time efficiency compared with the preamble mining algorithm based on AC algorithm.

Keywords: Protocol reverse · Frames segmentation · Preamble mining · Random probes

1 Introduction

Protocol reverse technology is a kind of technology being widely used in network traffic management and network security protection [1]. It uses bioinformatics [2] and natural language techniques [3] to analyze the protocol field locations based on network traffic, or to obtain the protocol field boundaries based on the information entropy [4].

As a class of important communication protocols, binary protocols are widely used in satellite links and underlying channels. Research on the reverse of binary protocols has get widespread attention in recent years. At present, most of the researches are on the basis of frames which have been segmented ideally. But actually, we can only intercept the bit-streams in the binary protocols on the network usually, especially in private link layer protocols and satellite communication protocols. Therefore, in many circumstances it is necessary to analyze the intercepted 01-bit stream and convert it into segmented frames before using the NLP technologies or bioinformatics to reverse the protocol format.

AC algorithm is a classic multi-mode mining and matching algorithm [5]. Up to now, most of the frame segmentation algorithms for binary protocols are based on the

J.-S. Pan et al. (eds.), *Advances in Intelligent Information Hiding and Multimedia Signal Processing*, Smart Innovation, Systems and Technologies 82,
DOI 10.1007/978-3-319-63859-1_39

AC algorithm. They use the AC algorithm to mine the preambles that appear between frames for padding and synchronization, after which the preambles are removed from the intercepted data stream to obtain frames that do not contain preambles. Jin Ling has proved that the AC algorithm has too much space overhead. The simple AC algorithm is only suitable for identifying short sequences [6]. Lei extract the beginning of the flag header sequence is by splicing short frequent sequence to construct long frequent sequences [7]. Wen records the short sequences appearing in the bit-stream by the AC algorithm and calculates the amount of information between the sequences to extract the frame boundary [8]. Article [9] analyses the starting sequence of frames by mining the candidate Preambles based on AC algorithm.

There are some problems with the preambles mining algorithms based on AC. First of all, these algorithms need to record most of the mode strings appearing in the entire data stream, which may cause larger space overhead and execution time. Secondly, these algorithms treat the preamble as a frequent pattern to mine, without making full use of the distribution characteristics and structural features of preambles. This brings a certain limit to the practical application of frame segmentation algorithms based on AC algorithm.

In view of this, a preamble mining algorithm based on random probes is presented in this paper. It finds the recurring frequent sequences by randomly inserting the probes into the source data so as to extract the preambles for each frame. This algorithm makes full use of the structural features of the preambles itself and only needs to analyze some of the data in the data stream, which has a high enough success rate in theory. As a result, this algorithm has a smaller time overhead and space overhead.

2 Preamble Mining Algorithm on Binary Protocol

At present, the core idea of most preamble mining algorithms based on AC algorithm is to regard the preamble as the most frequent long sequence in the data stream. They construct a trie tree containing all or part of the patterns by traversing the entire data stream, and then construct a long frequent sequence as a preamble. While in fact, the preamble is not simply a long frequent sequence, but has some characteristics. At first, since both of the idle period of the channel and the gap between the two frames need to be filled with the preamble, the length of the preamble is usually very long. What's more, the preamble is usually repeated by a shorter data unit of which the length is usually 8 or 16 bits, which is called preamble unit in this paper.

2.1 Candidate Preamble Units Mining

The process of the algorithm is divided into two steps. The first step is called mining, with the aim of obtaining some long Candidate Codes containing the preamble units from the probes which are set in the data stream. The second step is called "filtering", the purpose of which is to score different Candidate Codes to extract the correct preamble unit.

The process of mining step is as follows:

1. Set the number of probes as N:

$$N = L \cdot a - \left(\lg\frac{L}{b} - 1 \right) \tag{1}$$

Where L is the total length of the data. 'a' and 'b' is parameters related to the number of Probes, where a is a real number greater than 0 and less than 1, b is an integer greater than 1 and less than exp (10, 1-a).

2. Construct the probe position vector P [p_0, p_1, …, p_{N-1}], where p_i is a random integer less then L. Initializes a global Hash-Table List <string, integer>, which is used to record alternative preamble units and their numbers of repetitions.
3. Candidates mining: Excavate the candidate preamble units of length K from each probes, the pseudo code of which is as follows:

```
Algorithm 1.

for each probe located at p:
  Initial a local Hash-Table map<string, integer>;
  for i = 1 to S/K:
    buffer <- read(DataStream[p], K);
    if map. Empty = true:
      map. Insert (buffer, 1);
    else if: map.find(buffer) = it
      It.second <- it.second + 1;
    Else:
      map.clear();
    itmax = argmax_{it∈map} (it.second) ;
  List.insert(itmax.first);
```

4. Refine the candidate: The purpose of this step is to extract the longest sub-string from each unit in the candidate set as the real candidate preamble unit. The pseudo code aimed at output is as follows:

```
Algorithm 2.
```

```
For each candidate unit str:
  s <- str;
  while true:
    len <- len(s);
    if len <= 4:
      return s;
    bufpre <- s[0,...,len/2 - 1];
    bufnext <- s[len/2,...,len];
    if Bufpre = Bufnext:
      s <- bufpre;
    else:
      return s;
```

2.2 Candidate Preamble Units Filtering

For some special protocols, the number of preamble units extracted in previous step may be more than one. For example, if the real preamble unit is "10010110", the candidate units been found will include the following eight types: "00101101", "01001011", "01011010", "01101001", "10010110", "10100101", "10110100", "11010010". The reason for this is that different probes may locate at different locations within the preamble unit but not the boundaries. This situation makes the preamble unit extracted by this algorithm not unique, which makes it hard to be used for frame segmentation.

Thus, we designed a candidate filtering algorithm that attempts to analyze the unique correct preamble unit from candidate set. The basic principle of the algorithm is that the role of the preamble is to synchronize the frames. If the beginning of the preamble unit and the beginning of the frame are the same, it cannot be used for synchronization because of confusion. By the way, the length of the frame is often an integer multiple of 4, because the network communication need to carry out memory synchronization. Based on these two characteristics, we try to use the candidate preamble unit for frame segmentation and calculate the degree of similarity between the beginning of preamble unit and the frame, so as to find the correct preamble unit.

For each candidate preamble unit, the process of calculating its score is as follows:

1. Read a certain length of data into a large buffer, as long as we guarantee it contain at least 10 frames;
2. Use the KMP algorithm to get the list of locations Pos $[p_1, p_2, \ldots, p_n]$ where the candidate preamble unit appears in the buffer.
3. Set the score of the candidate unit by comparing the beginning of the unit and the beginning of the frame after it and then calculate the length of frames. The pseudo code of this step is as follows:

Algorithm 3.

```
For each candidate preamble unit s:
    s.score <- 0;
    len <- len(str);
    Pos[0,...,n-1] <- Kmp( str, buffer);
    reverse <- 0;
    flag <- End;
    pre <- 0;
    next <- 0;
    i from 0 to n-1:
    if flag = End:
      if Pos[i] - Pos[i - 1] == len:
        reverse <- reverse + 1;
      else if reverse>= 8:
        flag <- Continue;reverse <- 0;
        Pre <- Pos[i - 1];
      else:
        reverse <- 0;
    else if flag = Continue:
      if Pos[i] - Pos[i - 1] = len:
        reverse <- reverse + 1;
      else if reverse >= 4:
        next <- poslist[i - 1];
        if (Pos[next] - Pos[pre]) mod 4 = 0:
          buf1 <- buffer[ pre + len, pre + len + 2];
          buf2 <- str[0, 2];
          if buf1 != buf2:
            s.score <- s.score + 1;
        flag = END;
```

After calculating all the scores of candidate units, we select the output unit with the highest score:

$$\text{result} = \arg\max_{s \in CS}(s.score) \tag{2}$$

3 Precision Analysis

In the previous chapter, we propose a preamble mining algorithm based on random probes. After the probes being inserted into the data stream, we can extract the preamble correctly with a great probability by analyzing the repetitive sequences starting from the probes, as long as there is a probe locating within the effective area of the preamble. As a result, when the number of probes is sufficient, the probability that there is at least one probe locating within the effective area of the preamble is very large.

Assume that the total number of segments of the preamble is F (which means the number of frames is F-1), the preamble unit is a short string of length K, and the number of units in per preamble segment is M. So the total length of the preamble is n, $n = F \times K \times M$. Assume that the length of the data to be analyzed is L, the number of probes being inserting into the data is N, and the length of each probe is S. So the length of the data covered by all the probes is P, where P equals to $N \times S$.

Conservatively, when the probe appears in the first half of the preamble and the probe is at the beginning of one of the preamble units, we can identify the true preamble. So the length of the effective area of the probes is P/2 K. Here, the simplest assumption is that the distance from the starting positions of each of the two adjacent preamble segments follows the exponential distribution, and the probes positions follow a uniform distribution. So the effective probability of each probe is r:

$$r = \frac{P/2K}{L} = \frac{N \times S/2K}{L} = \frac{N \times S}{2L \times K} \tag{3}$$

As a result, the probability of the presence of at least one effective probe is p:

$$\begin{aligned} p &= 1 - (1 - r)^{N} \\ &= 1 - \left(1 - \frac{N \times S}{2L \times K}\right)^{N} \\ &= 1 - \{1 - \frac{\left[L \cdot a - \left(\lg\frac{L}{b} - 1\right)\right] \times S}{2L \times K}\}^{L \times a - \left(\lg\frac{L}{b} - 1\right)} \end{aligned} \tag{4}$$

4 Simulation and Analysis

In the previous chapter, we discuss the theoretical correctness of the algorithm under different data characteristics and different parameters. In this section, we use CCSDS-TC to evaluate the correct rate of the algorithm in actual use. At the same time, the efficiency difference between this algorithm and the preamble mining algorithm based on AC algorithm is compared.

CCSDS-TC protocol is a common used satellite link protocol. A CCSDS-TC message consists of a start sequence, a header and its data field, while the frames are padded with preamble. The preamble is constructed by several repeated “10101010”. We generate a total of 300 data stream files, in which there are 2 kinds of total length of data and 3 kinds of repeated times of preamble unit. There are 50 data streams in each combination of data length and repeated times. 2 groups of experiments were performed on the data set, through which we discuss the effect of the number of probes, the length of probes for the correctness of the algorithm, and compare the efficiency of the algorithm and the preamble mining algorithm based on AC algorithm at the same time.

1. The effect of the number of probes on the correct rate
In the first experiment, we select 50 files from the data set with 1000 frames and 336 K data in each file. The probe length S is fixed at 64, and the probes number parameter a is 0.001, 0.0005, 0.0001, 0.00005, 0.0004, 0.0003, 0.0002, 0.00001 respectively. For each a, 20 replicated experiments were performed on the 50 files to calculate the number of times of which the correct results were obtained. As it has a small effect on the number of probes, the value of probes number parameter b is not discussed here, but being fixed tat 5000. The results are shown in Table 1:

Table 1. Precision of this algorithm on different numbers of probes

Value of a	Probes number	Length of data (bit)	Test times	Correct times	Precision
0.0005	166	336 K	1000	1000	1
0.0001	32	336 K	1000	1000	1
0.00005	15	336 K	1000	1000	1
0.00004	11	336 K	1000	1000	1
0.00003	8	336 K	1000	508	0.508
0.00002	5	336 K	1000	159	0.159
0.00001	2	336 K	1000	52	0.052

As can be seen from Table 1, when the number of probes is more than 10, the precision of the algorithm can reach 100% in the experimental set. With the further reduction of the number of probes, the precision of the algorithm drops sharply. This shows that by setting an appropriate number of probes, we can ensure the precision and improve the efficiency at the same time.
2. Time efficiency comparison between 2 algorithms
The second experiment was performed on all of the 300 data streams in the entire data set. For each of these data stream files, we carried out 20 repeated experiments on both of the 2 algorithms, including preamble mining algorithm on AC and on random probes. The parameters of the algorithm based on random probes are as follows: a = 0.0005, b = 1000, S = 32; while the total length of scanned data of AC algorithm is 2000, and its window size is 32 [9]. The total execution time and precision of these two algorithms are shown in Table 2:
As can be seen from Table 2, although the two algorithms can both analyze the correct preamble unit, the algorithm based on random probes uses less time than the AC-based algorithm. And when the length of data is long, we can also choose a smaller number of probes so as to improve the efficiency of our algorithm.

Table 2. Precision and execution time comparison of preamble mining algorithms using AC and random probes

Data stream	AC precision	Probes precision	AC time(s)	Probe time(s)
1	1	1	257.4	9.7
2	1	1	202.3	12.1
3	1	1	216.8	26.3
4	1	1	189.8	32.8
5	1	1	214.8	83.8
6	1	1	195.4	105.2

5 Conclusion

In order to solve the problem that the frame segmentation algorithm based on AC has large space overhead and low efficiency, this paper proposes a preamble mining algorithm based on random probes. It extracts the correct preamble by randomly inserting some probes into the data stream, from which to find continuous short mode strings, after which extracting the most frequently repeated strings as the candidate units, and then filtering them with the help of structural characteristics of the preamble. What's more, we analyze the effect of different parameters and data distribution on the precision of the algorithm through theoretical derivation and some experiments on CCSDS-TC protocol, and comparing the different time efficiency on the preamble mining algorithm and our algorithm. The experiments show that, the efficiency of the algorithm is greatly improved compared with the preamble mining algorithm based on AC.

Acknowledgement. This work is supported by the National Natural Science Foundation of China (Grant Number: 61471141, 61361166006), Key Technology Program of Shenzhen, China, (No. JSGG20160427185010977) and Basic Research Project of Shenzhen, China (grant Number: JCYJ20150513151706561).

References

Narayan, J., Shukla, S.K.: A survey of automatic protocol reverse engineering tools. ACM Comput. Surv. **48**(3), 1–26 (2015)

Marshall, A.: Beddoe: Network Protocol Analysis using Bioinformatics Algorithms (2004)

Luo, J.-Z., Shun-Zheng, Yu.: Position-based automatic reverse engineering of network protocols. J. Netw. Comput. Appl. **36**, 1070–1077 (2013)

Zhang, Z., Zhang, Z.: Toward unsupervised protocol feature word extraction. IEEE J. Sel. Areas Commun. **32**(10), 1894–1906 (2014)

Aho, A.V., Corasick, M.J.: Efficient string matching: an aid to bibliographic search. Commun. ACM **18**(6), 333–340 (1975)

Ling, J.: Study on bit stream oriented unknown frame head. A Dissertation Submitted to Shanghai Jiao Tong University for the Master Degree of Engineering, January 2011

Hezhou, W., Kaiping, X.: An unknown link Protocol bit stream segmentation Algorithm based on frequent statistics and association rules. J. Univ. Sci. Technol. China **43**(7), 554–560 (2013)

Aixia, W.: The technology research of feature selection for unknown protocol in the form of bit stream. A Master Thesis Submitted to University of Electronic Science and Technology of China, May 2015

Dong, L., Tao, W.: Unknown protocol frame segmentation algorithm based on preamble. J. Comput. Appl. **37**(2), 440–444 (2017)

Particle Swarm Optimization-Based Time Series Data Prediction

Xiuli Ning(✉), Yingcheng Xu, Ying Li, and Ya Li

Quality Management Branch, China National Institute of Standardization, Beijing 10001, China
nxl_warm0908@163.com

Abstract. Time series data is one of the forms of product quality inspection data. It is significant for analyzing and processing big data of product quality inspection to research prediction method of time series data. In this paper, we focus on the problem of the existence of particle scattering and the problem of the lack of computational efficiency. Particle swarm optimization (PSO) is integrated into the standard particle filter algorithm, which improves the sampling process of the particle and optimizes the distribution of the sample, and accelerates the convergence of the particle set. Speed, and improve the performance of particle filter. On this basis, the similarity between particle filter and artificial fish swarm algorithm is analyzed. Based on this similarity, the foraging behavior and clustering behavior of artificial fish. The results show that the proposed algorithm can effectively analyze the time series data. The results show that the proposed algorithm can be used to analyze the residual life prediction of particle swarm optimization based on artificial particle swarm optimization.

Keywords: Particle filter · Optimization method · Sequence data analysis

1 Introduction

The basic ideas of particle filter is a set of particle collection with weight is used to approximate posterior probability distribution, and therefore can express the uncertainty of prediction results by PDF. Lithium battery cycle life prediction method based on PF algorithm according to different batteries can be data, using the PF algorithm state tracking ability, flexible experience in setting the parameters of the model, to improve the precision of prediction results can be obtained at the same time the battery cycle life prediction results of the probability density distribution, uncertainty expression ability, but the results rely too much on the experience of the battery model, makes the prediction error is relatively large, and have the disadvantages such as lack of particle filter itself. The particle filter algorithm is used to represent the post-test distribution by the set of particles with weights, so in theory the algorithm can represent the probability distribution of all forms. However, because the importance function of the standard particle filter is suboptimal, there are some disadvantages:

(1) Particle shortage.
(2) Computational efficiency issues.

J.-S. Pan et al. (eds.), *Advances in Intelligent Information Hiding and Multimedia Signal Processing*, Smart Innovation, Systems and Technologies 82,
DOI 10.1007/978-3-319-63859-1_40

In order to improve the above problems, in this paper, the time series data prediction method of particle swarm optimization particle filter (pso-pf) algorithm is proposed. Particle Swarm Optimization algorithm, Particle Swarm Optimization, PSO) and ant colony algorithm, is a kind of Optimization algorithm based on Swarm intelligence, the algorithm and the ant colony algorithm has a certain similarity, mainly through the simulation of birds feeding behavior and abstract and populations of the intelligent Optimization algorithm is proposed. The PSO algorithm has many similarities with the particle's updating process, so it can be optimized by introducing PSO into particle filter.

2 Proposed Method

(1) **Method**

In the basic particle swarm optimization algorithm, the particles constantly update their states, and the particles must follow the following three principles:

(1) Keep inertia;

(2) Change the status according to the best location searched;

(3) Constantly follow the best positions of the group while searching, and change their status according to the optimal location of the group.

(2) **Process**

The PSO algorithm diagram is shown in Fig. 1.

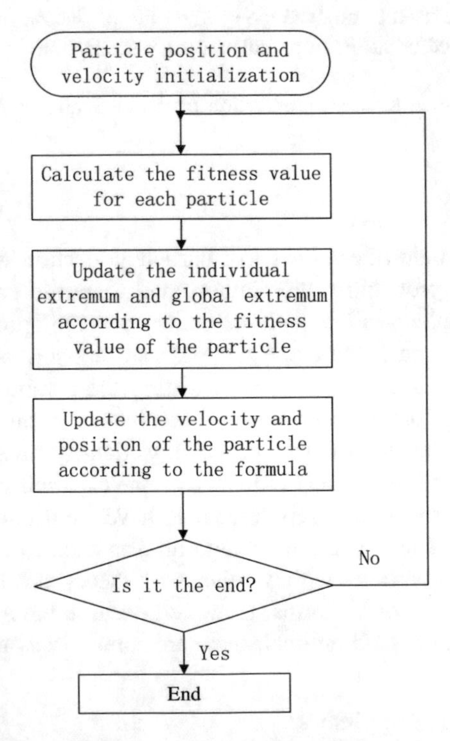

Fig. 1. The basic flow of particle swarm optimization algorithm

2.1 PSO - PF Algorithm

Particle filter algorithm and particle swarm optimization algorithm has a high similarity, both is the state of the particle, and the updated iteratively to obtain the optimal value or the best probability distribution, particle swarm optimization (PSO) algorithm is speed and position of particles more line to obtain the optimal solution of the problem, and the particle filter algorithm is to update the particle's position and weight to get the best a posteriori probability distribution, and based on this similarity, consider using PSO algorithm to improve the performance of standard PF algorithm.

The prediction algorithm flow diagram is shown in Fig. 2.

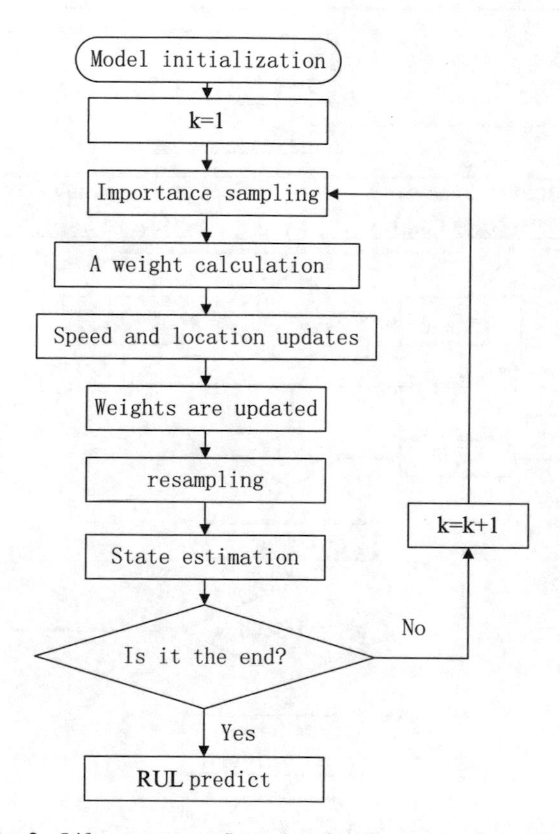

Fig. 2. Life expectancy flow chart based on pso-pf algorithm

2.2 AFSA-PF Algorithm

Artificial Fish Swarm Algorithm (AFSA) is based on research and imitation of Fish foraging, clustering and the pursuit of the tail. The behavior of biological groups is intelligent, so the algorithm is also a swarm intelligence algorithm. Artificial fish algorithm, the fish can treat optimization function in the search field of careful search for the optimal solution, this is a full representation of the group of smart ideas and specific applications.

The flow of the algorithm is shown in Fig. 3.

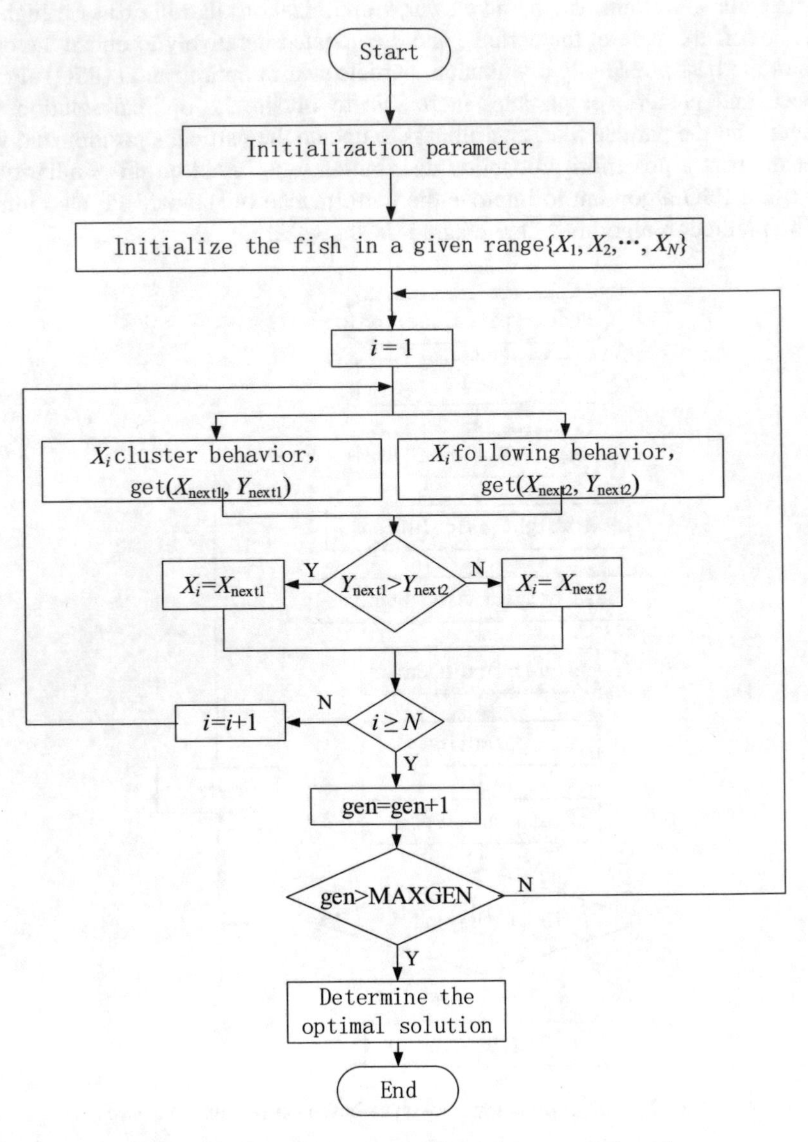

Fig. 3. The flowchart of the artificial fish swarm algorithm

2.3 AFSA-PF Algorithm Steps

The combination of AFSA and PF is mainly reflected in the selection and importance of the objective function (the measured function). To introduce AFSA cluster behavior and foraging behavior in the calculation of important weight, the following steps are:

(1) Initialize. At the moment k = 0, the importance function is sampled, the importance density function is:

$$x_k^i \sim q\left(x_k^i \middle| x_{k-1}^i, z_k\right) = p\left(x_k^i \middle| x_{k-1}^i\right) \tag{1}$$

(2) calculating right value

$$\begin{aligned}
\omega_k^i &= \omega_{k-1}^i p\left(z_k \middle| x_{k-1}^i\right) \\
&= \omega_{k-1}^i \frac{p\left(z_k \middle| x_k^i\right) p\left(x_k^i \middle| x_{k-1}^i\right)}{q\left(x_k^i \middle| x_{k-1}^i, z_k\right)} \\
&= \omega_{k-1}^i p\left(z_k \middle| x_k^i\right) \\
&= \omega_{k-1}^i \left[(2\pi)\sigma_v^2\right]^{-1/2} e^{-\frac{1}{2\sigma_v^2}\left[\left(z_k - \hat{z}_{k|k-1}^i\right)^2\right]}
\end{aligned} \tag{2}$$

Make the target function:

$$Y = \left[(2\pi)\sigma_v^2\right]^{-1/2} e^{-\frac{1}{2\sigma_v^2}\left[\left(z_k - \hat{z}_{k|k-1}^i\right)^2\right]} \tag{3}$$

Foraging behavior. When $Y_i < Y_j$, the particle keeps updating itself to the real state by comparing the target function.

$$\hat{x}_{k|k-1}^{i_m} = \hat{x}_{k|k-1}^{i_{m-1}} + r \cdot s \cdot \frac{\hat{x}_{k|k-1}^{j} - \hat{x}_{k|k-1}^{i_{m-1}}}{\left\| \hat{x}_{k|k-1}^{j} - \hat{x}_{k|k-1}^{i_{m-1}} \right\|} \tag{4}$$

(3) Weight renewal and normalization.

$$\begin{aligned}
\omega_k^i &= \omega_{k-1}^i p\left(z_k \middle| x_k^i\right) \\
\omega_k^i &= \omega_k^i \Big/ \sum\nolimits_{i=1}^{N} \omega_k^i
\end{aligned} \tag{5}$$

(4) Resampling. If the conditions ($N_{eff} = 1 \Big/ \sum_{i=1}^{N} \left(\omega_k^i\right)^2 < N_{threshold}$) for resampling are met:, then resampling.

(5) State estimation.

$$\hat{x}_k = \sum_{i=1}^{N_i} \omega_k^i x_k^i \tag{6}$$

(6) To judge whether to end.

3 Experiments and Analysis

The research results of this project will be used for time series data analysis, and the predictive algorithm is verified on the public data set. The public data set consists of two parts: (1) NASA PCOE test data; (2) Maryland CALCE test data.

The PSO-PF algorithm is used to predict lithium batteries. The number of particles is N = 100, and the training data is still 59 cycles and 117 cycles respectively. The simulation results are shown in Figs. 4 and 5. As can be seen from the graph, the training data is 59 cycles, the predicted results are 212 times and discharge cycles, as early as the actual charging and discharging times, the error is 2 times, although the training data is less, but the PSO-PF algorithm still can not achieve good forecasting results; the training data for 117 cycles of the predicted results are 217 cycles, later than the actual charge and discharge cycles, the error is 1. Comparing the results of this experiment with the predicted results of the standard PF algorithm of N = 500, we can see that the PSO-PF algorithm obtains almost identical prediction results in the case of

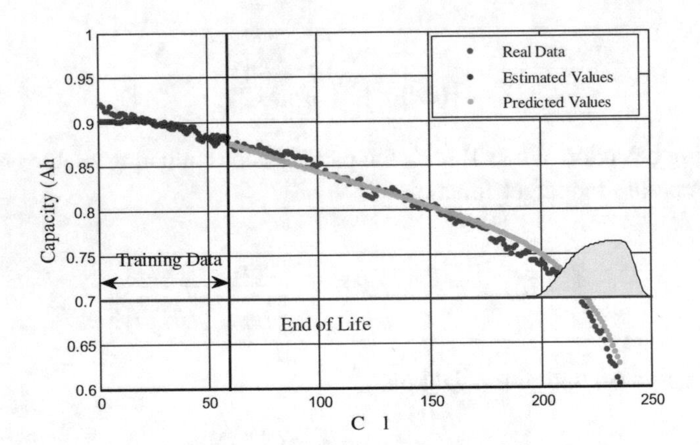

Fig. 4. Prediction with PSO-PF at 59 Cycles

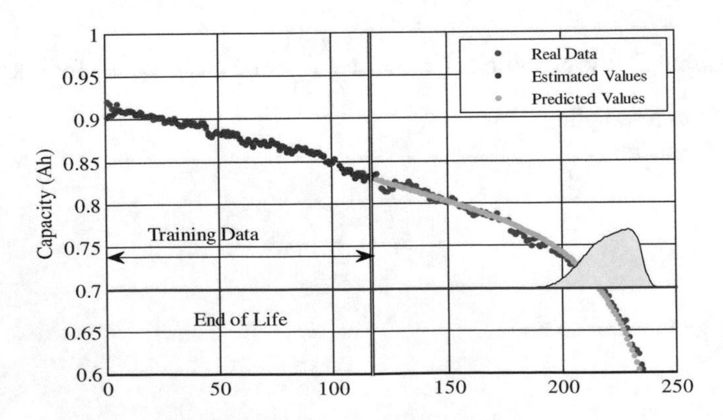

Fig. 5. Prediction with PSO-PF at 117 Cycles

fewer particles, and shows that the PSO-PF algorithm improves the accuracy and reduces the demand of particle number.

In order to make a more intuitive comparison, the results of the experiment are sorted out as shown in Table 1.

Table 1.

	PF (k = 59)	AFSA-PF (k = 59)	PF (k = 117)	AFSA-PF (k = 117)
Real life	216	216	216	216
Predictive life	196	228	225	218
Error	20	12	9	2

The K represents the forecast starting point in the table. The results show that the residual life prediction algorithm of lithium batteries based on artificial fish-swarm optimization is compared with the standard particle filter.

The prediction algorithm of residual life of lithium batteries has obviously improved in forecasting accuracy.

4 Conclusion

This paper proposes a time series forecasting algorithm based on particle swarm optimization, and proposes a particle filter algorithm for artificial fish swarm optimization. Firstly, this paper expounds the basic theory of particle swarm optimization algorithm and the specific process of the algorithm, describes the defects of particle filter and the feasibility of using particle swarm optimization algorithm to improve it. Secondly, the detailed steps of particle swarm optimization algorithm (PSO-PF) are introduced, and the PSO-PF algorithm is used to predict the residual life of the time series of lithium batteries, and the comparison simulation shows that the PSO-PF algorithm has higher performance than the standard particle filter algorithm.

On this basis, the principle of artificial fish swarm algorithm, the basic conception of artificial fish stocks and the four basic behaviors of individuals in the population are expounded. Then the algorithm is introduced into the particle filter algorithm, and the similarity of the two algorithms is analyzed. Based on this similarity, the foraging behaviors and clustering behaviors of artificial fishes are optimized for particle and weight calculation in particle filter algorithm. Finally, the framework of residual life prediction algorithm based on artificial fish swarm optimization is presented and the validity of the algorithm is proved by experiments.

Acknowledgment. This research is supported and funded by the National Key Research and Development Plan under Grant No. 2016YFF0202600 and No. 2016YFF0202604, the National Natural Science Foundation of China under Grant No. 71301152 and No. 91646122.

References

Chen, M.Y.: A hybrid ANFIS model for business failure prediction utilizing particle swarm optimization and subtractive clustering. Inf. Sci. **220**(1), 180–195 (2013)

Quan, H., Srinivasan, D., Khosravi, A.: Particle swarm optimization for construction of neural network-based prediction intervals. Neurocomputing **127**(6), 172–180 (2014)

Aladag, C.H., Yolcu, U., Egrioglu, E., et al.: An ARMA type fuzzy time series forecasting method based on particle swarm optimization. Math. Prob. Eng. **2013**(10), 3291–3299 (2013)

Ernawati, S.: Using particle swarm optimization to a financial time series prediction. In: International Conference on Distributed Framework and Applications, pp. 1–6 (2010)

Tian, Z., Wang, P., He, T.: Fuzzy time series based on k-means and particle swarm optimization algorithm. In: Man-Machine-Environment System Engineering (2016)

Time Series Data Analysis with Particle Filter-Based Relevance Vectors Machine Learning

Xiuli Ning(✉), Yingcheng Xu, Ying Li, and Ya Li

Quality Management Branch,
China National Institute of Standardization, Beijing 10001, China
nxl_warm0908@163.com

Abstract. Analyzing and processing big data of product quality inspection is the key to guarantee product quality and safety. Time series data is one of the most important forms of product quality inspection data. Therefore, it is significant to research time series data. In this paper, we focus on the time series prediction data to contain complex noise and uncertainties. We use the correlation vector machine to carry out regression modeling, find the regularity in this series of complex data, and return the RVM regression model with the largest information to establish the state Spatial model. And then use the particle filter method, the model is constantly updated, in order to achieve better prediction. The experimental results show that this method can effectively solve the problem of noise and uncertainty in time series data analysis, and obtain better performance of time series data analysis.

Keywords: Time series prediction · Relevance vector machine · Particle filter

1 Introduction

The Support Vector Machine (SVM) is a theory of statistical learning, which is mainly conducted in small sample sizes.

Relevance vector machine can be used for classification and regression, in view of the data contains the characteristics of the complex noise and uncertainty, the use of relevance vector machine for regression modeling, look for patterns in this a series of complex data, regression fitting out contains information about the biggest RVM regression model, to establish a state space model. Then the method of particle filtering is used to continuously update the model and realize the good prediction of the battery life. We know that by learning the relevance vector machine regression algorithm of classification standard is diverse, to the front, as defined by the relevant vector, through determine its existence difference, can be divided into two major kinds of regression algorithm.

One kind of decision condition is the superparameter threshold, which is called an iterative algorithm; Another type of decision is the improved algorithm for a particular operator, which is called a quick sequence sparse bayesian learning algorithm. One of the most basic method is based on super parameters threshold value by the size of the algorithm, this algorithm is the basic iterative relevance vector machine algorithm, is

J.-S. Pan et al. (eds.), *Advances in Intelligent Information Hiding and Multimedia Signal Processing*, Smart Innovation, Systems and Technologies 82,
DOI 10.1007/978-3-319-63859-1_41

applied to various fields. While the SVM classification and regression have achieved good results, but it still has many shortcomings, mainly reflected in the following aspects: (1) its nuclear function must satisfy the Mercer condition; (2) its predictions are not statistically significant, and there is no way to directly obtain the uncertain expression of the final forecast. (3) the number of basis functions grows with the linear growth of the training sample size, so the model's sparse is finite; (4) the method of cross verification is required to obtain the kernel function and parameters, which results in the increase of the training volume of the model parameters.

2 Algorithm

First of all, set the original training sample set for $\{x_n, y_n\}_{n=1}^{N}$, input values x_n and y_n target output value these two samples showed independent distribution is known, so you can get, x and y of the relationship between the two, said the following:

$$y = \phi w + \xi \tag{1}$$

Among them, the weight vector is the w parameter, $\mathbf{\Phi}$ is defined as a constitutive matrix of kernel functions:

$$\phi = [\phi(x_1), \phi(x_2), \ldots, \phi(x_n)] \tag{2}$$

$$\phi(x_n) = [k(x_n, x_1), k(x_n, x_2), \ldots k(x_n, x_N)]^T \tag{3}$$

$$\xi = [\xi_1, \xi_2, \ldots, \xi_n]^T, \xi_n \sim N(0, \sigma^2) \tag{4}$$

We can get the probability distribution of the target value by the iteration process:

$$p(y|w, \delta^2) = N(\phi_w, \delta^2) \tag{5}$$

In the RVM model, we can use bayesian learning theory to make our model more appropriate. We can use the theory of self-correlation decision to define the ARD prior distribution of the parameters above, which can be expressed as:

$$p(w|\alpha) = \prod_{i=0}^{N} N(w_i|0, \alpha_i^{-1}) \tag{6}$$

Among them, $\alpha = (\alpha_{0,}\alpha_1, \ldots, \alpha_N)$, this parameter is a very important hyperparameter, which is associated with the weight and determines the priori distribution.

To summarize relevance vector machine, the conclusion is as follows: relevance vector machine is a machine learning theory, which is based on the bayesian framework, with a relatively independent parameters represent the right value of each individual, it can make the model look more sparse. We have introduced this definition, the vectors and after model training, we're going to get some sample vector, the weights of sample vector is zero basis function of the basis functions, we put the sample of the sample vector is called a vector.

There are many parameters in our actual calculation, the posterior probability distribution of the weight is close to zero value, so can be neglected, and the right of the non-zero value is very important, so, we introduce the concept of the vectors, so when we study the relevance vector regression model, the nonzero vector and decision-making in the domain of the sample is not associated, it just represents the original samples, therefore, we can call these non-zero value relevance vector. Because the model is expected to predict the sample variable parameter is $p(y)$, then we can get a posteriori probability distribution of the predicted y_n corresponding to the x_n as follows:

$$p(y_n|y) = \int p(y_n|w, \alpha, \delta^2)p(w, \alpha, \delta^2|y)dwd\alpha d\delta^2 \tag{7}$$

Because we cannot get the posterior distribution $p(w, \alpha, \delta^2|y)$ of the model parameters directly through integrals, we decompose it into:

$$p(w, \alpha, \delta^2|y) = p(w|y, \alpha, \delta^2)p(\alpha, \delta^2|y) \tag{8}$$

$p(w|y, \alpha, \delta^2)$ has been asked, so we have transformed the learning problem of the correlation vector machine into the problem of the maximum value of α and δ^2 for the posteriori parameter, and then we can get it by calculating:

$$\alpha_i^{new} = \frac{1 - \alpha_i \sum_{ii}}{\mu_i^2} \tag{9}$$

$$(\delta^2)^{new} = \frac{\|y - \phi\mu\|^2}{N - \sum_{i=1}^{N} \alpha_i \sum_{ii}} \tag{10}$$

2.1 RVM-PF Algorithm

According to the above mentioned, we can learn the basic principle of the correlation vector machine. When a large time, the weight of this parameter corresponds to the associated vector of zero. In our practical application, we can also define the maximum value of some very important parameters, and analyze it according to the actual situation. Therefore, once $\alpha \geq \alpha_{\max}$, we can say that α is infinite, so this parameter associated with the vector parameter ω can approximate to 0. In this case, we can remove the corresponding base function and weight coefficient of α_i, and take the remaining samples as the correlation vectors.

Therefore, we can use this method to update the parameters and weights we need to repeat. This repeats several times, and after multiple updates, we can find the important vectors that are most needed in multiple samples, that is, the correlation vectors, and we can duplicate these vectors, and we can find many of their important parameters. According to this method, we can build the iterative algorithm of correlation vector machine. The specific iterative algorithms are as follows:

(1) Initialization: Initialize each parameter, select the training dataset we need, and then set the parameters.
(2) Calculating weight/
(3) Update the hyperparameter α and σ^2.
(4) Filter data.
(5) To determine the convergence. (2) (3) (4), and then according to the above definition of the parameters of the limit, to determine whether the number of RVM run to the standard, or the output of the accuracy of less than a certain limit until the meet the requirements.
(6) Output the result. Through the above steps, we can get the relevant vector and regression equation, and then you can output the results.

Relevant vector machine iterative regression algorithm has many advantages, the correlation vector machine regression iterative algorithm is very clear and very easy to implement, but it also has the disadvantage that the parameters are updated, we must calculate the posterior value Of the covariance inverse matrix, the computational complexity increases.

2.2 RVM-PF Algorithm Steps

The specific steps of the algorithm are:

(1) select multiple sets of data (three or more groups), and then the data preprocessing and other operations.
(2) initialize the RVM algorithm, set the parameters.
(3) RVM regression fit, the establishment of degradation model.
(4) According to the empirical model, the training data is attenuated by fitting the model to obtain the 95% confidence interval of the model parameters a, b, c and d, and the mean of each model parameter is taken as the initial value of the whole model parameter.
(5) In the particle filter, the initial model is set according to the model parameters obtained earlier. Need to set the data set C, the motion equation, the observation equation, the observed value Z, the initial sampling target, the predicted value, the failure threshold y_fail.
(6) According to the sampling particle set, the number of particle groups is defined. The setting of the number of particles needs to consider the accuracy of the prediction and the complexity of the calculation.
(7) The particles are sampled according to the sequence importance function.
(8) The weights of the particles are updated according to the observed observations and normalized.
(9) Resample the particle samples. Select the appropriate resampling method, followed by the process of particle resampling.

(10) If the forecast is not over, we will return to step (7) and continue to calculate the above process.
(11) According to our definition of the failure threshold, we can determine the time of failure, and then we can get the final life prediction results.

3 Experiments and Analysis

The results of this project will be used for time series data analysis, the prediction algorithm in the public data set has been verified. The public data set consists of two parts: (1) NASA PCOE test data; (2) Maryland CALCE test data.

The correlation vector machine is a learning method that can be classified and regressed. Therefore, we return the battery data A5, A8, A12 with RVM, and the result is shown in Fig. 1.

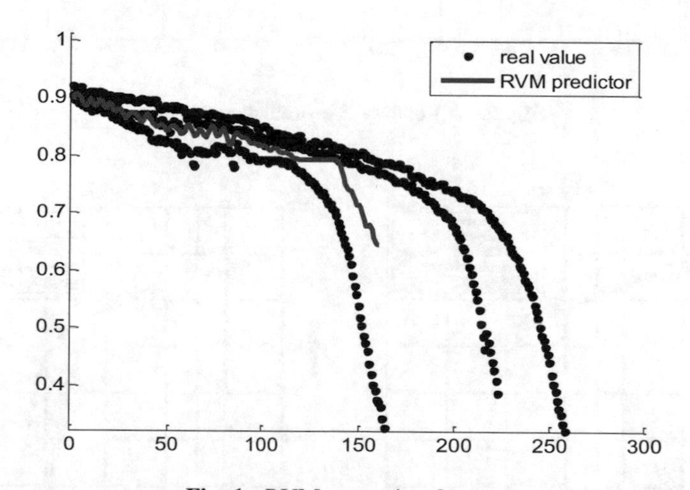

Fig. 1. RVM regression fit curve

The fitting data is fitted to the training data by the curve fitting tool in the MATLAB toolbox to obtain the 95% confidence interval of the model parameters, as shown in Table 1:

Table 1. Model parameter table

Parameter	Upper limit	Lower limit	Mean
a	–0.000002325	–0.0000007793	0.000000873
b	0.05738	0.04264	0.05001
c	0.9183	0.8978	0.908
d	–0.001314	–0.001086	–0.0012

The mean value of each model parameter is taken as the initial value of the whole model parameter. The initial values of a, b, c and d are: a = −0.000000873, b = 0.05001, c = 0.908d = −0.0012 (Figs. 2, 3, 4 and 5).

The initial parameter values is then applied to the particle filtering, prediction results as shown below:

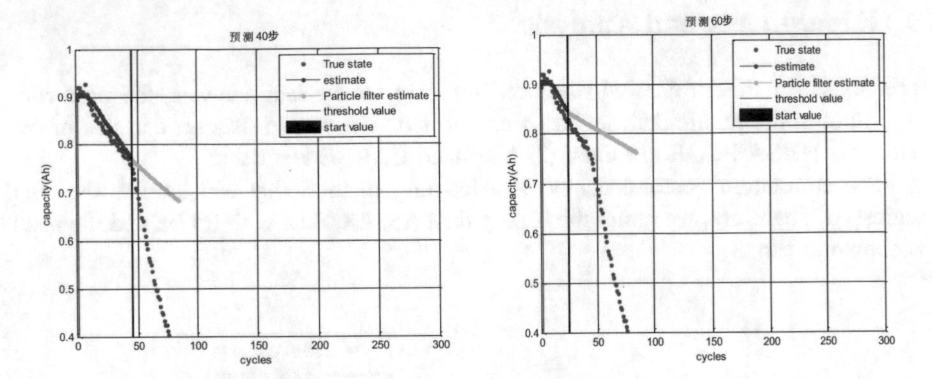

Fig. 2. A3 battery life prediction curve

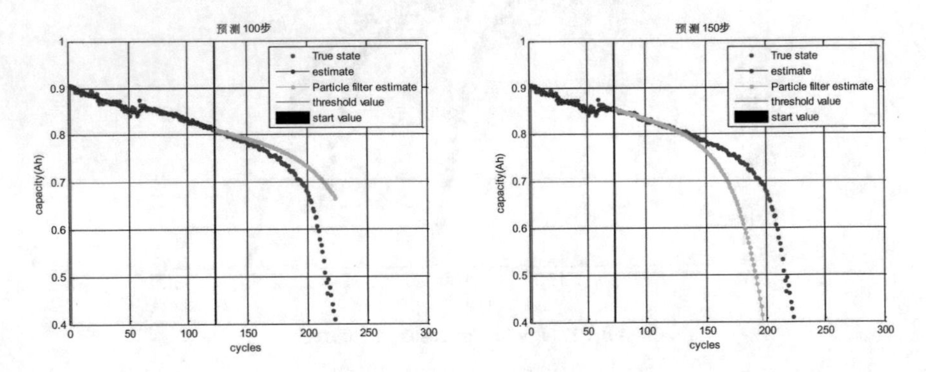

Fig. 3. A5 battery life prediction curve

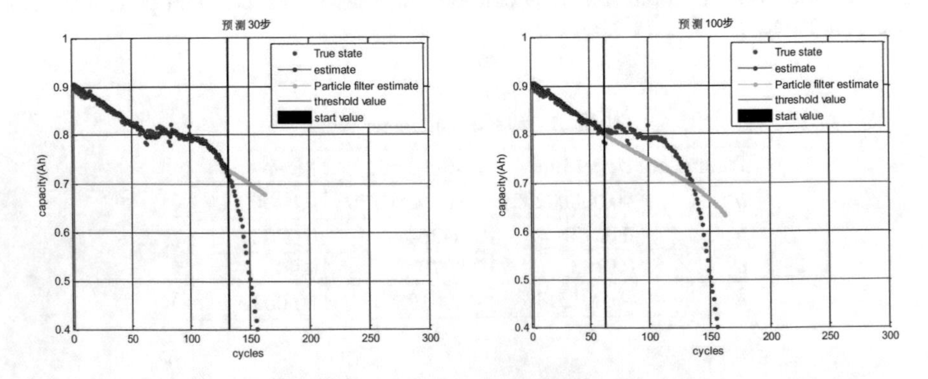

Fig. 4. A8 battery life prediction curve

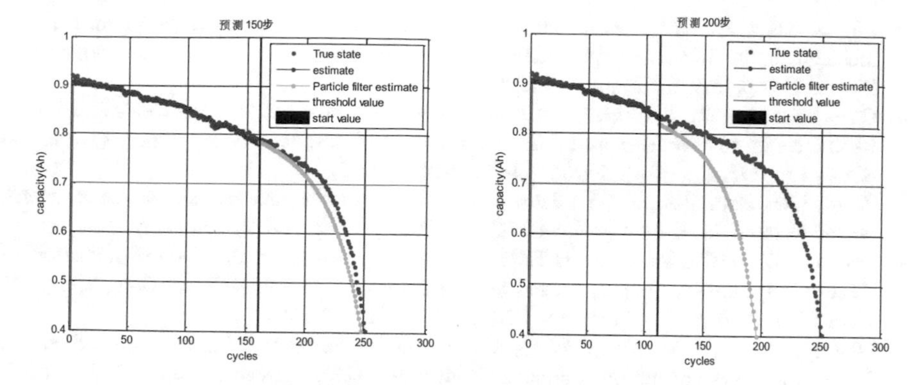

Fig. 5. A12 battery life prediction curve

Verified by the above forecast results, the feasibility of the method to predict the battery life is based on RVM the zanu-pf, and we can conclude that when the training data is small, the larger the error, when more training data, the more precise prediction.

4 Conclusion

For time series data contains complex noise and uncertain characteristics, is proposed based on particle filter related support vector machine (SVM) combining time series prediction method, first of all, the use of relevance vector machine for regression modeling, look for patterns in this a series of complex data, regression fitting out contains information about the biggest RVM regression model, to establish a state space model. Then the method of particle filtering is used to update the model to achieve better prediction. The experimental results show that the proposed method can be used to analyze the time series data effectively, and can be used to predict the big data.

Acknowledgment. This research is supported and funded by the National Key Research and Development Plan under Grant No. 2016YFF0202600 and No. 2016YFF020260, the National Natural Science Foundation of China under Grant No. 71301152 and No. 91646122.

References

1. Zhang, S., Liu, P.: Prediction of chaotic time series based on the relevance vector machine. In: IEEE Fifth International Conference on Advanced Computational Intelligence, pp. 314–318. IEEE (2012)
2. Nikolaev, N., Tino, P.: Sequential relevance vector machine learning from time series. In: IEEE International Joint Conference on Neural Networks. In: Proceedings of IEEE Xplore, IJCNN 2005, vol. 2, pp. 1308–1313 (2005)

3. Liu, F., Song, H., Qi, Q., et al.: Time series regression based on relevance vector learning mechanism. In: International Conference on Wireless Communications, Networking and Mobile Computing. IEEE, pp. 1–4 (2008)
4. Quinonero-Candela, J., Hansen, L.K.: Time series prediction based on the Relevance Vector Machine with adaptive kernels. In: IEEE International Conference on Acoustics, Speech, and Signal Processing. IEEE Xplore, pp. I-985–I-988 (2002)
5. Zhou, Z.Q., Zhu, Q.X., Xu, Y.: Time series extended finite-state machine-based relevance vector machine multi-fault prediction. Chem. Eng. Technol. **2017**, 40 (2017)
6. Fan, G., Deng-Wu, M.A., Ming-Hui, W.U., et al.: Condition time series prediction of electronic system based on optimized relevance vector machine. XI Tong Gong Cheng Yu Dian Zi Ji Shu/Syst. Eng. Electron. **35**(9), 2011–2015 (2013)
7. Islam, M.Z., Oh, C.M., Lee, C.W.: Real time moving object tracking by particle filter. In: International Symposium on Computer Science and ITS Applications, pp. 347–352. IEEE Computer Society (2008)

GUI of GMS Simulation Tool Using Fuzzy Methods

Yeonchan Lee[1], Jaeseok Choi[1(✉)], Myeunghoon Jung[2], and Junzo Watada[3]

[1] Department of Electrical Engineering, Gyeongsang National University, Jinju-si, Gyeongsangnam-do 52828, South Korea
{kkng1914, jachoi}@gnu.ac.kr
[2] Korea South-East Power Corporation, Jinju-si, Gyeongsangnam-do, South Korea
mhjeong@koenergy.kr
[3] Department of Computer and Information Sciences, 32610 Seri Iskandar, Perak Darul Ridzuan, Malaysia
junzo.watada@gmail.com

Abstract. This study used the fuzzy theory for creating preventive maintenance scheduling of generator (GMS). GMS simulation tool is developed in optimum by the CO_2 emission, probabilistic production cost, and reliability criteria LOLE (Loss of load expectation), EENS(Expected energy not served) in the objective functions. GMS has shown in graphical user interface (GUI) for the user-friendliness of the generator. It can be resulted in generator maintenance scheduling, reliability indices of power system, total probabilistic production costs and each power company probabilistic production costs.

Keywords: Simulation tools · Fuzzy methods · Generator maintenance scheduling

1 Introduction

In the past, the utility, Korea Electric Power Corporation (KEPCO), had taken charge of all power generation, transmission, and market, leading Korea's power policy direction. However, in the 21st century, the electric power market has been split into the transmission and distribution market operated by KEPCO, the power generation market in which various companies are involved, and the power exchange market that relays the above. As the electric power market has been re-organized as a competitive system, the establishment of generator maintenance scheduling (GMS) has also changed.

The current GMS plan are based on the schedule and maintenance plans of each company, and the companies apply for a schedule for preventive GMS every year. The Korea Power Exchange (KPX) decides on the preventive GMS plans by approving the request per preliminary power standard of the day. However, since GMS affects various factors such as the stability of electricity, the cost of settlement, and the yield of power generation companies, each company may have different opinions on GMS.

J.-S. Pan et al. (eds.), *Advances in Intelligent Information Hiding and Multimedia Signal Processing*, Smart Innovation, Systems and Technologies 82,
DOI 10.1007/978-3-319-63859-1_42

If the stability of electricity is secured and the cost of power generation is considered, it will be possible to establish a GMS plan that satisfies various conditions.

Therefore, this study aims to introduce a method for establishing an optimal GMS plan that utilizes the Fuzzy logic to consider power reserve, reliability and cost. In addition, this study introduces a GMS simulation tool that includes a graphic user interface (GUI) Probabilistic Generator Model

2 Probabilistic Generator Model

Model of the system, Fig. 1, consists of typical hydro generators and pumped storage power plant which are transformed to an equivalent generator of each type, conventional generators and load L is represented by the load duration curve of the Lpn peak load in n times [1–3].

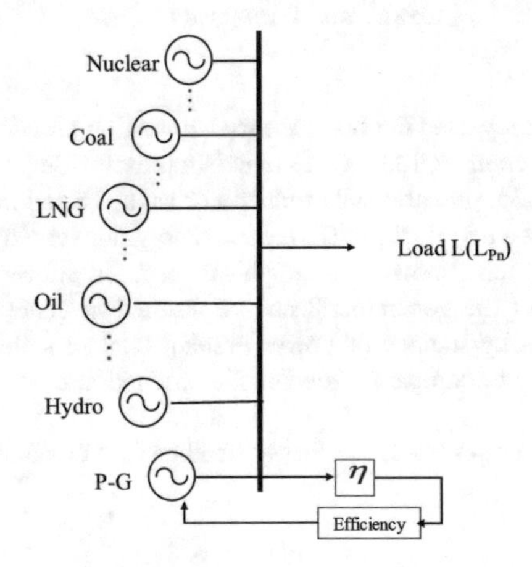

Fig. 1. Basic system model without transmission system

To obtain the Probabilistic power generation amount of the generators, the power generation to the #i generator is obtained as shown in Eq. (1) [4–7].

$$E_{in} = (1 - q_i)T \int_{ui-1}^{ui} \Phi_{in-1}(X)dX \tag{1}$$

Where, i: number of the economic order of generators

$u_i = C_1 + C_2 + \cdots + C_i$ [MW]
C_i: capacity of #i unit
$u_0 = 0$
ϕ_{in}: effective load duration curve
q_i: forced outage rate of #i unit

2.1 Establishment of Membership Functions

Membership function of fuzzy set for the production cost is defined as:

$$\mu_c\{X(t-1),u(t)\} = \begin{Bmatrix} 1 & :\Delta C(\cdot) \leq 0 \\ e^{-W_c \Delta C\{X(t-1),u(t)\}} & :\Delta C(\cdot) > 0 \end{Bmatrix}$$

Where, $\mu_c(\cdot)$: membership function of fuzzy set for production cost

$\Delta C(\cdot) = \{F(X(t))\text{-}Casp(t)\}/Casp(t)$
Casp(t): Aspiration level for production cost at #*t* stage
W_c: Weighting factor of the membership function for production cost

Membership functions of fuzzy set for the positive reliability (SRR) are defined as:

$$\mu_R\{X(t-1),u(t)\} = \begin{Bmatrix} 1 & :\Delta R(\cdot) \leq 0 \\ e^{-W_R \Delta R\{X(t-1),u(t)\}} & :\Delta R(\cdot) > 0 \end{Bmatrix}$$

Where, $\mu_R(\cdot)$: membership function of fuzzy sets for reliability

$\Delta R(\cdot) = \{SRR(X(t))\text{-}REQ_{SRR}(t)\}/REQ_{SRR}(t)$
$REQ_{SRR}(t)$: Aspiration level for reliability SRR at #*t*stage
W_{RSRR}: Weighting factor of the membership function for reliability SRR

Membership functions of fuzzy set for the negative reliability (LOLE) are defined as:

$$\mu_R\{X(t-1),u(t)\} = \begin{Bmatrix} 1 & :\Delta R(\cdot) \leq 0 \\ e^{-W_R \Delta R\{X(t-1),u(t)\}} & :\Delta R(\cdot) > 0 \end{Bmatrix}$$

Where, $\Delta R(\cdot) = \{LOLE(X(t))\text{-}REQ_{LOLE}(t)\}/REQ_{LOLE}(t)$
$REQ_{LOLE}(t)$: Aspiration level for reliability LOLE at #t stage
W_{RLOLE}: Weighting factor of the membership function for reliability LOLE

Membership functions of fuzzy set for the positive reliability, EIR are defined as:

$$\mu_R\{X(t-1),u(t)\} = \begin{Bmatrix} 1 & :\Delta R(\cdot) \leq 0 \\ e^{-W_R \Delta R\{X(t-1),u(t)\}} & :\Delta R(\cdot) > 0 \end{Bmatrix}$$

Where, $\Delta R(\cdot) = \{EIR(X(t))\text{-}REQ_{EIR}(t)\}/REQ_{EIR}(t)$
$REQ_{EIR}(t)$: Aspiration level for reliability EIR at #tstage
W_{REIR}: Weighting factor of the membership function for reliability EIR

2.2 Solution Procedure by the Fuzzy Search Method

Fuzzy decision set D can be formulated as:

$$D = C \cap R1 \cap R2 \cap R3 \tag{2}$$

Where, C: fuzzy set for production cost

R1: fuzzy set for reliability SRR
R2: fuzzy set for reliability LOLE
R3: fuzzy set for reliability EIR

Therefore, decision membership function can be formulated:

$$\mu_D(X(t)) = \max_{\min(t) \le u(t) \le u\max(t)} [\min\{\mu_c(\cdot), \mu_{R1}, (\cdot), \mu_{R2}(\cdot), \mu_A(\cdot), \mu_D(\cdot)(X(t-1))\}]$$

where, X(t) = X(t-1) + u(t)

$\mu_D(X(0)) = 1.0$
$\mu_D(\cdot)$: membership function of fuzzy set for decision function

3 Flow Chart

The flow chart in this paper is as follows. First, limit values of the generator, load and generator maintenance plan are set as input ones. To create the load, a virtual load duration curve is created using a load pattern and a peak load. In order to establish the generator maintenance plan of the generator, the solution of the generator maintenance plan of the generator is deducted.

If the generator maintenance scheduling satisfies the reserve, observes the limitations, and satisfies the optimal objective function, it should be established as the optimal generator maintenance plan. This is shown in Fig. 2.

4 GUI of GMS Simulation Tool

Figure 3 shows the control panel of GMS. Here, it is possible to set up a generator maintenance planning to each objective function (maximizing the supply reserve rate, maximizing the supply reserve power, minimizing the generation cost, minimizing the reliability, and Fuzzy function). You can also modify several parameters such as the minimum supply reserve rate, the price of CO_2, the capacity of hydropower and pumped storage. Figure 4 shows the execution of the simulation [10, 12].

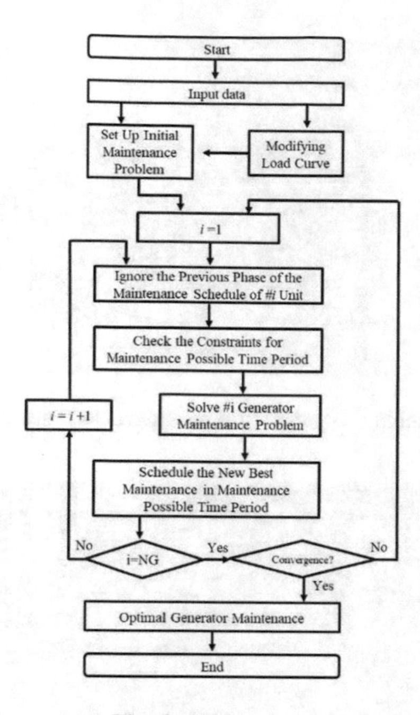

Fig. 2. Flow chart

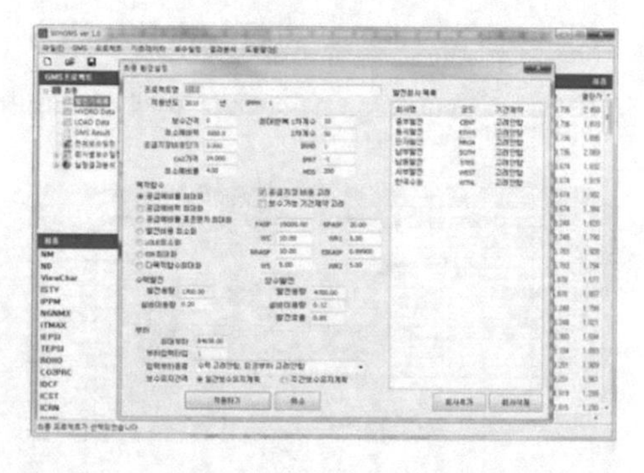

Fig. 3. Preferences panel (written in Koeran)

As seen in Fig. 5, generator maintenance scheduling can be shows. Figure 6 represents the result of power system. system`s reliability, costs, supply reserve rate, power amount and others can be checked in addition to the cost and power amount for each generation company.

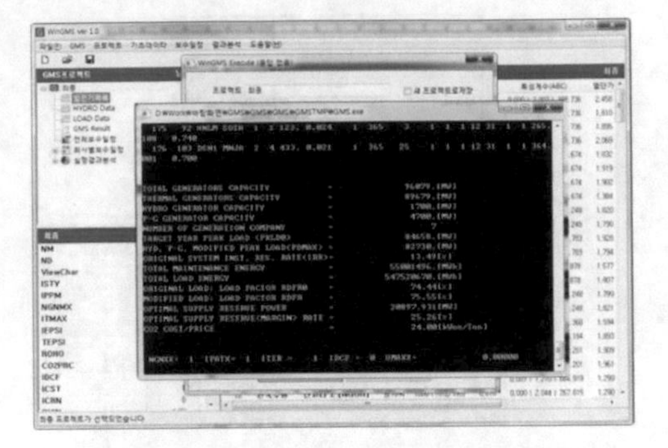

Fig. 4. Running to generation maintenance scheduling system

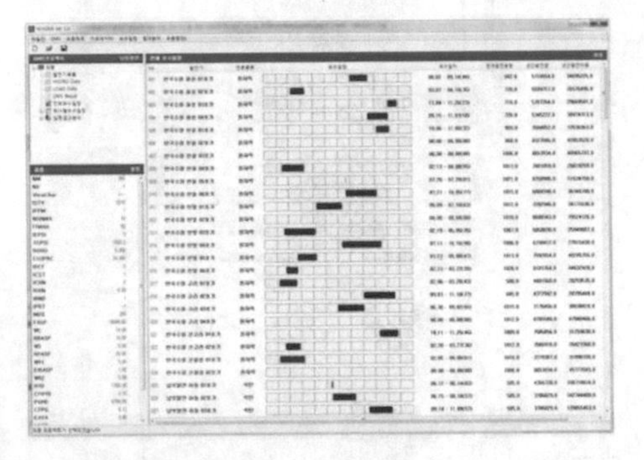

Fig. 5. Result of generation maintenance scheduling (total)

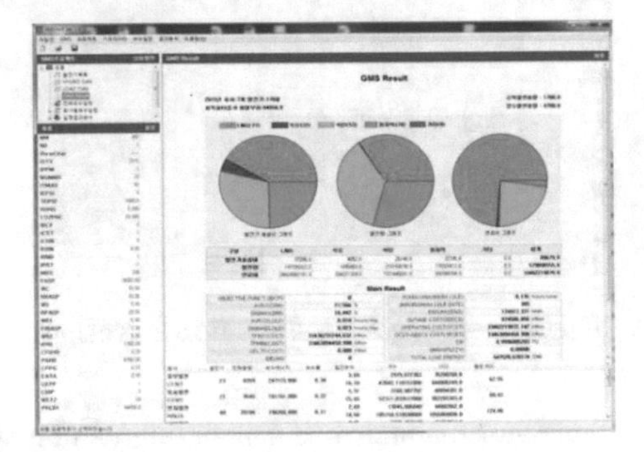

Fig. 6. Total system result

5 Case Study

In this study, it was input the scale in power generation system of 2015 and the peak load value. In case of the peak load is 80,637 MW and the total number of generators is set as 176. The number and capacity of generators are shown in Table 1. In addition, the scope of unit price on fuel of each generator is shown in Table 2.

Table 1. Generation capacity and CO_2 emissions

	Number of generators	Capacity [MW]	CO_2 emissions [Ton/MWh]
Nuclear	24	22,195	0
Coal	53	26,146	0.80 ~ 1.05
Oil	22	4,052	0.59 ~ 0.76
LNG(Liquefied Natural Gas)	77	37,286	0.36 ~ 0.49
Total	176	89,679	

Table 2. The calorie unit price per types of generators

	The calorie unit price
Nuclear power generation	1.290 ~ 2.175
Coal thermal power generation	15.483 ~ 17.086
Oil thermal power generation	162.756
LNG thermal power generation	71.273 ~ 79.452

Table 3. The case of this study

	Objective function
Case I	Maximization in minimum SRR
Case II	Maximization in minimum SRP
Case III	Minimization in total probabilistic cost
Case IV	Minimization in maximum LOLE
Case V	Fuzzy method

Table 4. Results of Model System

	Case I	Case II	Case III	Case IV	Case V
Average supply reserve rate [%]	34.231	34.18	34.336	33.772	34.345
Minimum supply reserve rate [%]	13.894	13.894	16.747	12.482	16.84
Minimum supply reserve power [MW]	8824	10939	3008	10939	10607
LOLE [hours/year]	0.47	0.368	2.935	0.068	2.935
EENS [MWh/year]	28844	23753	144388	9231	146703
EIR [PU]	0.99994	0.99995	0.99972	0.99998	0.99971
Production cost [10^6 Won]	29,593,199	29,582,156	29,482,114	29,525,719	29,538,406
CO_2 emissions [10^3Ton/year]	250,405	250,34,	250,614	250,191	250,607

This case study conducted simulations by dividing cases as shown in Table 3. The results of each simulated case are shown in Table 4

As verified in Table 4, the optimal GMS can be established according to each target function. The case 5 for Fuzzy is examined to find that GMS satisfying each target function can be achieved, though it cannot reach to optimal value.

6 Conclusions

This study simulated model system by introducing a GMS simulation tool and using it. The GMS simulation tool developed in this study established GMS by each target by using various target functions and exhibited it in GUI. The case study compared states of the whole systems through GMS optimizing each target function and using Fuzzy Method and also exhibited the GMS using Fuzzy Method.

This study suggests a method to establish GMS according to targets of many utilities along with changes in Electricity Market and shows the establishment of GMS, which can address their interests through GMS using the Fuzzy Method.

The fuzzy method is a function that can flexibly consider reliability, economy, and environment when considering the optimal objective function. It is a theory that can fuse different values wherein conflicts of interest. Therefore, It is expected to be useful forgenerator maintenance planning.

Subsequent studies on the generation maintenance scheduling embedded with various optimal objective functions such as SMP, are required. In addition, studies on the optimal generation maintenance scheduling configuration that considers a new renewable energy generator are strongly necessitated.

Acknowledgment. This work was supported by the Korean National Research Foundation (NRF) (No.#2012R1A2A2A01012803) and Korea South-East Power Co. (KOEN)

References

1. Choi, J.S.: A study on the generator maintenance scheduling considering load uncertainty & multi-criterion function. EESRI, Final Report (1995)
2. Lee, B.Y., Shim, K.B.: A comparative study on optimal generation maintenance scheduling with marginal maintenance cost and levelized risk method. KIEE **41**(1), 9–17 (1992) (written in Koeran)
3. Khatib, H.: Economic Evaluation of Projects in the Electricity Supply Industry. IEEE Power & Energy Series, vol. 44. MPG Books Limited (2003)
4. Kim, H.S., Moon, S.P., Do, D.H., Choi, J.S., Lee, S.Y., Shin, H.K., Kim, S.H., Kim, Y.S.: Generator maintenance scheduling considering air pollution. In: Proceedings on Autumn Meeting of Western Gyeongnam Area of KIEE, pp. 57–60 (1998) (written in Koeran)
5. Park, J., Wu, L., Choi, J., Baek, U.-K., Cho, W., Song, K., Cha, J., Lee, K.Y.: Fuzzy set theory based flexible generator maintenance scheduling. In: ISME 2009 (2009)

6. Park, J., Wu, L., Choi, J., Cha, J., Lee, K.Y.: Fuzzy set theory based on generator maintenance scheduling. In: ISME 2009. Dalian University of Technology, Dalian, China (2009)
7. Park, J., Wu, L., Choi, J., Baek, U.-K., Cho, W., Song, K., Cha, J., Lee, K.Y.: Fuzzy set theory based flexible generator maintenance scheduling. In: ISME 2009, August 5–7 (2009)
8. Oh, T., Park, J., Cho, K., Choi, J., Baek, U., El-Keib, A.A.: Generators maintenance scheduling using combined fuzzy set theory and GA. In: ICEE 2010, Paradise Hotel, Busan, Korea (2010)
9. Hwang, B., Lee, H., Park, K., Shin, Y.: A case study for determining system marginal price in Korea electricity market. In: KIEE 2011, Yongpyong Resort, Korea (2011) (written in Koeran)
10. Lee, Y., Lim, J., Oh, U., Do, D.P.N., Choi, J.: A study on the various objective functions based optimal generator maintenance scheduling in KOSEP. In: IGSC 2015, October 12–14 (2015)
11. Choi, J.: Visual system of probabilistic generator maintenance scheduling considering the various objective functions. In: PMAPS 2016, China (2016)

Discharge Summaries Classifier

Shusaku Tsumoto[1(✉)], Tomohiro Kimura[2], Haruko Iwata[3], and Shoji Hirano[1]

[1] Department of Medical Informatics, Faculty of Medicine, Shimane University, Matsue, Japan
{tsumoto,hirano}@med.shimane-u.ac.jp

[2] General Coordination Division, Faculty of Medicine, Shimane University, Matsue, Japan
t-kimura@med.shimane-u.ac.jp

[3] Center for Bed-Control, Shimane University Hospital, 89-1 Enya-cho, Izumo 693-8501, Japan
haruko23@med.shimane-u.ac.jp,
http://www.med.shimane-u.ac.jp/med_info/tsumoto/index.htm

Abstract. This paper proposes a method for construction of classifiers for discharge summaries. First, morphological analysis is applied to a set of summaries and a term matrix is generated. Second, correspond analysis is applied to the classification labels and the term matrix and generates two dimensional coordinates. By measuring the distance between categories and the assigned points, ranking of key words will be generated. Then, keywords are selected as attributes according to the rank, and training example for classifiers will be generated. Finally learning methods are applied to the training examples. Experimental validation shows that random forest achieved the best performance and the second best was the deep learner with a small difference, but decision tree methods with many keywords performed only a little worse than neural network or deep learning methods.

Keywords: Discharge summaries · Classifier · Deep learning · Random forest · Decision Tree · SVM

1 Introduction

Computerization of patient records enables to store “big unstructured text data” in a hospital information system. For example, our system in the university hospital, where about 1000 patients visit outpatient and about 600 patients stays inpatient, additionally stores about 200 GB text data per year, including patient records, discharge summaries and reports of radiology and pathology. Application of text mining to these resources is very important to discover useful

S. Tsumoto—This research is supported by Grant-in-Aid for Scientific Research (B) 15H2750 from Japan Society for the Promotion of Science(JSPS).

J.-S. Pan et al. (eds.), *Advances in Intelligent Information Hiding and Multimedia Signal Processing*, Smart Innovation, Systems and Technologies 82,
DOI 10.1007/978-3-319-63859-1_43

knowledge from text data, which can be viewed as a new type of support of clinical actions, researches and hospital management.

However, since standardization of terminology of medical terms has not yet been established, many varieties of vocabularies are used in medical records, which makes them difficult to analyze and extract patterns. In these resources, discharge summaries, which summarize the diagnosis and treatment for each patient during his/her stay, compactly describes the clinical course of the patient. Thus, compared with the other resources, the degree of redundancy of key words and the number of vocabularies are well controlled, which enables us to apply ordinary text mining methods [1,8,11].

Classification learning based on text mining usually generates a dataset from the texts and applies learning methods, such as decision tree and SVM, to generated data [9]. However, in ordinary text data analytics, the performance of such methods are not good because selection of keywords depends on frequencies of keywords, and connection of keywords and targets cannot be captured by frequencies. In order to solve the problem, Mesh [7] or other type of medical ontology are used to select keywords of interest and text mining is applied after the preprocessing [10].

This paper proposes a method for construction of classifiers for discharge summaries. First, morphological analysis is applied to a set of summaries and a term matrix is generated. Second, correspond analysis is applied to the classification labels and the term matrix and generates two dimensional coordinates. By measuring the distance between categories and the assigned points, ranking of key words will be generated. Then, keywords are selected as attributes according to the rank, and training example for classifiers will be generated. Finally learning methods are applied to the training examples. Experimental validation shows that random forest achieved the best performance and the second best was the deep learner with a small difference, but decision tree methods with many keywords performed only a little worse than neural network or deep learning methods.

The paper is organized as follows. Section 2 gives a proposed mining process. Section 3 shows the experimental results. Section 4 discusses the results obtained. Finally, Sect. 5 concludes this paper.

2 Methods

2.1 Motivation for Feature Selection

It is well known that feature selection is important even for deep learners. Although deep learners gain good performance in image analysis, in other cases, differences between deep learners and other classification method are very small. It may be due to that we have not yet suitable network structure and we may not use suitable features for classification. Since deep learners is good at recognition of images, here we propose a new feature selection method where two dimensional mapping of attributes is obtained by correspondence analysis.

2.2 Mining Process

Figure 1 shows the proposed total mining process, whose workflow is as follows. First, target discharge summaries are extracted from hospital information system. Next, morphological analysis, such as MeCab [3], is applied and a contingency table for keywords are generated. Thirdly, correspondence analysis is applied to a contingency table, and if dimension is set to 2, two dimensional coordinate is assigned to each key word and classification classes. Then, by using assigned coordinates, distances between classes and keywords are calculated and we use these distances values for ranking keywords: the smaller the distance is, the higher its ranking is. Fifth, ranking is used to select keywords for classification, and a classification table with a class and the selected keywords will be obtained. Finally, classification learning methods, such as decision tree, SVM, BNN or deep learning, are applied for construction of classifiers.

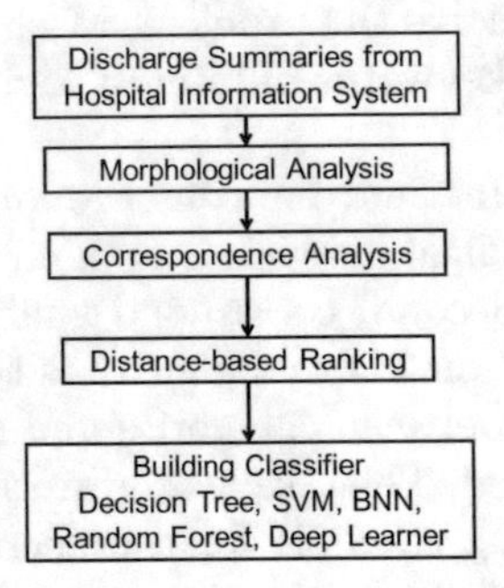

Fig. 1. Mining process

Table 1. DPC top ten diseases (fiscal year: 2015)

No	DPC	Cases	#Characters
1	Cataract (lateral)	445	270.12 ± 295.12
2	Cataract (bilateral)	152	287.87 ± 161.05
3	Type II Diabetes Mellitus (except for keto-acidosis)	145	6888.51 ± 1549.08
4	Lung Cancer (with surgical operation)	131	4535.57 ± 979.36
5	Uterus Cancer (without surgical operation)	121	1508.72 ± 1023.73
6	Lung Cancer (without surgical operation, chemotherapy)	120	2506.34 ± 1132.42
7	Uterus benigh tumor	111	2038.57 ± 910.91
8	Lung Cancer (without surgical operation, with chemotherapy)	110	3505.10 ± 1121.77
9	Shortage of Pregnancy	110	1182.35 ± 646.16
10	Injury of Elbow and Knee	99	1867.22 ± 639.75

2.3 Experiments

Classification. Table 1 shows the list of DPC codes the ranks of whose number of the patients admitted to Shimane University Hospital are top tens. First, we extracted discharge summaries of those diseases.

For morphological analysis, RMeCab [3] is used and the bag of keywords was generated. From the bag of keywords, a contingency table for these summaries are obtained. Then, correspondence analysis which is implemented in R3.3.2 was applied to the table and two dimensional coordinate was assigned to each keyword and each class. Next, the distances between the coordinate of a keyword and that of a class is calculated, and the ranking of keywords for each class was obtained. By using the ranking, a given number of keywords were selected to generate a classification table. Finally, decision tree (package: rpart [12]), random forest (package: randomForest [6], SVM (kernlab [4]), BNN(package: nnet [13]) and Deep Learner (darch [2]) were applied to the generated table. For parameters of Darch, the number of intermediate neurons are 10, (10,5), (40,10) and (100,10), whose epoch was 100. For all other packages, the default settings of parameters were used.

Evaluation Process. Evaluation process is based on repeated 2-fold cross validation [5]. First, a given dataset is randomly split into training examples and test samples half in half. Then, training examples is used for construction of a classifier, and the derived classifier is evaluated by remaining test samples. The above procedures were repeated for 100 times in this experiment, and the averaged accuracy was calculated.

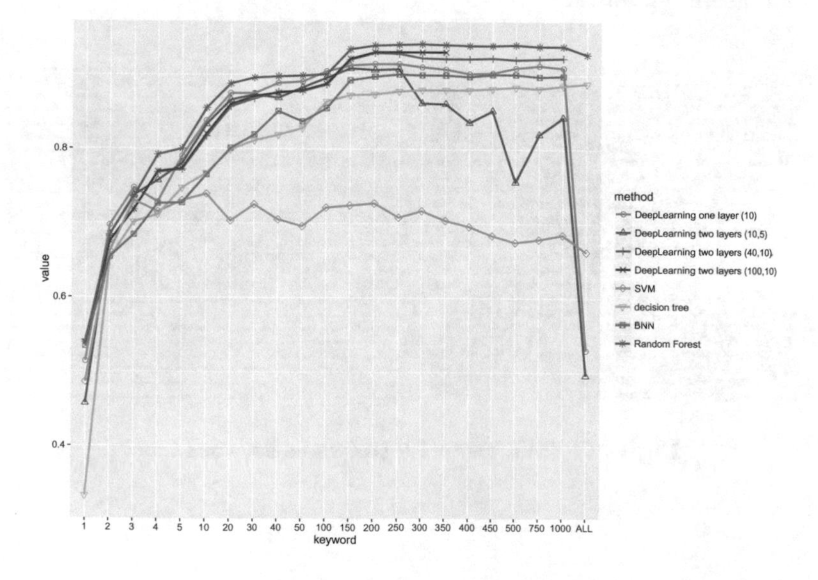

Fig. 2. Experimental results

The number of keywords varied from 1 to 1000, selected according to the rank given by correspondence analysis. For analysis, HP Proliant ML110 Gen9 (Xeon E5-2640 v3.2 2.6 GHz 8Core, 64 GBDRAM) was used.

3 Results

Figure 2 plots the evaluation results. The vertical axis denotes the averaged accuracy of a classifier, and the horizontal axis denotes the number of selected keywords. Until the number of keywords is 4, all the classifiers gained around 70% accuracy. However, larger than 5, SVM was rather decreasing and other methods were monotonically increasing and became plateau with the number of keywords being 200.

Random Forest achieved the best performance, compare with other classifiers. The second are Darch deep learners. Although the accuracy of BNN is worse than Darch (default setting) with 5 to 100 selected keywords, it approached Darch classifiers. Interestingly, the accuracy of decision tree monotonically increased and the highest value was obtained when all the keywords were used for analysis, although other classifiers performed worse than the achieved peak when the number of selected keywords was around 200.

Concerning deep learners, two intermediate layers with 10 and 5 neurons performed worse than one intermediate layers. In other cases, the averaged accuracies were a few percent better than BNN and approached that of random forest. In this experiment, the accuracy was getting higher when the number of intermediate layers and number of neurons increased. It will be our future work whether the change of intermediate neuron structure make a deep learner outperform random forest.

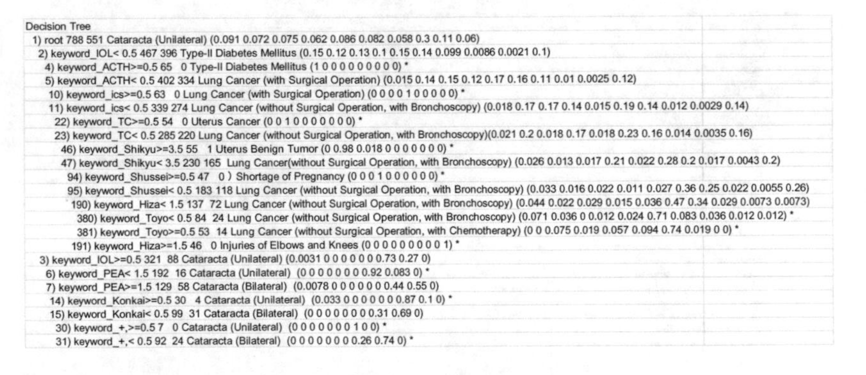

```
Decision Tree
 1) root 788 551 Cataracta (Unilateral) (0.091 0.072 0.075 0.062 0.086 0.082 0.058 0.3 0.11 0.06)
  2) keyword_IOL< 0.5 467 396 Type-II Diabetes Mellitus (0.15 0.12 0.13 0.1 0.15 0.14 0.099 0.0086 0.0021 0.1)
   4) keyword_ACTH>=0.5 65   0 Type-II Diabetes Mellitus (1 0 0 0 0 0 0 0 0 0) *
   5) keyword_ACTH< 0.5 402 334 Lung Cancer (with Surgical Operation) (0.015 0.14 0.15 0.12 0.17 0.16 0.11 0.01 0.0025 0.12)
    10) keyword_ics>=0.5 63   0 Lung Cancer (with Surgical Operation) (0 0 0 0 1 0 0 0 0 0) *
    11) keyword_ics< 0.5 339 274 Lung Cancer (without Surgical Operation, with Bronchoscopy) (0.018 0.17 0.17 0.14 0.015 0.19 0.14 0.012 0.0029 0.14)
     22) keyword_TC>=0.5 54   0 Uterus Cancer (0 0 1 0 0 0 0 0 0 0) *
     23) keyword_TC< 0.5 285 220 Lung Cancer (without Surgical Operation, with Bronchoscopy)(0.021 0.2 0.018 0.17 0.018 0.23 0.16 0.014 0.0035 0.16)
      46) keyword_Shikyu>=3.5 55   1 Uterus Benign Tumor (0 0.98 0.018 0 0 0 0 0 0 0) *
      47) keyword_Shikyu< 3.5 230 165  Lung Cancer(without Surgical Operation, with Bronchoscopy) (0.026 0.013 0.017 0.21 0.022 0.28 0.2 0.017 0.0043 0.2)
       94) keyword_Shussei>=0.5 47   0 ) Shortage of Pregnancy (0 0 0 1 0 0 0 0 0 0) *
       95) keyword_Shussei< 0.5 183 118 Lung Cancer (without Surgical Operation, with Bronchoscopy) (0.033 0.016 0.022 0.011 0.027 0.36 0.25 0.022 0.0055 0.26)
        190) keyword_Hiza< 1.5 137  72 Lung Cancer (without Surgical Operation, with Bronchoscopy) (0.044 0.022 0.029 0.015 0.036 0.47 0.34 0.029 0.0073 0.0073)
         380) keyword_Toyo< 0.5 84  24 Lung Cancer (without Surgical Operation, with Bronchoscopy) (0.071 0.036 0 0.012 0.024 0.71 0.083 0.036 0.012 0.012) *
         381) keyword_Toyo>=0.5 53  14 Lung Cancer (without Surgical Operation, with Chemotherapy) (0 0 0.075 0.019 0.057 0.094 0.74 0.019 0 0) *
        191) keyword_Hiza>=1.5 46   0 Injuries of Elbows and Knees (0 0 0 0 0 0 0 0 0 1) *
  3) keyword_IOL>=0.5 321  88 Cataracta (Unilateral) (0.0031 0 0 0 0 0 0 0.73 0.27 0)
   6) keyword_PEA< 1.5 192  16 Cataracta (Unilateral)  (0 0 0 0 0 0 0 0.92 0.083 0) *
   7) keyword_PEA>=1.5 129  58 Cataracta (Bilateral)  (0.0078 0 0 0 0 0 0 0.44 0.55 0)
    14) keyword_Konkai>=0.5 30   4 Cataracta (Unilateral)  (0.033 0 0 0 0 0 0 0.87 0.1 0) *
    15) keyword_Konkai< 0.5 99  31 Cataracta (Bilateral)  (0 0 0 0 0 0 0 0.31 0.69 0)
     30) keyword_+,>=0.5 7   0 Cataracta (Unilateral)  (0 0 0 0 0 0 0 1 0 0) *
     31) keyword_+,< 0.5 92  24 Cataracta (Bilateral)  (0 0 0 0 0 0 0 0.26 0.74 0) *
```

Fig. 3. Decision Tree with the whole keyword sets

Japanese	English	Cataract (Unilateral)	Cataracta (Bilateral)	Lung Cancer (with Surgical Operation)	Lung Cancer (without Surgical Operation, with Chemotherapy	Lung Cancer (without Surgical Operation, with Bronchoscopy)	Type-II DM	Uterus Cancer	Uterus Benign Tumor	Shortage of Pregnancy	Injuries of Elbows and Knees
IOL	IOL	111	108								
ACTH	ACTH						705				
ics	Uterus	486	486	129	815	604		339	323		508
TC	Uterus	244	228	569	398	403		10	566	774	977
Shikyu	Uterus	397	398	324	698	513		122	425	960	770
Shussei	Birth				625		442			375	
Hiza	Knee										20
Toyo	administer		965	866	146	275		795	877	21	
Hiza	Knee										20
IOL	IOL	111	108								
PEA	PEA	115	112								
Konkai	This time	173	173								
"+"	"+"	35	26								

Fig. 4. Ranking of keywords in Decision Tree

4 Discussion

4.1 Classification Accuracy of Decision Trees

The following two results were unexpected: one is that accuracy of decision trees monotonically increased. The other is that accuracy of random forest achieved the best performance. Since random forest can be viewed as refinement of decision tree, it can be said that representation by decision trees may give deep insights into hidden structure in the discharge summaries.

Figure 3 shows the decision tree obtained by all the keywords extracted by morphological analysis, where 11 attributes were used for description. We may have two observations. First, since the shape of the tree cannot be captured by linear combination, SVM, or linear combination of keywords may not gain classification accuracy. Second, the selection process based on correspondence analysis may not be appropriate for selection of keywords for SVM.

Figure 4 gives the location where the keywords used in the decision tree were located in the ranking in each classification class, which shows that all the keywords are not selected according to the ranking: it may be due to the differences in distances among the attributes are very small. In the future work, the nature of ranking should be examined.

However, medical experts gave the following comments after the review of the decision tree. First, the tree is very compact but reasonable. Second, selection of keywords are very interesting and reasonable. It may reflect that the description of disease summaries is completely different among the target diseases. Detailed examinations of discharge summaries should be needed for further evaluation, which will be one of our future researches.

4.2 Number of Keywords

Random Forest, SVM and Deep Learners reached their peak at 200 selected key words, where the averaged accuracy of SVM was 72.7%, whereas those of BNN, Darch, and Random Forest were 88.3, 91.5 and 93.2, respectively. It may due to structure of patterns hidden in a data set: complex network of keywords, maybe can be represented as a set of decision trees. Thus, since SVM is based on kernel function of linear combination of attributes, it may not be able to approximate hidden structures.

It is notable that the one intermediate layer achieved such performance, and the differences between one and two layers are very small. This observation gives a hypothesis that the hidden network structure is not so complex: rather a set of simple networks can be used, since a complex network may need more than two layers in the neural network structure. It will be our future work to validate this hypothesis.

4.3 Execution Time

Although the performance of BNN and Deep Learners were high, the problem is that they need more time for computation.

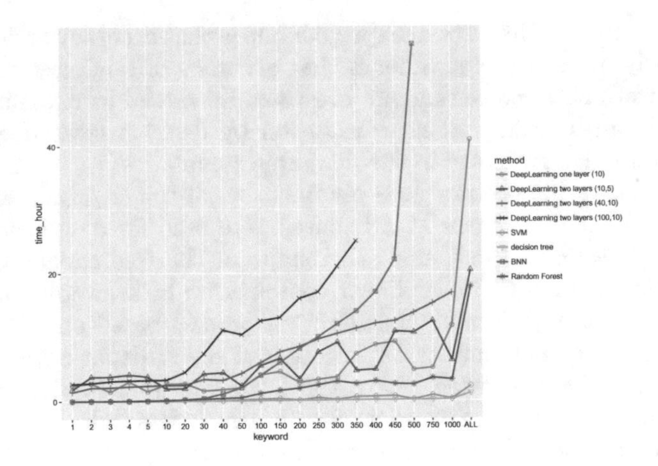

Fig. 5. Executed time for classification construction

Figure 5 shows empirical comparison of repeated 2-fold cross validation (100 Trials). The vertical and horizontal axes denote computational speed and the number of keywords respectively. Whereas the time need for SVM was 28 min in total, Deep Learners (10) needed 2 h 51 min (171 min), Random Forest and BNN were close to Deep Learners. BNN was fast when the number of keywords was small, but the accuracy grew exponentially with more than 200 keywords. In the case of Darch, the number of intermediate layers costed high computational time, although the growth rate was smaller than BNN.

5 Conclusion

It is well known that feature selection is important even for deep learners. Although deep learners gain good performance in image analysis, in other cases, differences between deep learners and other classification method are very small. It may be due to that we have not yet suitable network structure and we may not use suitable features for classification. Since deep learners is good at recognition of images, here we propose a new feature selection method where two dimensional mapping of attributes is obtained by correspondence analysis. Then, we propose a method for construction of classifiers for discharge summaries. First, morphological analysis is applied to a set of summaries and a term matrix is generated. Second, correspond analysis is applied to the classification labels and the term matrix and generates two dimensional coordinates. By measuring the distance between categories and the assigned points, ranking of key words will be generated. Then, keywords are selected as attributes according to the rank, and training example for classifiers will be generated. Finally learning methods are applied to the training examples. Experimental validation was conducted by discharge summaries stored in Shimane University Hospital in fiscal year of 2015. The results shows that random forest achieved the best performance around 93% classification accuracy. The second best was deep learners with a small difference, around 91%. But decision tree methods with many keywords performed only a little worse than neural network or deep learning methods. Furthermore, selected keywords and the tree structure was reasonable for domain experts.

This may be due to the hidden structure of knowledge in a data, which may be close to the structure approximated by a set of trees. It will be our future work to validate this hypothesis.

References

1. Aramaki, E., Miura, Y., Sotoike, M., Ohkuma, T., Masuichi, H., Ohe, K.: Construction of system on visualizing discharge summaries. In: 15h Annual Conference on Association of Natural Language Processing Society, pp. 348–351 (2009). (in Japanese)
2. Drees, M.: Implementierung und Analyse von tiefen Architekturen in R. Master's thesis, Fachhochschule Dortmund (2013)
3. Ishida, M.: Rmecab (2016). http://rmecab.jp/wiki/index.php?RMeCabFunctions
4. Karatzoglou, A., Smola, A., Hornik, K., Zeileis, A.: kernlab - an S4 package for kernel methods in R. J. Stat. Softw. **11**(9), 1–20 (2004). http://www.jstatsoft.org/v11/i09/
5. Kim, J.H.: Estimating classification error rate: repeated cross-validation, repeated hold-out and bootstrap. Comput. Stat. Data Anal. **53**(11), 3735–3745 (2009) http://dx.doi.org/10.1016/j.csda.2009.04.009
6. Liaw, A., Wiener, M.: Classification and regression by randomforest. R News **2**(3), 18–22 (2002). http://CRAN.R-project.org/doc/Rnews/
7. MeSH: Medical subject headings 2017, U.S. national library of medicine. https://meshb.nlm.nih.gov/#/fieldSearch

8. Miura, Y., Aramaki, E., Ohkuma, T., Sotoike, M., Sugihara, D., Masuichi, H., Ohe, K.: Automated extraction of side effects from electronic patient records (in Japanese). In: 16th Annual Conference on Association of Natural Language Processing Society, pp. 78–81 (2010)
9. Sebastiani, F.: Machine learning in automated text categorization. ACM Comput. Surv. **34**, 1–47 (2002)
10. Srinivasan, P.: Meshmap: a text mining tool for medline. In: Proceedings of AMIA Symposium, pp. 642–646 (2001)
11. Suzuki, T., Doi, S., Shimanda, G., Takasaki, M., Tamura, T., Fujita, S., Takabayashi, K.: Auto-selection of DRG codes from discharge summaries by text mining in several hospitals: analysis of difference of discharge summaries. Stud. Health Technol. Inform. **160**(Pt 2), 1020–1024 (2010)
12. Therneau, T.M., Atkinson, E.J.: An introduction to recursive partitioning using the RPART routines (2015). https://cran.r-project.org/web/packages/rpart/vignettes/longintro.pdf
13. Venables, W.N., Ripley, B.D.: Modern Applied Statistics with S, 4th edn. Springer, New York (2002). ISBN: 0-387-95457-0. http://www.stats.ox.ac.uk/pub/MASS4

Design and Implementation of Pseudo-Random Sequence Generator Based on Logistic Chaotic System and m-Sequence Using FPGA

Kai Feng[1,2,3] and Qun Ding[1,2,3(✉)]

[1] College of Electronic Engineering, Heilongjiang University, Harbin, China
vfengkai@126.com, qunding@aliyun.com
[2] Institute of Technology, Harbin, People's Republic of China
[3] College of Electronic Engineering, Heilongjiang University, Harbin, China

Abstract. Based on logistic chaotic system combined with m-sequence, a pseudo-random sequence generator is implemented on FPGA (Field-Programmable Gate Array) using DSP Builder design technology. The NIST (National Institute of Standard and Technology) standard called statistical test suite for random and pseudo-random number generators is used to test the pseudo-random sequence generated the generator. The results show that the sequence of the system has good random characteristics and can be applied to the actual stream key encryption system.

Keywords: Logistic chaotic system · m-sequence · FPGA · Pseudo-random sequence

1 Introduction

In the second half of the 20th century, the nonlinear system science has developed rapidly, and the study of chaos has occupied a great proportion. In 1963, Lorenz presented a deterministic nonperiodic flow model [1], and later raised the famous butterfly effect. In 1975, Chinese scholar Li Tianyan and American mathematician Yorke J published the famous article "Period three implies chaos" in the Mathematical Monthly [2], first introduced the concept of chaos.

To implement the chaotic system, there are two main ways, namely, analog and digital methods [3,4,10]. The former is usually by analog electronic circuits to implement, the typical technology is the chaotic masking. Cuomo K and Oppenheium A V constructed a chaotic masking system with Lorenz system [5,6]. The analog chaotic system is simple to construct but requires strict matching of circuit parameters between the transmitter and receiver. The effective way to solve these problems is based on the discretization and digitization processing technology, using FPGA technology to implement the chaos algorithm [7,8], but the digital chaotic system will degrade the chaotic characteristics of the system during the iterative process due to the finite precision effect, resulting in

J.-S. Pan et al. (eds.), *Advances in Intelligent Information Hiding and Multimedia Signal Processing*, Smart Innovation, Systems and Technologies 82,
DOI 10.1007/978-3-319-63859-1_44

the short period phenomenon. Although this phenomenon can be improved by improving the precision of the system, but it will greatly increase the cost of hardware resources. Under this premise, this paper describes the use of logistic chaotic system, at a relatively low precision, combined with the m-sequence, and with the use of FPGA technology to implement a pseudo-random sequence generator digital circuit design, and as a basis to generate pseudo-random sequence that can meet NIST test standard.

2 Logistic Chaotic System and M-Sequence

2.1 Logistic Chaotic System

Logistic chaotic system is the most widely used type of nonlinear dynamic discrete chaotic mapping system. This system is proposed by P E Verhulst in 1845 and extended by R May. The mapping equation is [9]:

$$x_{n+1} = \mu x_n (1 - x_n) \quad n = 0, 1, 2, 3 \cdots \tag{1}$$

Where the parameters $\mu \in (0, 4]$, $x_n \in (0, 1)$, when the system parameters change, the system's dynamic state also will change, when $3.569945672 < \mu \leq 4$, then the system is in a chaotic state. Of course, the system generated sequence will be affected by the initial value x_0. The state of logistic chaotic system changes with parameter μ, and the situation of change is shown in Fig. 1, we call Fig. 1 the bifurcation diagram of chaotic system.

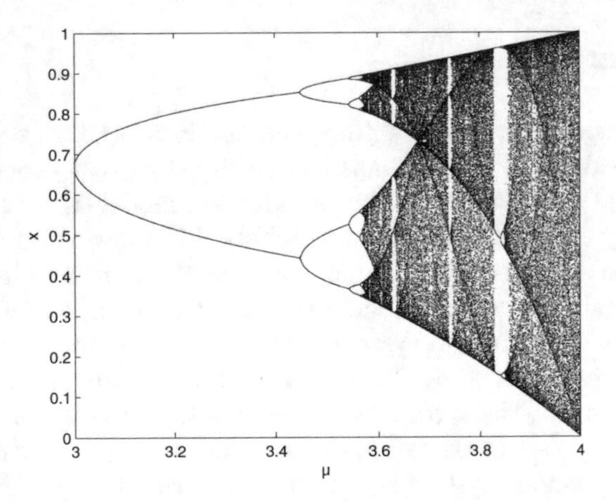

Fig. 1. Bifurcation diagram of logistic chaotic system.

2.2 m-Sequence

The m-sequence is the abbreviation of the longest linear feedback shifting register sequence. It is the longest sequence generated by linear feedback from multi-stage shift registers or other delay unit. As the m-sequence has many excellent performances, like easy to produce and the regularity, m-sequence has a wide range of applications in the spread spectrum communication.

There are 2^n possible different states of the n-stage shift register. In addition to the all 0 state, the remaining $2^n - 1$ states are available. After several shifts, it must be repeated a state appeared before, and then enter the loop state. The different positions of the feedback lines will result in different sequences of different periods. We want to find the position of the linear feedback lines, which can make the sequence of the longest, that is, the period

$$p = 2^n - 1 \tag{2}$$

The feedback polynomial of the m-sequence signal generator designed in this paper is

$$f(x) = x^{16} + x^{12} + x^3 + x + 1 \tag{3}$$

and the relationship of the feedback shift register is shown in Fig. 2.

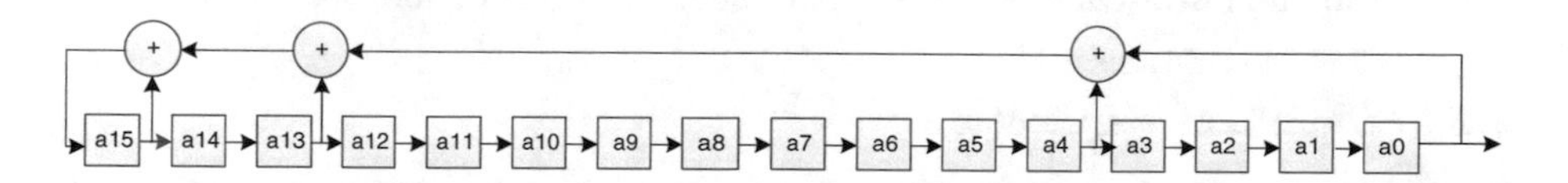

Fig. 2. 16-Stage linear feedback shifting register schematic diagram.

3 The Design of the Pseudo-Random Sequence Generator

3.1 Design of Pseudo-Random Sequence Generator Based on Logistic Chaotic System Combined with m-Sequence

The main design process of this paper is done in DSP Builder. According to the Eq. (1), it can be seen that the whole design should include the initial value module, the multiplication module, the adding module, the gain module, the delay module, the data selector module and the quantization module, AltBus module is here to implement this function and the system precision is set to 32bits. In addition to the need to note that in these modules, the data selector module and the delay module together to achieve the iterative function. And here the quantization function is implemented by shift module and extraction module. Quantization methods are varied. In this design, the discretization method is chosen to complete the quantization process, after the quantization, the 0–1 sequence is output from the logistic chaotic module, then the resulting logistic chaotic 0–1 sequence is XORed with the 16-stage m-sequence, and the schematic model of the system is shown in Fig. 3.

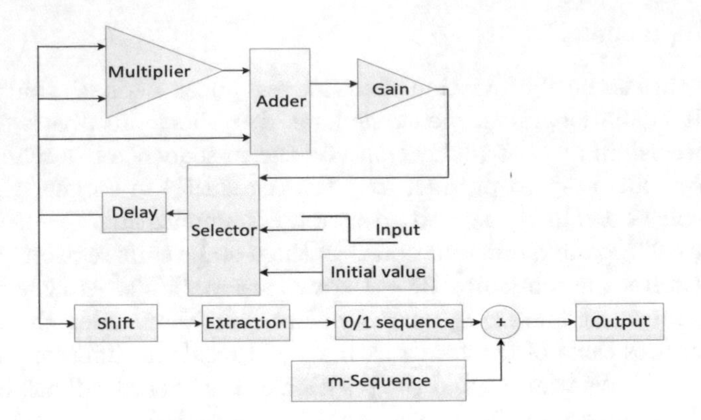

Fig. 3. 32-bit precision logistic chaotic system combined with 16-stage m-sequence pseudo-random sequence generator schematic diagram.

The first step, we input the high level, the initial value of the signal multiplexer is gated, then the input pulse is changed into low level, namely the iterative module consisting of a delay module, a multiplier, an adder and a gain module is selected. The initial value here of the circuit is 0.1, the parameters μ take 4,and the initial value of 16-stage m-sequence is 0000000000000001.

3.2 Simulation Verification

Simulink's graphical simulation verification is very powerful. Before Simulink simulation, you need to set the relevant simulation parameters of the design file, including simulation time and simulation mode. Figure 4 shows Simulink simulation results.

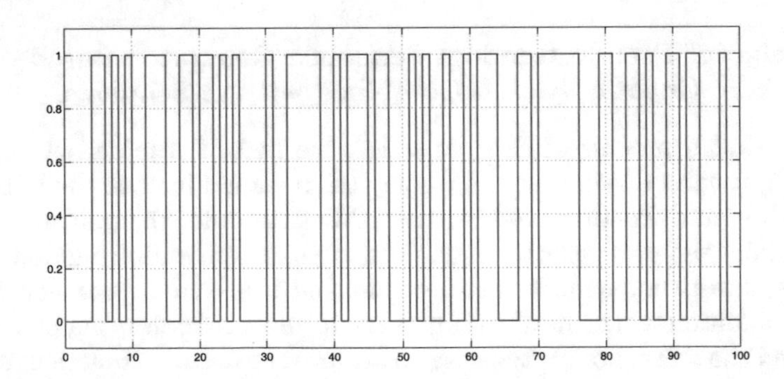

Fig. 4. Simulink simulation results.

After Simulink simulation, double-click the SignalCompiler module, click the "Analyze" button to analyze, and then click the "Compile" button to compile the .mdl model file and convert the .mdl file to VHDL code file. In the QuartusII

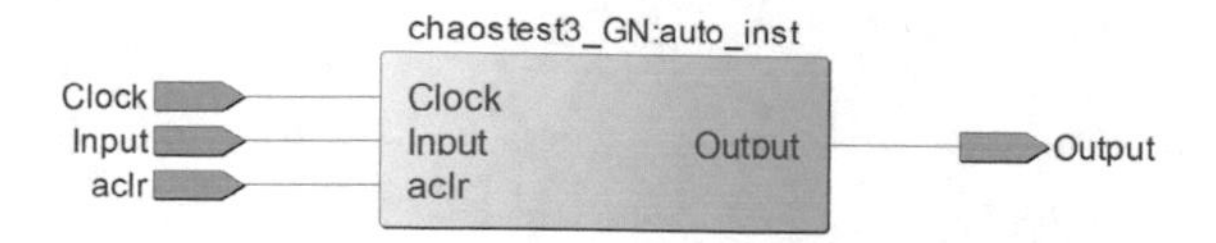

Fig. 5. Circuit diagram generated by Quartus II.

software the .qpf project file is compiled, and finally generate the circuit shown in Fig. 5.

After the above steps, the waveform simulation is followed. In contrast to the previous, the Simulink simulation is for the verification of the algorithm. However, the waveform simulation here is the verification that corresponds to the hardware structure. The simulation results are shown in Fig. 6.

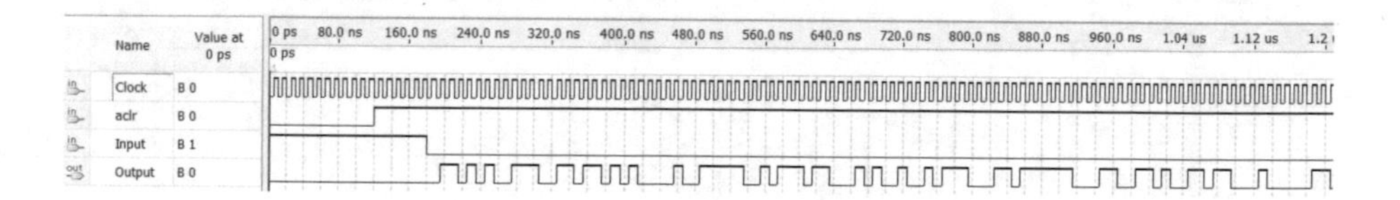

Fig. 6. Wave simulation.

3.3 Comparison of Logical Resource Usage of 32-Bit Precision Chaotic System Combined with 16-Stage m-Sequence and Precision Improved Chaotic System

With the development of digital circuit design, the research of chaotic system using digital circuit has been widely concerned. In this paper, the sequence autocorrelation characteristic of pseudo-random sequence of 32-bit precision logistic chaotic system and 32-bit precision logistic chaotic system combined with 16-stage m-sequence are tested by MATLAB.The results are shown in Fig. 7.

Obviously, the 32-bit precision chaotic system due to the finite precision effect, resulting in a clear periodic phenomenon, which is the short period variations of chaotic system in the digital system that need to overcome, but 32-bit precision logistic chaotic system combined with 16-stage m-sequence, in time when the sequence length reaches 1000000 bits, still has very good autocorrelation characteristics, and there is no significant periodic phenomenon. At the same time, after the experiment, it is found that the precision of the logistic chaotic system needs to be improved to 43 bits when the autocorrelation characteristic of the output 1000000 bits sequence can reach a good level on the logistic chaotic system. In order to compare the hardware resource consumption better, this paper presents the logical resource of 32-bit precision and 43-bit precision logistic chaotic system and the 32-bit precision logistic chaotic system combined with 16-stage m-sequence on EP4CE10F17C8 chip of Altera Cyclone IV series. The consumption is shown in Table 1.

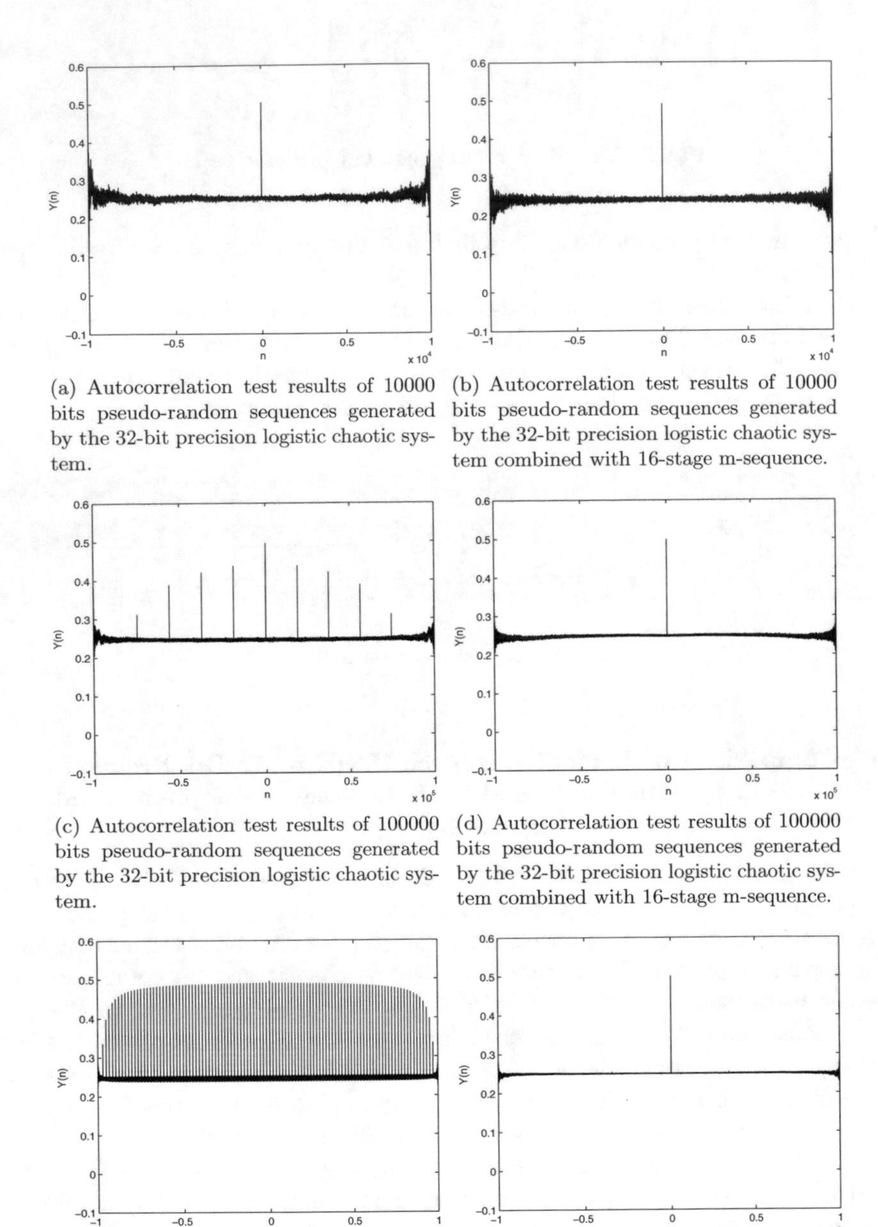

(a) Autocorrelation test results of 10000 bits pseudo-random sequences generated by the 32-bit precision logistic chaotic system.

(b) Autocorrelation test results of 10000 bits pseudo-random sequences generated by the 32-bit precision logistic chaotic system combined with 16-stage m-sequence.

(c) Autocorrelation test results of 100000 bits pseudo-random sequences generated by the 32-bit precision logistic chaotic system.

(d) Autocorrelation test results of 100000 bits pseudo-random sequences generated by the 32-bit precision logistic chaotic system combined with 16-stage m-sequence.

(e) Autocorrelation test results of 1000000 bits pseudo-random sequences generated by the 32-bit precision logistic chaotic system.

(f) Autocorrelation test results of 1000000 bits pseudo-random sequences generated by the 32-bit precision logistic chaotic system combined with 16-stage m-sequence.

Fig. 7. Autocorrelation test.

Table 1. The Hardware Resource Consumption

Name of system	Total logic elements	9-bit multiplier
32-bit precision logistic chaotic system	152	8
43-bit precision logistic chaotic system	308	17
32-bit precision logistic chaotic system combined with 16-stage m-sequence	169	8

It can be seen from the data in Table 1 that the 32-bit precision logistic chaotic system combined with 16-stage m-sequence achieves better performance while the consumption of the logic elements is only 17 more than 32-bit precision logistic chaotic system, and in the case of using the method to improve the precision of the chaotic system, the logic resources consumed by the same performance are more than double. Obviously, the chaotic system combined with the m-sequence is more economical and practical in the digital implementation.

4 Randomness Test of Pseudo-Random Sequence

The randomness test of this paper is based on the "statistical test suite for random and pseudo-random number generators" published by NIST. The NIST Test Suite is a statistical package consisting of 15 tests that were developed to test the randomness of binary sequences. These tests focus on a variety of

Table 2. Part of Test Results of 1000000 bits Sequence

Item	P value
Frequency	0.869731
Block frequency	0.434884
Cumulative sums	0.910261
	0.987173
Runs	0.926719
Longest run	0.543965
Rank	0.569129
FFT	0.992678
Overlapping template	0.268615
Universal	0.881875
Approximate entropy	0.366645
Serial	0.451909
	0.668384
Linear complexity	0.416879

different types of non-randomness that could exist in a sequence. Some tests are decomposable into a variety of subtests, for example, some has one subtest and some has more than one. Each subtest has its own p value. The test results are measured by the p value, if the computed p value is < 0.01, then the sequence is considered to be non-random, otherwise, the sequence is considered to be random, the higher the p value, the better the performance. The 1000000 bits 0–1 sequence generated by the pseudo-random sequence generator designed in this paper was tested. In the 15 items and total of 188 subtests, the sequences were tested, showing good randomness, for convenience, we only show part of the test results in Table 2.

5 Conclusion

The output of the chaotic system has the characteristics of initial sensitivity, intrinsic randomness, orbit of motion trajectory to meet the requirements of pseudo-random sequence design. With the popularization of digital system and the improvement of digital circuit design, the digital chaotic system is in the field of confidential communication will have a broad application prospects. In this paper, 32-bit precision logistic chaotic system combined with 16-stage m-sequence pseudo-random sequence generator based on FPGA is implemented. Compared with improving the precision of digital chaotic system, it will use less logic resources to improve the short periodic phenomenon. The performance of the pseudo-random sequence generator designed in this paper is tested by NIST test. The test sequence length is 1000000 bits, and NIST test results show that the generated chaotic pseudo-random sequence has good performance and can be used directly in practical encryption applications. The work of this paper provides a good technical support for the application of chaotic system in the field of confidential communication.

Acknowledgment. Project supported by the National Natural Science Foundation of China (Grant Nos. 61471158), and Project supported by the Specialized Research Fund for the Doctoral Program of Higher Education of China (Grant No. 20132301110004).

References

1. Lorenz, E.N.: Deterministic nonperiodic flows. J. Atmos. Sci. **20**, 130–141 (1963)
2. Li, T.Y., Yorke, J.A.: Period three implies chaos. J. Am. Math. Monthly **80**(10), 985–992 (1975)
3. Wang, G.Y., Zheng, Y., Liu, J.B.: A hyperchaotic Lorenz attractor and its circuit implementation. J. Acta Phys. Sinca **56**(6), 3113–3120 (2007)
4. Liu, Y.Z.: A new hyperchaotic Lü system and its circuit realization. J. Acta Phys. Sinca **57**(3), 1439–1443 (2008)
5. Cuomo, K.M., Oppenheim, A.V., Strogatz, S.: Synchronization of Lorenz-based chaotic circuits with applications to communications. J. IEEE Trans. CAS-II **40**(10), 626–632 (1993)

6. Cuomo, K.M., Oppenheim, A.: Circuit implementation if synchronized chaos with application to communication. J. Phys. Rev. Lett. **71**, 65–68 (1993)
7. Zhou, W.J., Yu, S.M.: Chaotic digital communication system based on field programmable gate array technology - design and Implementation. J. Acta Phys. Sinca **58**(1), 113–119 (2009)
8. Wang, G.Y., Bao, X.L., Wang, Z.L.: Design and FPGA implementation of a new hyperchaotic system. J. Chin. Phys. B **17**(10), 3596–3602 (2008)
9. May, R.: Simple mathematical model with very complicated dynamic. J. Nature **261**, 459–467 (1976)
10. Zhao, B.S., Qi, C.Y.: Chaotic signal generator design based on discrete system. J. Inf. Hiding Multimedia Sig. Process. **7**(1), 50–58 (2016)

Cryptanalysis of a Random Number Generator Based on a Chaotic Circuit

Salih Ergün(✉)

TÜBİTAK-Informatics and Information Security Research Center,
P.O. Box 74, 41470 Gebze, Kocaeli, Turkey
salih.ergun@tubitak.gov.tr

Abstract. This paper introduces an algebraic cryptanalysis of a random number generator (RNG) based on a chaotic circuit using two ring oscillators coupled by diodes. An attack system is proposed to discover the security weaknesses of the chaotic RNG. Convergence of the attack system is proved using master slave synchronization scheme where the only information available are the structure of the RNG and a scalar time series observed from the chaotic circuit. Simulation and numerical results verifying the feasibility of the attack system are given. The RNG does not fulfill Big Crush and Diehard statistical test suites, the previous and the next bit can be predicted, while the same output bit sequence of the RNG can be reproduced.

1 Introduction

People have needed to keep their critical data secure since they began to communicate with each other. Over the last decades there has been an increasing emphasis on using tools of information secrecy. Certainly, random number generators (RNGs) have more prominently positioned into the focal point of research as the core component of the secure systems. Although many people are even unaware that they are using them, we use RNGs in our daily business. If we ever obtained money from a bank's cash dispenser, ordered goods over the internet with a credit card, or watched pay TV we have used RNGs. Public/private key-pairs for asymmetric algorithms, keys for symmetric and hybrid crypto-systems, one-time pad, nonces and padding bytes are created by using RNGs [1].

Being aware of any knowledge on the design of the RNG should not provide a useful prediction about the output bit sequence. Even so, fulfilling the requirements for secrecy of cryptographic applications using the RNG dictate three secrecy criteria as a "must": 1. The output bit sequence of the RNG must pass all the statistical tests of randomness; 2. The previous and the next random bit must be unpredictable and; 3. The same output bit sequence of the RNG must not be able to be reproduced [1].

An important principle of modern cryptography is the Kerckhoff's assumption [1], states that the overall security of any cryptographic system entirely depends on the security of the key, and assumes that all the other parameters

J.-S. Pan et al. (eds.), *Advances in Intelligent Information Hiding and Multimedia Signal Processing*, Smart Innovation, Systems and Technologies 82,
DOI 10.1007/978-3-319-63859-1_45

of the system are publicly known. Cryptanalysis is the complementary of cryptography. Interaction between these two branches of cryptology form modern cryptography which has become strong only because of cryptanalysis revealing weaknesses in existing cryptographic systems.

There are four fundamental random number generation methods out of all RNG designs reported in the literature: 1. Amplification of a noise source [2,3]; 2. Jittered oscillator sampling [4]; 3. Discrete-time chaotic maps [5] and; 4. Continuous-time chaotic oscillators [6,8]. Although the use of discrete-time chaotic maps in the realization of RNG has been widely accepted for a long period of time, it has been shown during the last decade that continuous-time chaotic oscillators can also be used to realize RNGs [6,8]. In particular, a "true" RNG based on continuous-time chaos has been proposed in [6].

In this paper we target the RNG reported in [6] and further propose an attack system to discover the security weaknesses of the targeted system. The strength of a cryptographic system almost depends on the strength of the key used or in other words on the difficulty for an attacker to predict the key. On the contrary to recent RNG design [8], where the effect of noise generated by circuit components was analyzed to address security issue, the target random number generation system [6] pointed out the deterministic chaos itself as the source of randomness.

The organization of the paper is as follows. In Sect. 2 the target RNG system is described in detail; In Sect. 3 an attack system is proposed to cryptanalyze the target system and its convergence is proved; Sect. 4 illustrates the numerical results with simulations which is followed by concluding remarks.

2 Target System

Chaotic oscillators are categorized into two groups: discrete-time or continuous-time, respectively regarding on the evolution of the dynamical systems. In comparison with RNGs based on discrete-time chaotic sources it appears that RNGs based on continuous-time chaos can be implemented using less complex and more robust structures, particularly due to the absence of successive sample-and-hold and multiplier stages.

In target random number generation system [6], a simple continuous-time chaotic circuit is utilized as the core of the RNG. This chaotic system is derived from two ring oscillators coupled by diodes [6].

Using the normalized quantities: $x_n = v_n/V_{th}$, $y_d = i_d R_d/V_{th}$, $t = T/RC$, $\alpha = G_m R$, $\beta = C/(C+C_1)$, $\gamma = R/R_1$, $\delta = R/R_d$, and $\varepsilon = R/R_2$, the equations of the chaotic circuit transform into the following normalized equation Eq. 1:

The equations in 1 generate chaos for different sets of parameters. The chaotic attractor shown in Fig. 1 is obtained from the numerical analysis of the system with $\alpha = 3.7$, $\beta = 0.1$, $\gamma = 1$, $\delta = 100$ and $\varepsilon = 2.5$.

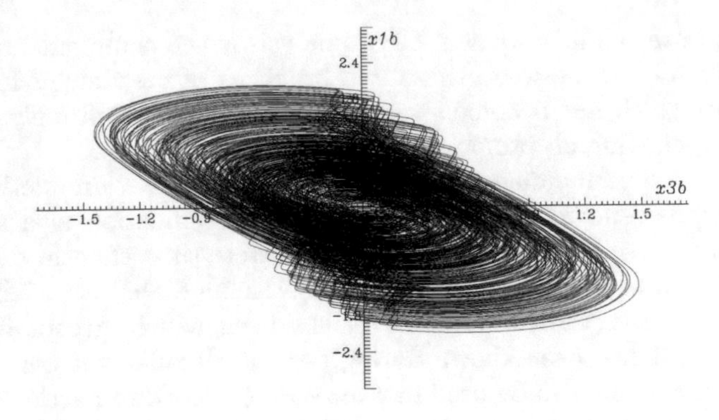

Fig. 1. Numerical analysis results of the chaotic system for $\alpha = 3.7$, $\beta = 0.1$, $\gamma = 1$, $\delta = 100$ and $\varepsilon = 2.5$.

$$
\begin{aligned}
\dot{x_{1a1}} &= -x_{1a1} - \alpha x_{3a1} \\
\dot{x_{2a1}} &= -x_{2a1} - \alpha x_{1a1} \\
\dot{x_{3a1}} &= -\beta(\gamma + 1)x_{3a1} - \alpha\beta x_{2a1} - \beta\delta y_d \\
\dot{x_{1b1}} &= -x_{1b1} - \alpha x_{3b1} \\
\dot{x_{2b1}} &= -x_{2b1} - \alpha x_{1b1} \\
\dot{x_{3b1}} &= -(\varepsilon + 1)x_{3b1} - \alpha x_{2b1} + \delta y_d \\
&where \\
y_d &= \begin{cases} x_{3a1} - x_{3b1} - 1 \; for \; x_{3a1} - x_{3b1} > 1 \\ 0 \qquad\qquad\quad for \; |x_{3a1} - x_{3b1}| \leq 1 \\ x_{3a1} - x_{3b1} + 1 \; for \; x_{3a1} - x_{3b1} < -1 \end{cases}
\end{aligned}
\tag{1}
$$

Target random number generation mechanism is depicted in Fig. 2 where bit generation method is based on jittered oscillator sampling technique. As depicted in Fig. 2 output of a fast oscillator is sampled on the rising edge of a jittered slower clock using a D flip-flop where the jittered slow clock is realized by chaotic ring oscillator circuit.

In this design, if the fast and the slower clock frequencies are known as well as the starting phase difference ΔT, the output of the fast oscillator, sampled at the rising edge of the jittered slower clock, can be predicted. It can be shown that the output bit sequence $S_{(bit)i}$ is the inverse of least significant bit of the ratio between the total periods of the jittered slower clock and period of the fast clock:

$$S_{(bit)i} = \left(\frac{\lfloor \frac{(\sum_{j=1}^{i} T_{slow\ j}) - \Delta T}{T_{fast}/2} \rfloor mod2}{(2d_{fast})}\right)' \tag{2}$$

where $T_{fast} = \frac{1}{f_{fast}}$, f_{fast}, d_{fast} are the period, frequency and the duty cycle of the fast clock, respectively, and the periods of the jittered slower clock $T_{slow\,j}$ are obtained at times t satisfying:

$$s(t) = x_{3a1}(t) = Q\, with\ \frac{ds}{dt} > 0 \tag{3}$$

where $x_{3a1}(t)$ is the chaotic signal, and Q is the logic threshold of the D flip-flop. We have numerically verified that, for high $\frac{f_{fast}}{f_{slow\ center}}$ ratios, the effect of

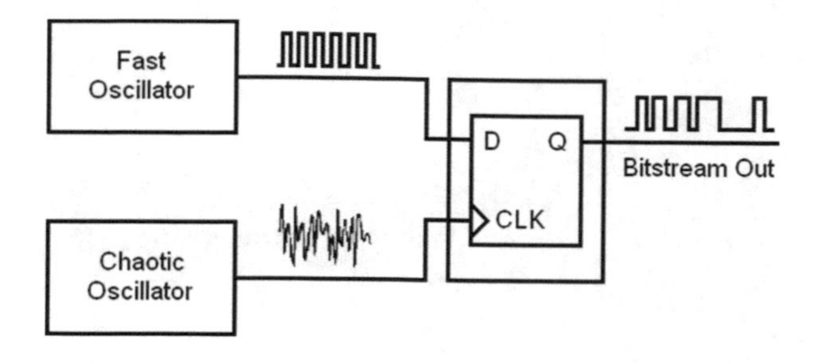

Fig. 2. Target random number generation system.

ΔT becomes negligible and the mean value (m_{output}) of the output sequence S_{bit} approaches the fast clock duty cycle d_{fast} where frequency of the chaotic signal, corresponding to mean frequency of the jittered slower clock $f_{slow\ center}$, determines the throughput data rate (f_{rng}). It should be noted that, anyone who knows the chaotic signal output can reproduce the same output bit sequence.

The authors of [6] have preferred to use NIST 800-22 [9] statistical test suite in order to analyze output randomness of their RNG design. However, Big Crush [10] and Diehard [11] statistical test suites which are available at the publication date of target paper [6] weren't applied to output bit stream of the RNG [7]. It should be noted that, the target random number generation system [6] doesn't satisfy the first secrecy criteria, which states that "RNG must pass all the statistical tests of randomness."

3 Attack System

After the seminal work on chaotic systems by Pecora and Carroll [12], synchronization of chaotic systems has been an increasingly active area of research [13]. In this paper, convergence of attack and target systems is numerically demonstrated using master slave synchronization scheme by means of feedback method [13]. In order to provide an algebraic cryptanalysis of the target random number generation system a attack system is proposed which is given by the following Eq. 4:

$$
\begin{aligned}
\dot{x}_{1a2} &= -x_{1a2} - \alpha x_{3a2} \\
\dot{x}_{2a2} &= -x_{2a2} - \alpha x_{1a2} \\
\dot{x}_{3a2} &= -\beta(\gamma + 1)x_{3a2} - \alpha\beta x_{2a2} - \beta\delta y_d \\
\dot{x}_{1b2} &= -x_{1b2} - \alpha x_{3b2} \\
\dot{x}_{2b2} &= -x_{2b2} - \alpha x_{1b2} \\
\dot{x}_{3b2} &= -(\varepsilon + 1)x_{3b2} - \alpha x_{2b2} + \delta y_d + c(x_{3b1} - x_{3b2}) \\
&where \\
y_d &= \begin{cases} x_{3a2} - x_{3b2} - 1 & for\, x_{3a2} - x_{3b2} > 1 \\ 0 & for\, |x_{3a2} - x_{3b2}| \leq 1 \\ x_{3a2} - x_{3b2} + 1 & for\, x_{3a2} - x_{3b2} < -1 \end{cases}
\end{aligned}
\tag{4}
$$

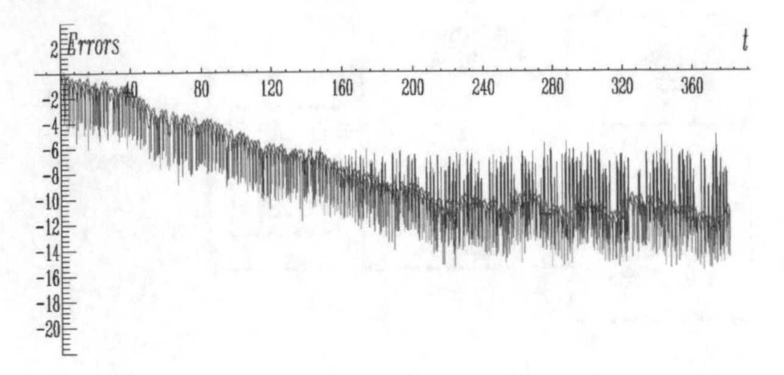

Fig. 3. Synchronization errors: Log $|e_{x1a}(t)|$ (red line), Log $|e_{x2a}(t)|$ (blue line), and Log $|e_{x3a}(t)|$ (green line).

where e is the coupling strength between the target (master) and attack (slave) systems and the only information available are the structure of the target random number generation system and a scalar time series observed from x_{3b1}.

In this paper, we are able to construct the attack system expressed by the Eq. 4 that synchronizes ($x_{3b2} \to x_{3b1}$ for $t \to \infty$) where t is the normalized time. We define the error signals as $e_{x1a} = x_{1a1} - x_{1a2}$, $e_{x2a} = x_{2a1} - x_{2a2}$, and $e_{x3a} = x_{3a1} - x_{3a2}$ where the aim of the attack is to design the coupling strength such that $|e(t)| \to 0$ as $t \to \infty$.

The master slave synchronization of attack and target systems is verified by the conditional Lyapunov Exponents, and as firstly reported in [12], is achievable if the largest conditional Lyapunov Exponent is negative. When e is greater than 1.05 then the largest conditional Lyapunov Exponent is negative and hence identical synchronization of target and attack systems starting with different initial conditions is achieved and stable [12]. (Largest conditional Lyapunov Exponent is -0.00102704 for $e = 1.1$). However for e is equal to or less than 1.05, largest conditional Lyapunov Exponent is positive and identical synchronization is unstable.

Log $|e_{x1a}(t)|$, Log $|e_{x2a}(t)|$, and Log $|e_{x3a}(t)|$ are shown in Fig. 3 for $e = 2$, where the synchronization effect is better than that of $e = 1.1$ (Largest conditional Lyapunov Exponent is -0.0262916 for $e = 2$), which indicate that the identical synchronization is achieved in less than $380t$.

4 Numerical Results

We numerically demonstrate the proposed attack system using a 4th-order Runge-Kutta algorithm with adaptive step size and its convergence is illustrated in Fig. 3. Numerical results of $x_{1a1} - x_{1a2}$, $x_{2a1} - x_{2a2}$, and $x_{3a1} - x_{3a2}$ are also given in Figs. 4, 5, and 6, respectively illustrating the unsynchronized behavior and the synchronization of target and attack systems.

It is observed from the given figures that, master slave synchronization is achieved and stable. As shown by black lines in these figures, no synchronous

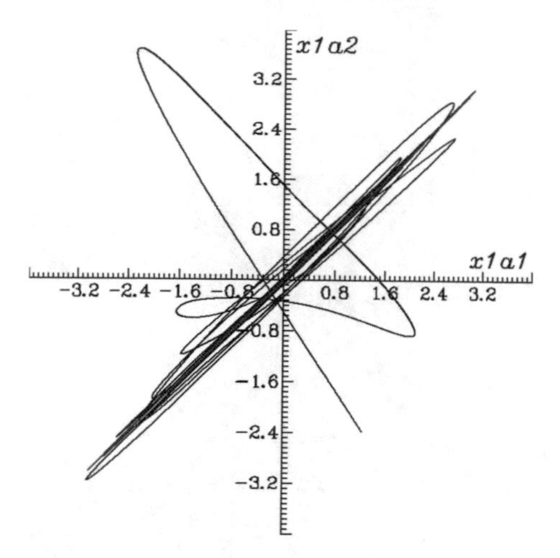

Fig. 4. Numerical result of $x_{1a1} - x_{1a2}$ illustrating the unsynchronized behavior and the synchronization of target and attack systems.

phenomenon is observed before 380t. In time, the proposed attack system converges to the target system and identical synchronization is achieved where colored lines depict synchronized behaviors of chaotic states in Figs. 4, 5, and 6, respectively.

Since the identical synchronization of attack and target systems is achieved ($x_{3b2} \rightarrow x_{3b1}$) in 380$t$, the estimated value of $S_{(bit)i}$ bit which is generated according to the procedure explained in Sect. 2 converges to its fixed value.

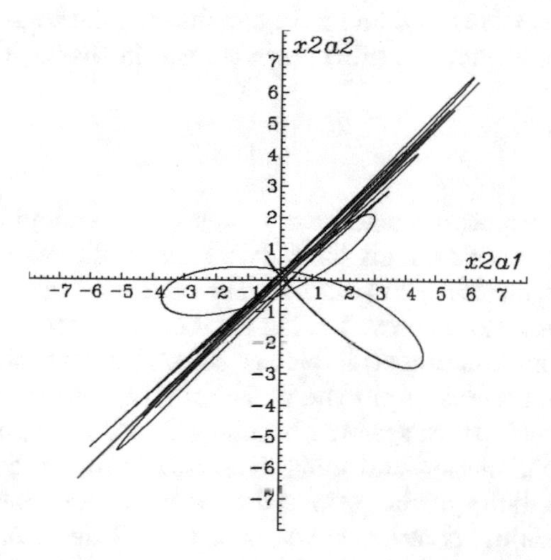

Fig. 5. Numerical result of $x_{2a1} - x_{2a2}$ illustrating the unsynchronized behavior and the synchronization of target and attack systems.

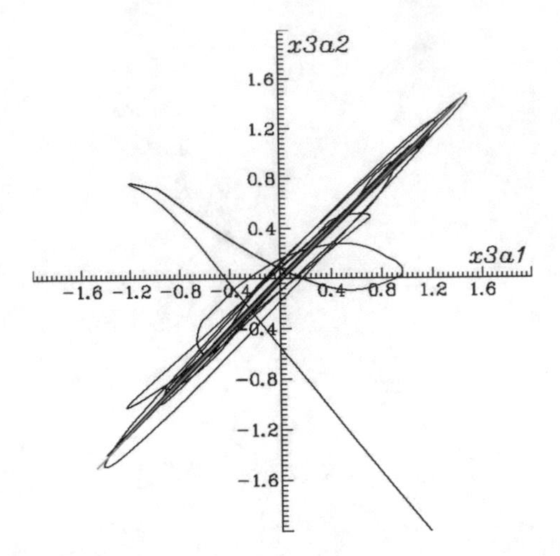

Fig. 6. Numerical result of $x_{3a1} - x_{3a2}$ illustrating the unsynchronized behavior and the synchronization of target and attack systems.

As a result, it is obvious that identical synchronization of chaotic systems is achieved and hence output bit streams of target and attack systems are synchronized.

It is clearly shown that master slave synchronization of proposed attack system is achieved. Hence, output bit sequences of target and attack systems are synchronized. As a result, cryptanalysis of the target random number generation system not only predicts the previous and the next random bit but also demonstrates that the same output bit sequence of the target random number generation system can be reproduced. In conclusion, the target system [6] satisfies neither the second, nor the third secrecy criteria that a RNG must satisfy.

5 Conclusions

In this paper, we propose an algebraic attack on a random number generator (RNG) based on a chaotic circuit using two ring oscillators coupled by diodes. An attack system is introduced to discover the security weaknesses of the chaotic RNG and its convergence is proved using master slave synchronization scheme. Although the only information available are the structure of the target RNG and a scalar time series observed from the target chaotic system, identical synchronization of target and attack systems is achieved and hence output bit streams are synchronized. Simulation and numerical results presented in this work not only verify the feasibility of the proposed attack but also encourage its use for the security analysis of the other chaos based RNG designs. Proposed attack, renders generated bit streams predictable, thereby qualifying the target RNG to be used as a not true but pseudo random source.

References

1. Menezes, A., van Oorschot, P., Vanstone, S.: Handbook of Applied Cryptology. CRC Press, Boca Raton (1996)
2. Göv, N.C., Mıhçak, M.K., Ergün, S.: True random number generation via sampling from flat band-limited Gaussian processes. IEEE Trans. Circ. Syst. I **58**(5), 1044–1051 (2011)
3. Petrie, C.S., Connelly, J.A.: A noise-based IC random number generator for applications in cryptography. IEEE Trans. Circ. Syst. I **47**(5), 615–621 (2000)
4. Bucci, M., Germani, L., Luzzi, R., Trifiletti, A., Varanonuovo, M.: A high speed oscillator-based truly random number source for cryptographic applications on a smart card IC. IEEE Trans. Comput. **52**, 403–409 (2003)
5. Stojanovski, T., Pihl, J., Kocarev, L.: Chaos-based random number generators-part II: practical realization. IEEE Trans. Circ. Syst. I **48**(3), 382–385 (2001)
6. Çicek, İ., Dündar, G.: A chaos based integrated jitter booster circuit for true random number generators. In: Proceedings of the European Conference Circuit Theory and Design (ECCTD 2013), pp. 1–4, August 2013
7. Çicek, İ.: Design aspects of discrete time chaos based true random number generators, Ph.D. thesis. http://www.ee.boun.edu.tr/Research/RecentPhD Disserta-tions.aspx
8. Ergün, S., Güler, Ü., Asada, K.: A high speed IC truly random number generator based on chaotic sampling of regular waveform. IEICE Trans. Fundam. Electron. Commun. Comput. Sci. **E94–A**(1), 180–190 (2011)
9. National Institute of Standard and Technology: A Statistical Test Suite for Random and Pseudo Random Number Generators for Cryptographic Applications, NIST 800-22, May 2001. http://csrc.nist.gov/rng/SP800-22b.pdf
10. L'Ecuyer, P.: Universit'e de Montr'eal: Empirical Testing of Random Number Generators (2002). http://www.iro.umontreal.ca/~lecuyer/
11. Marsalgia, G.: Diehard: A Battery of Tests of Randomness (1997). http://stat.fsu.edu/~geo/diehard.htm
12. Pecora, L.M., Carroll, T.L.: Synchronization in chaotic systems. Phys. Rev. Lett. **64**(8), 821–824 (1990)
13. Hasler, M.: Synchronization principles and applications. In: Toumazou, C. (ed.) Tutorials IEEE International Symposium on Circuits and Systems (ISCAS 1994), London, England, pp. 314–327 (1994)

Training Method for a Feed Forward Neural Network Based on Meta-heuristics

Haydee Melo[1](✉), Huiming Zhang[1], Pandian Vasant[2], and Junzo Watada[3]

[1] Graduate School of Production, Information and Systems, Waseda University, Tokyo, Japan
melo.haydee@asagi.waseda.jp, huimingde@gmail.com

[2] Fundamental and Applied Sciences Department, Universiti Teknologi PETRONAS, Seri Iskandar, Malaysia

[3] Deparment of Computer and Information Sciences, Universiti Teknologi PETRONAS, Seri Iskandar, Malaysia
junzow@osb.att.ne.jp

Abstract. This paper proposes a Gaussian-Cauchy Particle Swarm Optimization (PSO) algorithm to provide the optimized parameters for a Feed Forward Neural Network. The improved PSO trains the Neural Network by optimizing the network weights and bias in the Neural Network. In comparison with the Back Propagation Neural Network, the Gaussian-Cauchy PSO Neural Network converges faster and is immune to local minima.

Keywords: Particle Swarm Optimization · Neural network · Training algorithm · Gaussian distribution · Cauchy distribution

1 Introduction

The Back Propagation (BP) algorithm is the most widely implemented algorithm for training an ANN [1]. However, BP has disadvantages such as slow convergence and easily getting trapped in local minima [2]. Many learning techniques have been developed for the BP such as gradient descendant, resilient, BGFS quasi-Newton, one-step secant, Levenberg-Marquad and Bayesian regularization whose objective is to minimize the error. These learning techniques do not overcome the BP algorithm disadvantages generated from poor network structure specification and tuning parameters. The Particle Swarm Optimization (PSO) has demonstrated to have better results as a training algorithm [3]. Even though PSO has a good performance in global optima search [3,4], parameters tuning has influence in its efficiency [5]. In addition, PSO has no momentum that is a serious disadvantage; causing to be difficult to reach the global optimum. In this paper a PSO with Gaussian and Cauchy random variables is proposed. Gaussian random variables can improve the search performance and the information sharing among the particles by allowing more stable search for an optimal solution.

J.-S. Pan et al. (eds.), *Advances in Intelligent Information Hiding and Multimedia Signal Processing*, Smart Innovation, Systems and Technologies 82,
DOI 10.1007/978-3-319-63859-1_46

The use of Cauchy random variables is proposed for ensuring escape from local minima. In contrast with other methods the combination of Gaussian-Cauchy probability distributions introduces learning into the algorithm by faster convergence and escape from the local minima.

2 Gaussian-Cauchy Particle Swarm Optimization Neural Network

2.1 Particle Swarm Optimization

In PSO [6], particles are initialized in uniform randomness over a searching space D. At each iteration, each particle updates velocity and position based on its best position B_p and the neighbors' best position G_b. For the number of particles n, let us denote the position of the ith particle by $X_i = (x_{i1}, x_{i2}, \cdots, x_{iD})$; the best position vector as $B_{ip} = (b_{i1}, b_{i2}, \cdots b_{iD})$, the best position of all the swarm $G_b = (g_{b1}, g_{b2}, \cdots, g_{bD})$ and the velocity by $V_i = (v_{i1}, v_{i2}, \cdots, v_{iD})$, respectively. Eberhart [6] introduced an internal weight q. Furthermore, velocity limits $[v_{min}, v_{max}]$ and position limits $[x_{min}, x_{max}]$ are proposed. The new position and velocity at $(t+1)$ can be calculated as follows:

$$v_{id}(t+1) = q \times v_{id}(t) + c_1 \times r_1[b_{id}(t) - x_{id}(t)] + c_2 \times r_2[g_{bd}(t) - x_{id}(t)] \quad (1)$$

$$x_{id}(t+1) = x_{id}(t) + v_{id}(t+1) \quad (2)$$

$$1 \leq i \leq n 1 \leq d \leq D \quad (3)$$

where c_1, c_2 are the social and cognitive acceleration factors and r_1, r_2 are randomly generated numbers. In the beginning of iterations the weight decreases rapidly, but after a certain number of iterations the weight decreases slowly by using the following formula:

$$q = \begin{cases} q_0 - ((q_0 - q_1)/f_1) \times f & 1 \leq f \leq f_1 \\ (q_1) \times e^{(f_1 - f)/k}, & f_1 \leq f \leq f_2 \end{cases} \quad (4)$$

where q_0 and q_1 are the inertial weights that are used in the initial stage and in the linear reduction respectively, the f_2 is the total number of generations in the algorithm, f_1 is the number of generations for the linear reduction strategy. In the reduction weight strategy k adjusts the non-linear reduction and the different values in k modify the curve [7].

2.2 Gaussian Random Variables

According to statistical theory, a particle from the swarm without initial velocity, the position of the particle can be considered to be inside of a parallelogram (uniform probability). After many interactions we observe that the parallelograms overlap as shown in Fig. 1. Using the central limit theorem we assume a

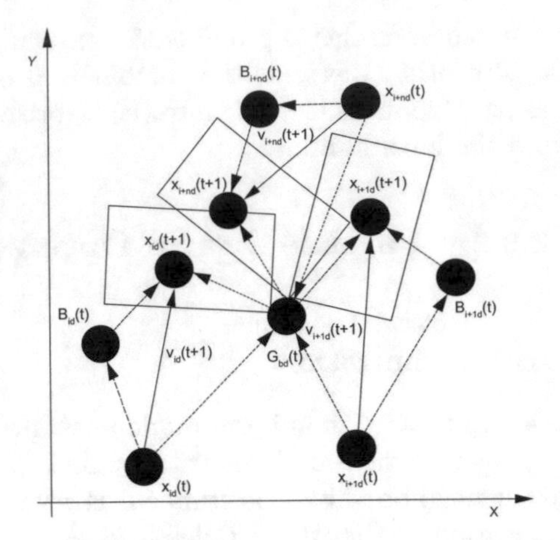

Fig. 1. Visualization of the swarm

Gaussian distribution. Consequently, the uniformly distributed random variables can be replaced for Gaussian random variables $G(\mu, \sigma^2)$:

$$v_{id}(t+1) = q \times v_{id}(t) + c_1 \times G(\mu, \sigma^2)[b_{id}(t) - x_{id}(t)] + c_2 \times G(\mu, \sigma^2)[g_{bd}(t) - x_{id}(t)] \tag{5}$$

$$x_{id}(t+1) = x_{id}(t) + v_{id}(t+1) \tag{6}$$

$$1 \leq i \leq n \tag{7}$$

$$1 \leq d \leq D \tag{8}$$

2.3 Ziggurat Algorithm

The Ziggurat algorithm generates a point x from a probability distribution which is made up of n-equal-area regions; $n - 1$ rectangles can cover the desired distribution, on top of a rectangular base that includes the tail of the distribution. Then the right hand edge of each rectangle is placed to cover the distribution as the gray area in each rectangle. Therefore, some of the area in the top right of each rectangle is outside of the distribution (points with $y > f(x)$). However R_0 is an exception and is entirely within the probability distribution function (pdf) [8]. The V size of each rectangle area can be then calculated with the following equation:

$$x_2[f(x_3) - f(x_2)] = x_1[f(x_2) - f(x_1)] = x_1 f(x_1) + \int_{x_1}^{\infty} f(x)\,\mathrm{d}x = V \tag{9}$$

where r is the rightmost x_1 and C denotes the number of blocks that is partitioned.

$$V(r) = rf(r) + \int_{r}^{\infty} f(x)\,\mathrm{d}x \tag{10}$$

where the value x_i can be found with the substitution from:

$$x_i = f^{-1}(f(x_i - 1) + v(r)/x_{i-1}), i = 2, \ldots, C - 1. \quad (11)$$

2.4 Cauchy Random Variables

Cauchy distribution prevents the PSO from fast trapping in local minima. By analyzing the trajectory of a particle x_{id} in the swarm (Fig. 1), we observed that the particle converged to a weighted mean between of B_p and G_b. The Cauchy distribution is characterized for the lack of a mean that could be useful in PSO for escape from the local minima.

$$v_{id}(t+1) = v_{id}(t) + c_1 \times C(\mu, \sigma^2)[b_{id}(t) - x_{id}(t)] + c_2 \times C(\mu, \sigma^2) \times [g_{bd}(t) - x_{id}(t)] \quad (12)$$

$$x_{id}(t+1) = x_{id}(t) + v_{id}(t+1) \quad (13)$$

$$1 \leq i \leq n1 \leq d \leq D \quad (14)$$

2.5 Artificial Forward Neural Network

An ANN is principally formed of different layers: input layer, hidden layer and output layer as depicted in Fig. 2. These weights connect the input layer with the hidden layer and the hidden layers with the output layers, respectively.

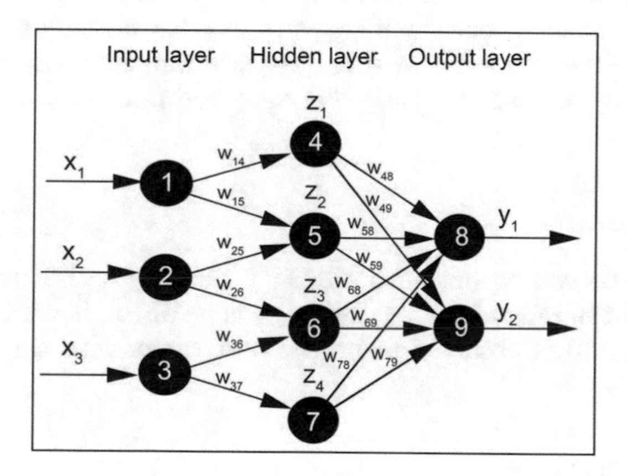

Fig. 2. Neural network structure

2.6 Feed Forward Neural Network

Considering a network with n neurons, o hidden neurons, and m output neuron. All the input units ($x_i, i = 1, ..., n$) calculate the transfer the value to all the hidden units ($z_j, j = 1, ..., H$) as follows:

$$z_j = \frac{1}{1 + exp(-(\sum_{i=0}^{n} x_i w_{ij} - \theta_j))} \quad (15)$$

where the weight is denoted as w_{ij}, the bias is denoted as θ_j. Output units ($y_k, k = 1, ...O$) are calculated as follows:

$$y_k = \frac{1}{1 + exp(-(\sum_{j=0}^{H} z_j w_{jk} - \theta_k))} \quad (16)$$

All the output units ($y_k, k = 1, ...O$) are used to calculate the error E as follows:

$$E = \frac{1}{m} \sum_{i=0}^{m} (d_i - Y_i)^2 \quad (17)$$

Then the fitness function can be formulated as

$$fitness(X_i) = E(X_i) \quad (18)$$

3 Encoding Strategy and Parameter Selection

3.1 Encoding Strategy

In this paper a matrix strategy for encoding the weights is used having the advantage of understating the weight structure in the network. For example, using a 2-3-2 network structure, the first column is constituted by w_{31}, w_{41}, w_{51}. Then the second column consists of w_{32}, w_{42}, w_{52}. To simplify the encoding, one matrix configuration is used so w_{63}, w_{64}, w_{65} are weights from the hidden units to the output unit y_1, and w_{73}, w_{74}, w_{75} are weights from the hidden units to the output unit y_2.

3.2 Parameters

The experiment is carried out with a $c_1 = 1.2$ and $c_2 = 1.2$ that achieved the best results and the range for v_{min} and v_{max} is [−10,10]. Then the search space range of (−100,100) is observed to be the best range with a swarm size of 25 particles.

3.3 Pseudo Code

The GCPSO-NN procedure is composed of the following steps:

Step 1: Initialize the training values, target values and parameters in the FFNN.

Step 2: The particles' positions and velocities are initiated randomly and the iteration t is set to 1.

Step 3: The particle's fitness is determined by introducing the weight values into the FFNN.

Step 4: The best position B_p and the global best G_b are determined from the current particles' positions.

Step 5: The velocity and position of each particle are calculated with Gaussian random variables. The boundaries for velocity $[v_{min}, v_{max}]$ and position $[v_{min}, v_{max}]$ are checked if the new position or velocity is beyond the boundaries then the new value is set to be the minimum or maximum.

Step 6: Check for change in the G_b. If there is no change in the global best, then the velocity and position of each particle are calculated with Cauchy random variables.

Step 7: Repeat from step 3 to 6 until the stop criteria are met or the maximum iterations is reached.

Step 8: Output the trained weights.

4 Simulation Results

In this paper, the optimization goal is to minimize the objective function. The accuracy of the network base on MSE on the training set is used as objective function by minimizing the network rate. For initial values in the network, the weights were generated within in the range of [−10,10], and the bias was set to 0. The initial inertial weight was set to 1.8. The initial population was set to 100.

4.1 Function Approximation Problem

For the approximation problem a 1-S-1 network structure was trained for approximating the function $f = sin(2x)e^x$, where $S = 3, 5, 7$, and the x_i was obtained from the range $[0, \pi]$, and the sampling interval was 0.03 for the training set and 0.1 for the testing set in the range $[0.02, \pi]$. Table 1 shows the performance comparison for the three algorithms. Different hidden unit numbers performance is shown in Fig. 3 (Table 2).

Table 1. The MSE of the function approximation problem for the training methods.

Hidden unit	GCPSO	PSO	BP
3	3.97e-009	1.39e-008	1.23e000
4	2.05e-011	6.21e-003	−1.91e001
5	6.56e-011	1.24e-008	−1.91e001
6	7.56e-011	3.17e-008	−1.19e001
7	6.65e-010	8.75e-008	−1.91e001

4.2 Iris Data Classification Problem

The data set is composed of 150 instances divided into 3 classes of Iris: Setosa, Versicolour and Virginica. Each class has 50 instances with 4 attributes: petal width, petal length, sepal width and sepal length. The problem is to classify the

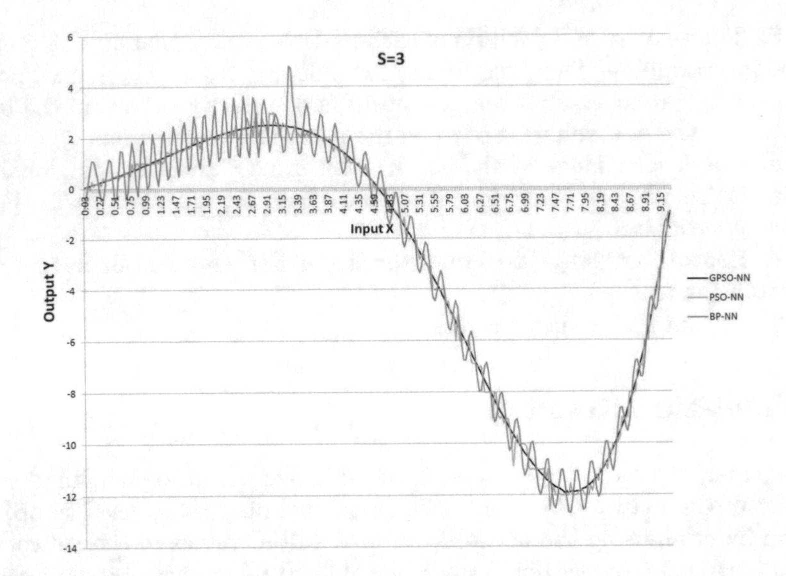

Fig. 3. The generated curves from the test data set for the function $f = sin(2x)e^x$ for training algorithms: Gaussian-Cauchy PSO, PSO-NN and Back Propagation algorithm respectively for different values of S.

Table 2. The CPU time for function approximation problem for the three training methods.

Hidden unit	GCPSO	PSO	BP
3	3.354 s	9.332 s	9.332 s
4	3.085 s	8.165 s	8.165 s
5	3.364 s	8.883 s	8.542 s
6	3.766 s	9.273 s	7.952 s
7	4.322 s	9.246 s	9.354 s

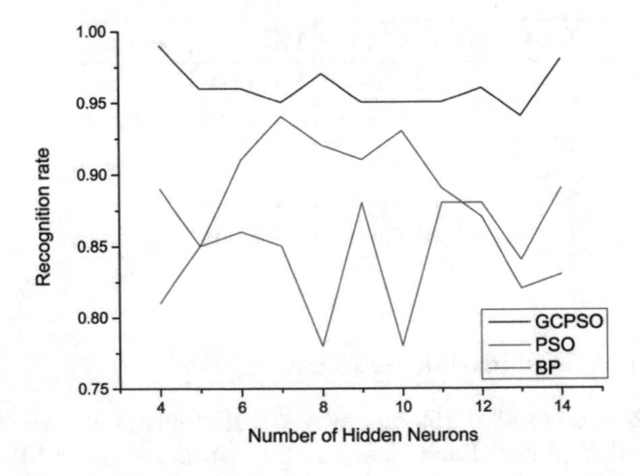

Fig. 4. Best recognition rate

data according to these 4 attributes. For this problem a 4-S-3 FFNN structure was used, where $S = 4, 5, 6, 7, 8, ..., N$ is the number of neurons in the hidden layer. Figure 4 shows the performance of the three training algorithms. The best recognition rate that the BP algorithm achieved was 0.85 while the best recognition rate attained by the PSO algorithm was 0.88 and 0.97 for the GCPSO with any number of hidden nodes. The proposed CGPSO-NN achieved best mean recognition rate 0.9185.

5 Conclusions

A Gaussian-Cauchy Particle Swarm optimization for training a Feed Forward Neural problem has been presented in this paper. The experiment was pursed with different numbers of hidden units configuration for analyzing the performance of three training methods: Back-propagation algorithm, PSO and the Gaussian-Cauchy PSO. The performance of both algorithms was better when the number of hidden units increased. Future work is to apply this algorithm to a real scenario and to incorporate this training algorithm into recurrent neural networks and recurrent fuzzy network.

Acknowledgments. This work was supported by Petronas Corporation, Petroleum Research Fund (PRF) No. 0153AB-A33.

References

1. Cerqueira, J.J.F., Palhares, A.G.B., Madrid, M.K.: A simple adaptive back-propagation algorithm for multilayered feedforward perceptrons. In: Proceedings of the IEEE International Conference on Systems, Man and Cybernetics, vol. 3, p. 6, 6–9 October 2002
2. Gudise, V.G., Venaayagamoorthy, G.K.: Comparison of particle swarm optimization and back propagation as training algorithms for neural networks. In: Proceedings of the IEEE Swarm Intelligence Symposium (SIS 2003), pp. 110–117 (2003)
3. Wang, W., Cao, J., et al.: Reservoir parameter prediction of neural network based on particle swarm optimization. J. Southwest Pet. Univ. **29**(6), 31 (2007)
4. Juang, C.F.: A hybrid genetic algorithm and particle swarm optimization for recurrent network design. IEEE Trans. Syst. Man Cybern. **32**, 997–1006 (2004)
5. Settles, M., Raylander, B.: Neural network learning using particle swarm optimizers. In: Advances in Information Science and Soft Computing, pp. 224–226 (2002)
6. Eberhart, R.C., Shi, Y.: Comparison between genetic algorithms and particle swarm optimization. In: Proceedings of the 7th Annual Conference on Genetic and Evolutionary Computation, pp. 611–616. Springer, Berlin (1998)
7. Zhang, J.-R., et al.: A hybrid particle swarm optimization-back propagation algorithm for feed forward neural network training. Appl. Math. Comput. **185**, 1026–1037 (2007)
8. George, M., Tsang, W.W.: The Ziggurat method for generating random variables. J. Stat. Softw. **5**(8), 1–7 (2000)

Hamming Code Protection Method of FPGA-Based Space System

Dan Wang, Baolong Guo(✉), Yunyi Yan, and Haibin Xu

Mechanical and Electrical Engineering, Xidian University, Xi'an 710071, China
13572975876@163.com, blguo@xidian.edu.cn

Abstract. FPGA is susceptible to single events upset (SEU) effect in space. SEU will lead to SRAM-based FPGA memory cell's content changes which may result in system instability or even collapse. In this paper, the soft error probability model is established. For Hamming code protection, we have formulated an exact reliability model and estimated the consumption of system after conducting protection. Simulation test of ISE14.4 version XC4VSX55 models of FPGA is given. By the model simplification, the costs in power, area and speed of the protective methods are estimated.

Keywords: SEU · Soft error · Hamming code · Protection costs

1 Introduction

FPGA devices based on SRAM technology have been widely used in the aerospace electronic systems. There are a lot of high-energy particles in space. It may happen Single Event Upset (SEU), when high-energy particles impinging on FPGA devices. SEU will lead to SRAM-based FPGA memory cell's content changes which may result in system instability or even collapse [1]. Therefore, the protection methods relating to SRAM-based FPGA SEU have become a hotspot in recent years [2].

Hamming code protection is an effective fault-tolerant technology to SEU. The coding and decoding principle of Hamming code is clearly explained by Hamming [3]. Tian Kai and He Li change parallel coding of the Hamming code into serial coding, optimizing timing and saving resources in the basic of pipeline [4]. Wang Aizhen adds a parity check to the overall encoding, forming (8, 4) Hamming code [5]. It can correct one-bit error and detect two-bit error, improving the reliability of data transmission. The Hsiao code in the literature [6] has smaller area, lower power consumption and higher probability of 4-bit error detection than the Hamming code, but there is no effective algorithm to select the optimal matrix.

Most of the recent studies have improved the algorithm of Hamming code itself, but the effective model of Hamming code combined with FPGA protection have not been established. In this paper firstly, soft error of system caused by the space energetic particles bombarding is defined. Then Hamming code protection method for SEU has been analyzed. At last the performance and costs of Hamming code protection method have been estimated through the model that may provide guidance in engineering applications.

J.-S. Pan et al. (eds.), *Advances in Intelligent Information Hiding and Multimedia Signal Processing*, Smart Innovation, Systems and Technologies 82,
DOI 10.1007/978-3-319-63859-1_47

2 Modeling

2.1 Soft Error Rate of FPGA

S is defined as the area of the sensitive area, f is particle radiation flow rate in the area of S in the period of time T, p is the probability of each arrived particle caused the device flipping, the frequency of SEU is

$$R_p = \frac{fpS}{N} \quad (1)$$

N represents number of times of SEU. R_p represents frequency of SEU. So the error probability due to the single-particle radiation is

$$P_{SEU} \approx 1 - P\{N_{SEU} = 0\} \approx 1 - e^{-R_p NT} \quad (2)$$

SEU of one-bit in the FPGA memory cells does not necessarily cause the output error in the system, depending on the bit that occurs SEU is whether a status bit which can affect the correct execution of system, that is ACE bit [7]. If ACE bit occurs flip, it will affect the validity of the system. The definition of soft error robust factor (soft error robust factor, *SERF*) to represent visible error probability which caused by each bit flip and has a value of ACE-bit proportion of devices, that is $SERF \in [0, 1]$. For unprotected system, the probability of error in theory is

$$P_0 \approx \left(1 - e^{-R_p NT}\right) \times SERF \quad (3)$$

E_0 is defined as the number of SEU system occurs in unit time, which is error frequency. After time T_t, there will be a mistake in the system. That is to say after time T_t, the exception of error frequency is 1. For error frequency-time function $p_1(t)$

$$\int_0^{T_t} p_1(t)dt = E_0 \times T_t = 1 \quad (4)$$

By Formula (3), we know $p_1(t)$ is an increasing function of time. To simplify the model, we use a function to replace approximatively probability-time function in (4). That is $p_1(t) = k \times t$ and $k > 0$, so before the protection the probability of error is

$$p_1(t) = 2E_0^2 t \quad (5)$$

2.2 Hamming Code Protection

We analysis data encoding and decoding of Hamming code by assuming m-bit data, n-bit Hamming code, the code length $l = m + n$. According to Hamming code matrix, we can obtain correspondence between m data bits and n bits Hamming code. Hamming encoding circuit diagram is shown in Fig. 1.

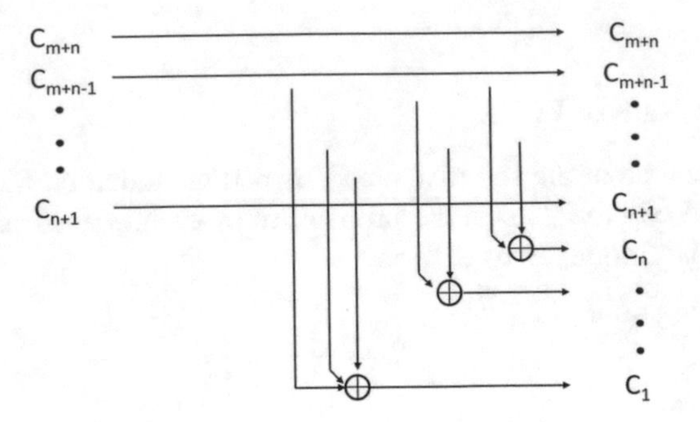

Fig. 1. Encoding circuit of Hamming code

According to principle of Hamming encoding, the generation of parity check matrix is varies based on the different generated matrix. According to the generated matrix, parity check matrix can be obtained, which can be obtained logic parity check matrix encoding, resulting in calibration factor. Hamming code's decoding circuit diagram is shown in Fig. 2.

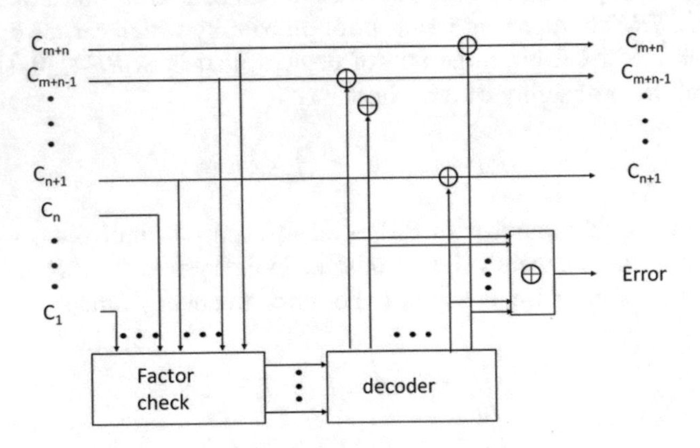

Fig. 2. Decoding circuit of Hamming code

Assuming without protection, area occupied by data resource is D_0 bits. The probability of error occurred of each-bit data in module is same at the same point in time. But the error probability will increase over time. The function of data error rate $p_1(t)$ and error rate $p_3(t)$ of every bit data in module is

$$p_3(t) = 1 - \sqrt[D_0]{1 - p_1(t)} \tag{6}$$

One bit data error can be corrected with Hamming code protection. Only all data bits correct or just one-bit occurs error, the data is correct. The error probability of module data is

$$p_a(t) = p_1(t) - D_0\left[(1-p_1(t))\frac{D_0-1}{D_0} - (1-p_1(t))\right] \tag{7}$$

We make integration of $p_a(t)$ in time period $[0, T_t]$, the error probability of module after protection is

$$E_p = 1 + \frac{D_0S_0}{2E_0(2D_0-1)}\left[1-\left(1-\frac{2D_0E_0}{S_0}\right)^{2-\frac{1}{D_0}}\right] + \frac{2D_0S_0 - D_0^2E_0}{S_0} - \frac{D_0}{S_0} \tag{8}$$

2.3 Cost Estimation of Protection Methods

In resource overhead, power consumption is considered as P_0, and unit is watts (W), which referred to P_s as static power, dynamic power denoted P_d. Static power consumption associated with the current work environment. Dynamic power is related to clock, programmable logic cells, I/O ports and memory. The power related to clock is P_1. The power related to programmable logic cells is P_2. The power related to I/O ports is P_3. The power related to memory is P_4. The power formula is

$$P_0 = P_s + P_d = P_s + P_1 + P_2 + P_3 + P_4 \tag{9}$$

The speed in resource overhead is V_0 and unit is MHz. According to analysis of identification of engineering resources of FPGA, the speed formula is

$$V_0 = \frac{1}{T_{co} + T_{\log ic} + T_{route} + T_{su}} \approx \frac{1}{T_{co} + T_{su} + 3 \times n \times T_{LUT}} \tag{10}$$

The output trigger delay is T_{co}. The logic delay is $T_{\log ic}$. The routing delay is T_{route}. The lookup table delay is T_{LUT}. The trigger setup time is T_{su}.

The area of resource overhead is S_0 and its unit is bit. According to analysis of identification of engineering resources of FPGA, the formula is

$$S_0 = \alpha \times N_{LUT} + \beta \times N_{BRAM} \tag{11}$$

The number of LUT is N_{LUT}. The number of BRAM is N_{BRAM}. $\alpha = 2^4 = 16$. $\beta = 18 \times 1024 \times 18 = 147456$.

The protection of Hamming code only protect data. The redundancy bit of error correction code varies with the different data bits that need protection. In actual operation, the number of data bits needs protected may vary, and it is difficult to accurately forecast changes in the data space. In order to ensure sufficient resources in practice, it is assumed that each data requires 10 bits redundant storage space (8 redundant bits to 247 bits in length data protection). The area is about

$$S_p \approx S_0 + 10N_k + S_x \tag{12}$$

The number of variables that need protection is N_k. The additional area due to coding matrix and decoding matrix is S_x.

For power, only programmable logic unit power P_2 and memory power P_4 changes. After protection the power of module is

$$P_p = P_s + P_d \approx P_s + P_1 + P_3 + \varepsilon(P_2 + P_4) + P_x \tag{13}$$

Change factor that the storage space and the programmable logic unit regarding power is ε. Additional power due to encoding circuit decoding circuit is P_x.

Error correction codes not only changes the storage and routing logic, but also take up additional execution time of the module during the execution of the coding and decoding. Logic delay $T_{\log ic}$ and routing delay T_{route} will increase. The speed after protection is

$$V_p = \frac{1}{T_{co} + T_{\log ic} + t_1 + T_{route} + t_2 + T_{su}} \tag{14}$$

3 Data and Result

For Hamming code, 32-bit data requires six bits Hamming codes, which are located 1, 2, 4, 8, 16, 32 bits, a total of $32 + 6 = 38$-bit data. Hamming encoder encode data according to the rules. Decoder decodes 38 bits including data and Hamming code, and filtered 6 bits Hamming code, output 32-bit data. To determine whether an error occurs, correct the inverted position, and outputs the result of 38 bit which over-turned. The simulation test is based ISE14.4 version for XC4VSX55 model FPGA. Voting only designed for a vote of one bit in testing. Count the number of LUT, Slice Registers (SR), LUTRAM, flip-flops (FF), the input I/O count, the output I/O count, BRAM and detect the change of system before protection and after protection. With the mapping between these parameters and the module cost, we estimate the approximate costs.

Static consumption power P_s and clock frequency power P_1, programmable logic unit power P_2, input and output I/O ports power P_3 and memory consumption power P_4 of original module can be obtain by ISE. By test results in the Table 1, we can approximate the additional power P_x by voter and delayer, the additional area S_x and reduced speed V_x. The power, area, speed after protection is approximately calculated in Table 2. These can provide references to protection methods in engineering applications.

Clock frequency is f_0. Bits of voter is N_d.

Table 1. Resource testing of Hamming code

Parameter	Encoder	Decoder	Sum
LUT	71	115	186
SR	48	85	133
LUTRAM	0	0	0
FF	48	54	102
IO	0	0	0
BRAM	1	1	2

Table 2. Power, area and speed of Hamming code protection methods

Method	Hamming code
ε	1.184
$P_x(W)$	$6.8 \times 10^{-5} f_0$
$S_x(bit)$	2.98×10^{-5}
$V_x(MHz)$	9.6×10^{-6}

4 Conclusion

For Hamming code method, it needs the encoder and the decoder. The encoder and the decoder need LUT, SR, FF and BRAM, this will lead the number of LUT, SR, FF and BRAM increase. Its power is influenced by the clock frequency while area and speed are not affected. In conclusion, this paper discusses the mechanism of soft error and analysis Hamming code of protection, simplified theoretical model, which was used to predict the resource consumption in engineering application. Based on ISE14.4 version, we conduct simulation test for XC4VSX55 model FPGA and obtained increased resources of several protection methods, estimating power, speed, area cost to provide a reference for engineering analysis and design of FPGA.

Acknowledgment. This work was supported by the National Natural Science Foundation of China under Grants No. 61571346.

References

1. Violante, M., Ceschia, M., Reorda, M.S., Paccagnella, A., Bernardi, P., Rebaudengo, M., Candelori, A.: Analyzing SEU effects is SRAM-based FPGAsb. In: 9th IEEE on-Line Testing Symposium, IOLTS 2003, pp. 119–123 (2003)
2. Zheng, D., Vladimirova, T.: Application of programmable system-on-a-chip for small satellite computers. Chin. Space Sci. Technol. **24**(1), 37–44 (2004)
3. Hamming, R.W.: Error detecting and error correcting codes. Bell Labs Tech. J. **29**(2), 147–160 (1950)
4. Tian, K., Li, H.E., Tian, F., Liang, T.: Improved design of hamming code decoder and FPGA verification. Video Eng. **37**(17), 232–235 (2013)

5. Wang, A.: Design of expansion hamming coder/decoder and its FPGA realization. Mod. Electron. Tech. **31**(19), 187–188 (2008)
6. Zhao, Y., Hua, G.: A method of fault tolerant design for memory. Aerosp. Control Appl. **35**(3), 61–64 (2009)
7. Ostler, P.S., Caffrey, M.P., Gibelyou, D.S., Graham, P.S.: SRAM FPGA reliability analysis for harsh radiation environments. IEEE Trans. Nucl. Sci. **56**(6), 3519–3526 (2009)

Optimization of AES and RSA Algorithm and Its Mixed Encryption System

Jia Liu, Chunlei Fan, Xingyu Tian, and Qun Ding(✉)

Electronic Engineering College, Heilongjiang University, Harbin, China
qunding@aliyun.com, 18846089521@163.com

Abstract. An improved key expansion method is proposed to improve the security performance in AES key expansion. There is large difference of operation time between Mixcolumns and Inverse Mixcolumns, we propose the simplest form of MixColumn and InvMixColumn operation on finite field $GF(2^N)$ which consumes same computing resources in the process of encryption and decryption. In terms of the defection of RSA operation efficiency, traditional double prime number is replaced by four prime number, Chinese remainder theorem combined with Montgomery modular multiplication is also presented to optimize modular exponentiation. On this basis, we adopt message digest, digital signature, digital envelope and other technologies to build a mixed encryption system which encompasses convenient key management and high-efficiency encryption and decryption, combined with the advantages of AES and RSA. The experimental results show that optimized algorithm has high speed and feasibility.

Keywords: AES algorithm · Chinese remainder theorem · RSA algorithm · Digital signature

1 Introduction

With the development of Internet and communication technology, information security problem has received much attention, information security technology based on data encryption has developed rapidly. Data encryption technology in terms of secret key type divided to symmetric encryption and asymmetric encryption. AES is one of the most popular symmetric algorithms and a new generation of data encryption standard after DES, this algorithm has the highest security and the fastest operation speed standard. RSA is one of the most widely used asymmetric algorithms [1], the characteristics of RSA are high security and easy to carry out, it can not only encrypt data but also be authenticated. This paper analyses research status of AES and RSA and improves existing problem, on this basis, we adopt message digests, digital signature [2], digital envelope and other security technologies to build a mixed encryption system, which transmits information effectively and securely and authenticate data resources efficiently

J.-S. Pan et al. (eds.), *Advances in Intelligent Information Hiding and Multimedia Signal Processing*, Smart Innovation, Systems and Technologies 82,
DOI 10.1007/978-3-319-63859-1_48

to prevent denial situation. AES is a new generation data encryption standard and encompasses a lot of advantages, such as, high security, high performance, flexible and easy to use. However, there is increasingly unsafe factors and high demand of operation efficiency today.

2 AES Algorithm Optimization

2.1 AES Key Expansion Analysis and Improvement

Based on the table look-up method [3], this paper improves the key expansion method and solves the problem of unequal time-consuming in the process of encryption and decryption, and further improves the computing efficiency under the premise of ensuring security.

AES key expansion process shown in Fig. 1 (10 rounds of AES as an example). The first key is 128 bits, four words w_0, w_1, w_2 and w_3. Then w_4 is generated by w_0 and w_3, w_5 is generated by w_1 and w_4, w_6 is generated by w_2 and w_5, and w_7 is generated by w_3 and w_6. w_4, w_5, w_6, and w_7 are called the first round of expansion keys, and then push down as the basis, until all the keys are generated. In which the first word of each round key to carry out a complex operation f, which consists of an S-box function SubWord, a shift operation RotWord and a round of constant XOR composition. Each round of operation depends on the previous round, followed by pushing down to get the desired arbitrary round key. This kind of key expansion method has advantage of high efficiency and immediacy, but if attacker gets one round key, the whole key can be cracked. Because each word w_i is related to w_{i-1} and w_{i-4}, in other words, if any two of them are known, the third will be deduced.

In view of the above security risks, this paper proposes an improved key expansion algorithm from the aspects of anti-attack strength and execution time of the program, the initial key w_0, w_1, w_2, w_3 is unchanged, the first round expansion key w_4, w_5, w_6, w_7 is a new set of expansion key, there is no connection between w_4, w_5, w_6, w_7 and w_0, w_1, w_2, w_3, on the basis of new key, AES inherent algorithm is used for key expansion until all the sub-key is generated. The principle of this method is given in Fig. 2. After this change, since there is no relationship between the initial key and the extended key, the attacker cannot deduce the entire key from a round key. If we use exhaustive key attack, we assume that the seed key length is k bit, the best case of exhaustive key attack is 1 and the worst case is 2^k. Since the probability of each case is equal, the average complexity is

$$\sum_{i=1}^{2k} \frac{1}{2^k} \times i = \frac{1}{2^k} \times \sum_{i=1}^{2k} i = \frac{1}{2^k} \times \frac{(1+2^k) \times 2^k}{2} = \frac{1}{2} + 2^{k-1} \approx 2^{k-1} \tag{1}$$

For 10 rounds of AES algorithm, the attacker needs to try 2^{127} possible keys on average, and the key expansion algorithm in this paper makes the attacker need to try 2^{255} possible keys on average. In terms of current computing power,

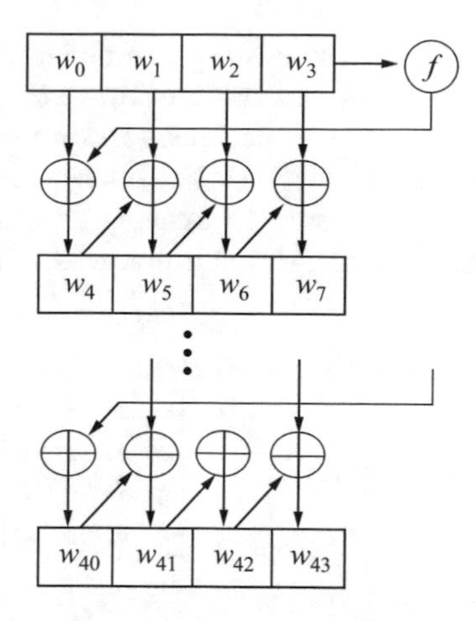

Fig. 1. AES key expansion process

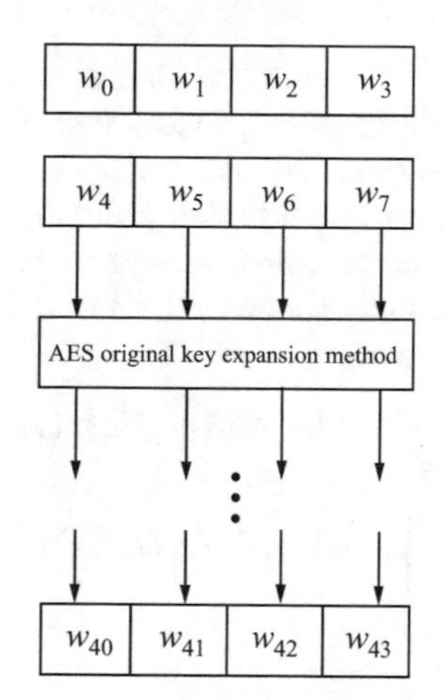

Fig. 2. Improved key expansion algorithm

completing this exhaustive search will take at least hundreds of millions of years. Therefore, we made a small change in original method, the improved key expansion method overcomes the original security risks and ensures the efficiency of the program.

2.2 Analysis and Optimization of Mixcolumns and Inverse Mixcolumns

AES algorithm encryption and decryption operation time-consuming is different, the reason is that the algorithm complexity of Mixcolumns and Inverse Mixcolumns is distinct. In a Mixcolumns transformation, each column of the state is treated as a polynomial over $GF(2^8)$ and associated with a fixed polynomial $C(x) = \{03\}x^3 + \{01\}x^2 + \{01\}x + \{02\}$, and mod the modulo polynomial $x^4 + 1$. The matrix is expressed as follows:

$$\begin{bmatrix} 02 & 03 & 01 & 01 \\ 01 & 02 & 03 & 01 \\ 01 & 01 & 02 & 03 \\ 03 & 01 & 01 & 02 \end{bmatrix} \begin{bmatrix} a_{00} & a_{01} & a_{02} & a_{03} \\ a_{10} & a_{11} & a_{12} & a_{13} \\ a_{20} & a_{21} & a_{22} & a_{23} \\ a_{30} & a_{31} & a_{32} & a_{33} \end{bmatrix} = \begin{bmatrix} b_{00} & b_{01} & b_{02} & b_{03} \\ b_{10} & b_{11} & b_{12} & b_{13} \\ b_{20} & b_{21} & b_{22} & b_{23} \\ b_{30} & b_{31} & b_{32} & b_{33} \end{bmatrix} \quad (2)$$

Inverse Mixcolumns process can also be expressed as the matrix multiplication:

$$\begin{bmatrix} 0E & 0B & 0D & 09 \\ 09 & 0E & 0B & 0D \\ 0D & 09 & 0E & 0B \\ 0E & 0D & 09 & 0E \end{bmatrix} \begin{bmatrix} b_{00} & b_{01} & b_{02} & b_{03} \\ b_{10} & b_{11} & b_{12} & b_{13} \\ b_{20} & b_{21} & b_{22} & b_{23} \\ b_{30} & b_{31} & b_{32} & b_{33} \end{bmatrix} = \begin{bmatrix} a_{00} & a_{01} & a_{02} & a_{03} \\ a_{10} & a_{11} & a_{12} & a_{10} \\ a_{20} & a_{21} & a_{22} & a_{23} \\ a_{30} & a_{31} & a_{32} & a_{33} \end{bmatrix} \quad (3)$$

It is obvious that decryption process is much more complex than encryption process. In encryption process, Mixcolumns transform needs to perform four

XOR additions and two times xtime multiplications. Inverse Mixcolumns transform in decryption process requires nine XOR additions and twelve times xtime multiplications [4]. As multiplication consumes more time and space resources, resulting in there is delay during decryption process relative to the encryption process, in practice, the process is often difficult for users to accept. In this paper, we use theorem 1 to find Mixcolumns and Inverse Mixcolumns with the simplest form.

Theorem 1. *In finite field $GF(2^8)$, if there is a linear matrix A, $A = \begin{bmatrix} a & b & c & d \\ d & a & b & c \\ c & d & a & b \\ b & c & d & a \end{bmatrix}$, $a, b, c, d \in GF(2^8)/\{0\}$, if $A^{-1} = A$, then $A = \begin{bmatrix} a & b & c & b \\ b & a & b & c \\ c & b & a & b \\ b & c & b & a \end{bmatrix}$, $a^2 + c^2 = 1$.*

It is proved that g is a generator in finite field $GF(2^8)$, α, β, γ, ρ are orders of elements a, b, c, d respectively. The following equation can be constructed by $AA^{-1} = 1$:

$$\begin{bmatrix} a & b & c & d \\ d & a & b & c \\ c & d & a & b \\ b & c & d & a \end{bmatrix} \begin{bmatrix} a & b & c & d \\ d & a & b & c \\ c & d & a & b \\ b & c & d & a \end{bmatrix} = \begin{bmatrix} 1 & 0 & 0 & 0 \\ 0 & 1 & 0 & 0 \\ 0 & 0 & 1 & 0 \\ 0 & 0 & 0 & 1 \end{bmatrix} \tag{4}$$

We can get $\begin{cases} a^2 + c^2 = 1 \\ b^2 + d^2 = 0 \end{cases}$, which is $\begin{cases} g^{2\alpha} + g^{2\gamma} = 1 \\ g^{2\beta} + g^{2\rho} = 0 \end{cases}$, since $g^{2\beta} + g^{2\rho} = 0$, $1 \leq \beta, \rho \leq 225$, then $2\beta = 2\rho$ or $2\beta = 2\rho \pm 255$, but 2β is even number, $2\beta = 2\rho \pm 255$ is invalid. So $\beta = \rho$, $b = d$. Above all, $A = \begin{bmatrix} a & b & c & b \\ b & a & b & c \\ c & b & a & b \\ b & c & b & a \end{bmatrix}$, $a^2 + c^2 = 1$. According to above theorem, if $M = \begin{bmatrix} 2 & 1 & 3 & 1 \\ 1 & 2 & 1 & 3 \\ 3 & 1 & 2 & 1 \\ 1 & 3 & 1 & 2 \end{bmatrix} = M^{-1}$, we use this matrix to replace Mixcolumns and Inverse Mixcolumns in the original algorithm, so Mixcolumns and Inverse Mixcolumns consume the same computing resources, which solves the time delay problem of decryption relative to encryption with high practical value.

3 Improvement of RSA Algorithm and Its Application in Digital Signature

Modular exponential operation of large number factorization in RSA algorithm is time-consuming, which has been restricted the development of RSA. Aiming at solving this problem, many scholars have proposed different optimization algorithms, among them, it is effective obviously that Chinese residual theorem (CRT) is efficient for decryption or signature. It has been proved that if we consider the computational cost of the Chinese remainder theorem, the operational speed of the dual primes CRT-RSA is respectively 3.32 times than the original algorithm times (modulo 1024 bits) and 3.47 times (the model is 2048 bit) [5]. Although the speed is satisfactory, there is security problem. Therefore, the original double prime numbers RSA algorithm is changed to four prime numbers, and then applied to the digital signature [6,7], the specific ideas are as follows.

3.1 Fundamentals of Four Prime RSA Algorithm

On the basis of traditional double-prime RSA cipher algorithm [8], the number of primes is taken as 4, the algorithm is still established, which is described as follows:

(a) Select four different large prime p, q, r, s randomly, calculate $n = pqrs, \varphi(n) = (p-1)(q-1)(r-1)(s-1)$.
(b) Take the encryption key e which satisfies certain conditions, calculate the private key d satisfied $de \equiv 1 \bmod \varphi(n)$.
(c) Encryption and decryption process is same as the traditional algorithm, which is, encryption algorithm: $c = E(m) = m^e \bmod n$, decryption algorithm: $m = D(c) = c^d \bmod n$.

3.2 Application of RSA Algorithm with Four Prime Numbers in Digital Signature

First of all, we introduce hash function, it refers to any length of the message is mapped to a fixed-length message of a function, generally used for message integrity testing and certification. Using the hash function and RSA algorithm with four primes, the signature process is as follows:

(a) User A sends the message M generated by hash function H to produce message digest $D = H(M)$.
(b) User A signs the message digest with private key d, $S = D^d \bmod n$.
(c) User A sends message M and signature S to user B.
(d) After receiving the message and signature, user B decrypts the signature S with public key of A, and then computes the message digest D' with the hash function to determine whether D' is equal to D. If they are same, then the message does come from A, and has not been tampered during transmission; Otherwise, it is likely that the message is not come from A and has been tampered by others.

3.3 The Montgomery Module Exponentiation Algorithm and Chinese Remainder Theorem Are Used to Optimize the Signature Process

The Montgomery Module Exponentiation Algorithm. Montgomery modular multiplication is a method of changing division operation to shift operation to simplify two-modulus multiplication algorithm. In this paper, we use an improved Montgomery multiplication CISO(coarsely integrated operand scanning) algorithm [9]. $m = (m_{n-1} \cdots m_1 m_0)_b$ is an integer, b is a binary number, $x = (x_{n-1} \cdots x_1 x_0)_b$, $y = (y_{n-1} \cdots y_1 y_0)_b$, $\gcd(m, b) = 1$, $R = b^n$, $m' = -m^{-1} \bmod b$ (m^{-1} is the multiplicative inverse of m modulo b). We define calculation $mont(u, v)$ is $uvR^{-1} \bmod m$, and then we get computing method of $x^e \bmod m$ by using above-mentioned CIOS algorithm combining exponential algorithm, x is integer, $1 \leq x < m$, $e = (e_t \cdots e_1 e_0)_2$, $e_t = 1$. The algorithm is described as follows:

(a) $x' = \mathrm{mod}(x, R^2 \bmod m)$, $A = R \bmod m$.
(b) Performing i from t to 0: $A = mont(A, A)$, and if $e_i = 1$ then $A = mont(A, x')$.
(c) $A = mont(A, 1)$.
(d) $A = x^e \bmod m$.

It is known that if only modulo multiplication is used, the Montgomery modular multiplication does not improve speed, because the algorithm uses general modular arithmetic and modular inversion in pre-processing. But if it is modular exponentiation, which means $(3 \log_{10} e)/2$ times modular multiplication, the pre-processing can be carried out only once without division of Montgomery multiplication, which greatly improve the speed of modular exponentiation operation.

Chinese Remainder Theorem. Chinese Remainder Theorem is a method of solving a congruence group and an important theorem in number theory [10]. Let r integers $m_1, m_2 \cdots, m_r$, two of which are prime, $a_1, a_2 \cdots, a_r$ is any integer r, then the modular $M = m_1 m_2 \cdots m_r$ of $x \equiv a_i \bmod m_i (1 \leq i \leq r)$ has a unique solution, the expression is $x = \sum_{i=1}^{r} a_i M_i y_i (\mathrm{mod} M)$, where $M_i = M/m_i, y_i M_i \equiv 1 \bmod m_i, 1 \leq i \leq r$.

It can be seen that Chinese Remainder Theorem can convert modular exponentiation of high-order large numbers into modulo exponentiation of relatively small numbers of lower bits.

Application of Chinese Remainder Theorem. Using Chinese Remainder Theorem, the digital signature of Message Digest D can be transformed into the following operations:

(a) Calculate $m_p = D \bmod p, m_q = D \bmod q$, $m_r = D \bmod r$, $m_s = D \bmod s$;
(b) Calculate $d_p = d \bmod (p-1)$, $d_q = d \bmod (q-1), d_r = d \bmod (r-1)$, $d_s = d \bmod (s-1)$;

(c) Calculate $M_1 = m_p^{d_p} \bmod p$, $M_2 = m_q^{d_q} \bmod q$, $M_3 = m_r^{d_r} \bmod r$, $M_4 = m_s^{ds} \bmod s$;
(d) $S = (M_1(qrs)^{p-1} + M_2(prs)^{q-1} + M_3(pqs)^{r-1} + M_4(pqr)^{s-1}) \bmod n$, then signature S is calculated.

In the above calculation, traditional signature algorithm $S = D^d \bmod n$ is transformed to solve four congruences: $S \equiv D^d \bmod p$, $S = D^d \bmod q$, $S = D^d \bmod r$ and $S = D^d \bmod s$. In this paper, we use Fermat's little theorem instead of the extended Euclidean algorithm, for any integer A that is not divisible by prime p, there are $A^{p-1} \equiv 1 mod p$, $A^{-1} \equiv A^{p-2} \bmod p$, which can be solved by a polynomial operation instead of one of the inverse elements, which further improves computational efficiency.

4 Hybrid Encryption System

In this paper, based on optimization algorithm, we propose a hybrid encryption system that transmit data securely and authenticate identically combined with the advantages of two algorithms. Information transmission process is shown in Fig. 3, the specific steps are as follows: Sender A performs hash function operation on original message M to obtain a message digest D. Then, Sender A adapts its own private key PVA to signature message digest D using RSA algorithm to obtain S, encrypts original message M and digital signature S with AES algorithm. Encrypted information E is obtained. After that, A uses public key PBB of receiver B and RSA algorithm to encrypt AES key SK to form digital envelope DE, which like symmetric key SK is loaded into a envelope with public key encryption of recipient. At last, A sends encrypted information E with digital envelope DE to receiver B.

Receiver B receives digital envelope DE, firstly he removes symmetric key SK with its own private key PVB and decrypts digital envelope. Then, receiver B decrypts encrypted information E with symmetric key SK and AES algorithm, original message M and digital signature S are obtained. B verifies digital signature, firstly B uses sender's public key PBA to decrypt digital signature S to obtain the message digest D. After that, B uses original information with

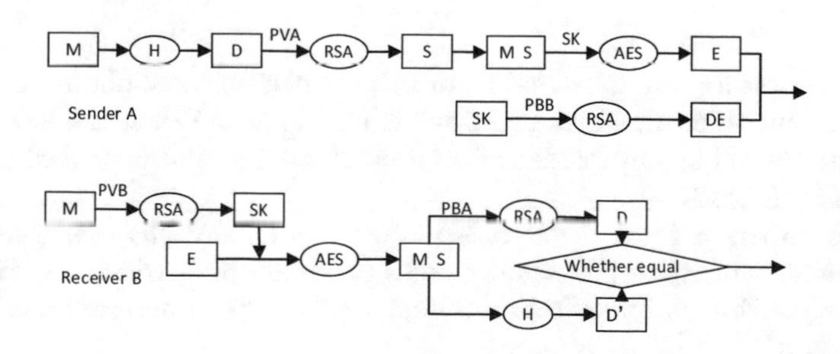

Fig. 3. AES and RSA hybrid encryption system.

hash function to calculate a new message digest D ' and compares two message digests, if they are equal, then data has not been tampered, it is confidential transmission, the signature is true; if they are not equal, data has been tampered, information is not confidential transmission, B refuse signature.

5 Experimental Results and Comparative Analysis

This paper uses 10 rounds of AES algorithm in order to take advantages of AES efficiency, we use modulus of 2048bit RSA algorithm taking into account the security. For hashing function, because MD5 and SHA1 have been cracked, this paper uses SHA512 [11]. The experimental operating system is Windows7; development tool is Visual C++ 6.0.

5.1 AES Diffusion Confusion Test

Diffusion and confusion are two basic ways that Shannon has proposed to design a cryptosystem to counter adversary's statistical analysis. In this paper, during encryption and decryption process, we use same matrix, which will affect original algorithm of diffusion and confusion characteristics, this article illustrate characteristics through experiments.

Firstly, we will test the spread of 128 bit string AES encryption to ensure that key is unchanged and record the number of ciphertext bits when plaintext changes one bit, due to space reasons, here are only three changes in plaintext test results. When plaintext changes, changes caused by ciphertext of original algorithm and improved algorithm are shown in Table 1.

Table 1. The number of ciphertext bits changes when key is unchanged

Plaintext changes	Original algorithm	Improved algorithm
1 bit	65 ± 5 bits	64 ± 7 bits
2 bits	63 ± 7 bits	63 ± 5 bits
3 bits	63 ± 7 bits	64 ± 5 bits

Then, we test its confusion and record the impact of the ciphertext when key changes one bit to ensure that plaintext is unchanged. When the key changes, the changes caused by cipher text of original algorithm and improved algorithm are shown in Table 2.

In this paper, a total of 30 bits plaintext and key changes were tested, and the number of ciphertext changes was about 64 bits, which indicated that improved algorithm had no effect on diffusion aliasing characteristics of the original algorithm.

Table 2. The number of ciphertext bits changes when plaintext is unchanged

Key changes	Original algorithm	Improved algorithm
1 bit	63 ± 7 bits	63 ± 5 bits
2 bits	64 ± 7 bits	64 ± 7 bits
3 bits	64 ± 5 bits	63 ± 7 bits

5.2 AES Encryption and Decryption Rate Test

In this paper, we encrypt and decrypt 10000 times on five groups of 128 bit string in order to make sure experimental results more obvious, and record the time-consuming situation, Table 3 shows that the acceleration of improved AES algorithm is increased by 22%, decryption rate can be increased by 26% compared to the original algorithm, encryption and decryption time-consuming is equal. It shows that improved AES algorithm has some advantages over the original one, and it solves the problem of delay in decryption and encryption in original algorithm.

Table 3. Time-consuming of encryption and decryption

Time/ms	Original algorithm		Improved AES algorithm	
	Encrypt	Decrypt	Encrypt	Decrypt
The first	470	491	348	351
The second	450	501	340	341
The third	461	490	351	353
The fourth	471	500	358	361
The fifth	460	481	349	351
Average	462.7	486.3	355.2	358.4

5.3 RSA Signature Rate Test

In this paper, based on C++, we build a complete large number libraries and use unsigned long integer to store large numbers array from low bit to high bit, that is, a large number is represented as 2^{32}, in this way can we minimize the basic number of cycles in operation and improve operational efficiency. Then, we use improved Miller-Rabin algorithm [12], two 1024-bit primes and four 512-bit prime numbers are randomly generated. Traditional double-prime RSA algorithm, double prime RSA blend CRT and four primes CRT-RSA algorithm is used separately to sign three 512-bit message digests and records their time-consuming behavior. The experimental results are shown in Table 4.

From the table, we know that the signature efficiency of dual-prime number RSA optimize with CRT is 3.36 times compared with traditional signature algorithm, which is close to theoretical value 3.47. And four prime RSA signature

Table 4. Time-consuming of three signature algorithm

Time/ms	Traditional double-prime RSA signature	Double-prime RSA signature	Four primes CRT-RSA signature
The first	3432	996	311
The second	3621	1107	388
The third	3568	1059	329
Average	3541	1054	326

algorithm with CRT and Montgomery modular exponential algorithm is about 10.86 times as efficient as traditional algorithm, which is 3.23 times compared with double-prime CRT-RSA. It shows that improved RSA signature can greatly improve the signature speed.

6 Conclusion

This paper mainly studies popular symmetric encryption algorithm AES and asymmetric encryption algorithm RSA, we analyzes its research status and improves the shortcomings, and proves that improved algorithm does have some advantages through experiments. And then combined with respective advantages, we build a hybrid encryption system, next phase we will study the application of hybrid encryption system in specific devices.

Acknowledgment. Project supported by the National Natural Science Foundation of China (Grant Nos. 61471158), and Project supported by the Modern sensor technology of Universities in Heilongjiang (Grant No. 2012TD007).

References

1. Wenling, W., Dengguo, F.: Research status of block cipher mode. Chin. J. Comput. **29**, 22–25 (2006)
2. Qiuyu, Z., Pengfei, X., Yibo, H., Ruihong, D.: An efficient speech perceptual hashing authentication algorithm based on wavelet packet decomposition. J. Inf. Hiding Multimed. Sig. Process. **6**, 311–322 (2015)
3. Guihua, C., Xuemei, Q., Yonglong, L.: Polynomial modulo operation in AES algorithm and its performance analysis. Comput. Technol. Dev. **20**, 115–118 (2010)
4. Wiener, M.J.: Cryptoanalysis of short RSA secret exponents. IEEE Inf. Theory Soc. **36**, 553–558 (1990)
5. Boneh, D., Durffe, G.: Cryptanalysis of RSA with private key d less than $N^{0.292}$. IEEE Inf. Theory Soc. **46**, 1339–1349 (2000)
6. Keying, H.: Research on Improved RSA Algorithm. University of Electronic Science and Technology, Chengdu (2010)
7. Yang, B.: Modern Cryptography. Tsinghua University Press, Beijing (2007)

8. Xiaofei, F., Huanying, H.: Security analysis of CRT - RSA algorithm. Microcomput. Inf. **25**, 36–38 (2009)
9. An, W.: A Fast Implementation of RSA Public Key Cryptography. Shandong University, Jinan (2008)
10. Gongliang, C.: Information Security Mathematical Basis. Tsinghua University Press, Beijing (2011)
11. Stalling, W.: Cryprography and Network Security Principles and Practice. Publishing House of Eletronics Industry, Beijing (2011)
12. Couveigne, J.M., Ezome, T., Lercier, R.: A faster pseudo-primality test. Rendiconti del Circolo Matematico di Palermo **61**, 261–278 (2012)

Fast Coding Unit Depth Decision for HEVC Intra Coding

Yueying Wu[1,2,3], Pengyu Liu[1,2,3](✉), Zeqi Feng[1,2,3], and Kebin Jia[1,2,3](✉)

[1] Beijing Advanced Innovation Center for Future Internet Technology, Beijing University of Technology, Beijing 100124, China
{liupengyu, kebinj}@bjut.edu.cn

[2] College of Electronic Information and Control Engineering, Beijing University of Technology, Beijing, China

[3] Beijing Laboratory of Advanced Information Networks, Beijing University of Technology, Beijing, China

Abstract. To reduce the complexity of High Efficiency Video Coding (HEVC), this paper focuses on innovative works for fast coding tree unit (CTU) depth decision in HEVC intra coding based on a proposed CTU depth range possibility mechanism. First, the depth correlation factor (DCF) is computed by analyzing the temporal-spatial correlation of the co-located CTU and its spatial adjacent CTUs in previous frame. Second, the most possible depth range (MPDR) of current CTU is predicted according to DCF and the known coding depth information of neighboring CTUs in current frame. Experimental results show that the proposed method can achieve 34.06% coding time saving on average with 1.36% BDBR increase and 0.12 dB BDPSNR decrease compared with HM15.0. The proposed method is expected to be applied in the real-time environments.

Keywords: HEVC · Intra coding · CTU · Temporal-spatial correlation · MPDR

1 Introduction

Motivated by the demand of improving the compression performance for high definition (HD) and ultra-high definition (UHD) videos, the emerging high efficiency video coding (HEVC) standard [1] was established by the Joint Collaborative Team on Video Coding (JCT-VC). HEVC has shown significant advances in compression efficiency and outperforms the existing H.264/AVC standard [2] by 50% coding bit-rate reduction with equivalent visual quality [3]. The flexible quad-tree coding structure is the main reason for the improvement in HEVC coding performance [4]. HEVC broke 16×16 sized macroblock (MB) coding structure in H.264/AVC and adopted the quad-tree structured CTU depth traversal strategy, which allowed each CTU recursively splitting into four coding unit (CU) with depths from 0 to 3. Flexible quad-tree coding structure of HEVC is drawn in Fig. 1.

J.-S. Pan et al. (eds.), *Advances in Intelligent Information Hiding and Multimedia Signal Processing*, Smart Innovation, Systems and Technologies 82,
DOI 10.1007/978-3-319-63859-1_49

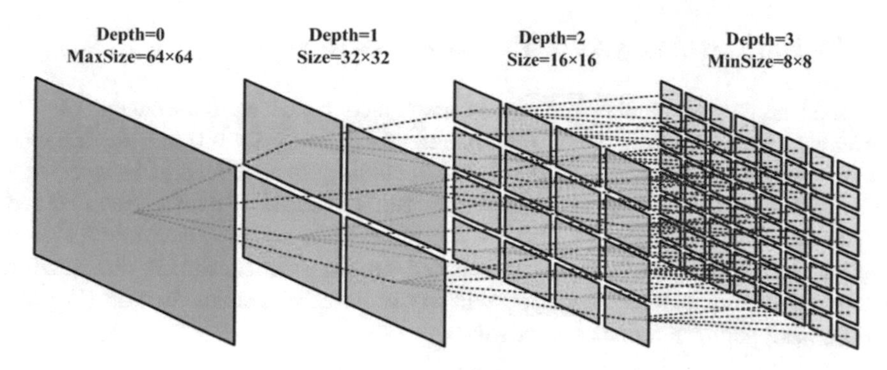

Fig. 1. The quad-tree coding structure in HEVC

However, the performance improvement is at the expense of the increased coding complexity. Especially in intra coding, the complexity caused by CTU depth range traversal dramatically accounted for about 90% of the total computation cost. Therefore, as a relatively time consuming part of HEVC encoder, computational complexity reduction with negligible performance loss in CTU depth decision has been deeply researched.

At present, research on improving HEVC efficiency by reducing quad-tree complexity can be divided into three categories. First, many works have been focused on low complexity CTU depth range decision method based on rate-distortion (RD) cost [5–7]. Kim [5] and Ma [6] utilized the early termination strategy in CTU splitting by comparing the RD cost with its corresponding thresholds. Cen [7] skipped small sized CU by analyzing the RD cost in neighbor CUs. However, the coding time saving might decrease when the to-be-encoded image contained abundant detail information. Second, some fast CTU depth decision methods were proposed based on the texture complexity [8, 9]. As stated in [8, 9], some unnecessary CTU depth candidates were removed to reduce the coding burden by analyzing the texture complexity. However, the computation of texture features will bring more complexity. Finally, spatial correlation between adjacent CTUs also provided a support for low complexity CTU depth decision algorithm [10, 11]. Cheng [10] and Shen [11] utilized the depth of spatial adjacent CTUs to predict the optimal depth of current CU. However, the above algorithms lacked the comprehensive utilization of temporal-spatial correlation.

Thus, to overcome the above difficulties, a CTU depth range possibility mechanism is proposed in this paper. First, depth correlation factor (DCF) is established to measure the temporal-spatial correlation between current CTU and its neighboring CTUs. Second, a concept of most possible depth range (MPDR) is designed for predicting the possible depth range of current CTU.

The rest of the paper is organized as follows. Section 2 describes the motivation for the proposed method. Section 3 provides the proposed method in detail. Experiment results are stated and discussed in Sect. 4. Finally, Sect. 5 draws the conclusion.

2 Motivations Analysis of Proposed Method

As stated above, the coding process is performed based on the concept of CTU. However, HEVC adopts the fixed CTU depth search range from 0 to 3 for the whole video sequences. But actually, larger depths including 2 and 3 tend to be selected for CTUs with abundant textures and complicated motions. Smaller depths such as 0 and 1 are selected for CTUs in homogeneous and motionless region. Thus the CTU depth range should be adaptively determined based on the characteristic of the quad-tree structure. The computation complexity will be decreased if some rarely used CU depths could be skipped or terminated in advance.

Motivation 1: Early skipping or terminating unnecessary CU depths can decrease coding complexity and save encoding time.

As shown in Fig. 2, the CTU depth range can be ultimately divided into four types, including [0,1], [1,2], [1,3] and [2,3]. Encoding time saving is calculated when current CTU depth range is conditioned to the above 4 types in Table 1. It can be observed that the encoding time saving can achieve up to 60.90% when current CTU adopted depth range [0,1].

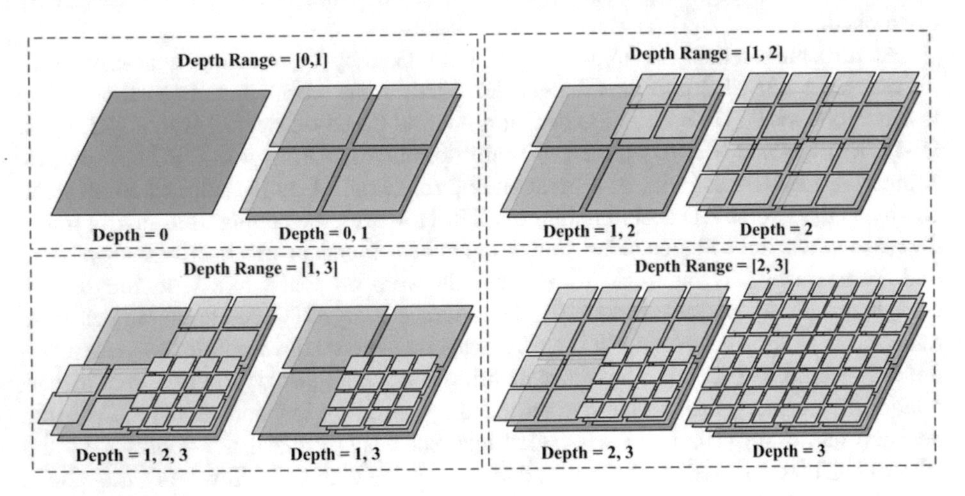

Fig. 2. Four types of possible depth range and its corresponding quad-tree structure

Further, how to early decide the most possible depth range (MPDR) of current CTU become the key to the fast depth decision method.

Motivation 2: Optimal depth range of current CTU has a strong correlation with its spatial adjacent CTUs in current frame and its co-located CTU in previous frame. Figure 3 shows the diagram of current CTU and its spatial adjacent CTUs.

As everyone knows, temporal and spatial correlation is the essential to remove the data redundancy. And such correlation also exists in the CTU depth decision process. The correlation between current CTU, its neighboring CTUs and its co-located CTU is demonstrated by the experimental results in Table 2.

Table 1. Encoding time saving under different depth range (DR)

Class \ DR	[0,1]	[1,2]	[1,3]	[2,3]
Class A	60.51%	49.88%	6.56%	20.96%
Class B	60.90%	51.38%	10.05%	18.64%
Class C	59.32%	48.70%	10.00%	22.60%
Class D	60.50%	49.36%	8.94%	19.48%
Average	60.31%	49.83%	8.89%	20.42%

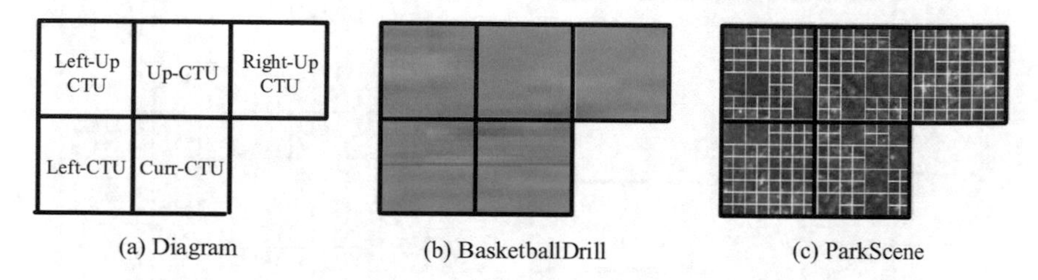

(a) Diagram (b) BasketballDrill (c) ParkScene

Fig. 3. Diagram of current CTU and its spatial adjacent CTUs

Table 2. DR comparison between current CTU and its temporal-spatial adjacent CTUs

Class	Probability				
	Co-located CTU	Up CTU	Left CTU	Left-Up CTU	Right-Up CTU
Class A	96.64%	94.52%	96.32%	91.21%	91.55%
Class B	95.05%	91.68%	94.88%	87.49%	88.10%
Class C	91.14%	84.65%	90.53%	77.16%	77.13%
Class D	82.74%	65.11%	80.80%	52.10%	53.86%
Average	91.69%	83.99%	90.63%	76.99%	77.66%

As stated above, optimal depth range of current CTU has strong correlation with its temporal-spatial adjacent CTUs. Thus, the depth range of current CTU can be early predicted by its temporal and spatial adjacent CTU. And it is noteworthy that there are differences among the correlation degree. But in traditional fast CTU depth decision method, the correlation was expressed by fixed weight. Hence, the correlation weight will be analyzed to solve the problem on correlation description.

3 Fast CTU Depth Range Decision Method

3.1 Depth Correlation Factor (DCF) Establishment

As stated above, the temporal-spatial adjacent CTUs have significant correlation in depth decision, which can be used as a foundation to evaluate the correlation weights.

Therefore, to accurately describe the correlation between current CTU and its spatial adjacent CTUs, an adaptive DCF is defined according to the temporal-spatial correlation. The specific coding process is as follows:

Step 1. Recording CTU depth partition information
In HEVC, the size of CTUs perform as 64 × 64. Therefore, the CTUs are divided into 256 4 × 4 units and the depth partition information of each CTU is stored with a 16 × 16 sized matrix Ω. Then, the matrix Ω_{Corr}, Ω_{Up}, Ω_{Left}, $\Omega_{Left\text{-}Up}$ and $\Omega_{Right\text{-}Up}$ are used to record the depth partition information of co-located CTU and its spatial adjacent CTUs in previous frame respectively. Figure 4 shows an example of matrix Ω.

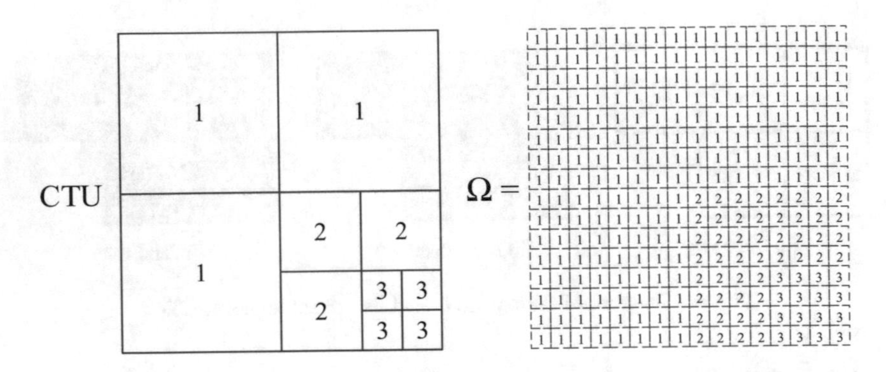

Fig. 4. An example of matrix Ω

Step 2. Calculating depth correlation
Supposing matrix $\Omega_{Corr} = (x_{ij})_{16 \times 16}$ stores depth partition information of co-located CTU and matrix $\Omega_{*} = (y_{*ij})_{16 \times 16}$ stores depth information of its corresponding spatial adjacent CTUs where $* \in \{Up, Left, Left\text{-}Up, Right\text{-}Up\}$. 16 × 16 represents the element numbers of matrix Ω and max depth minus minimum depth equals to 3. Then, the correlation factor $\boldsymbol{r}_{*}$ between Ω_{Corr} and Ω_{*} is defined and calculated by formula (1).

$$r* = 1 - \frac{1}{3 \times 16 \times 16} \sum_{i=1}^{16} \sum_{j=1}^{16} \left| x_{ij} - y_{*ij} \right| \tag{1}$$

Step 3. Establishing depth correlation factor (DCF)
From the perspective of the temporal correlation in CTU depth decision, the DCF $\boldsymbol{R}_{*}$ between current CTU and its spatial adjacent CTUs can be approximatively expressed based on $\boldsymbol{r}_{*}$ which is calculated and obtained from its previous frame.

$$R* = r* \tag{2}$$

So far, the establishment process of DCF for current CTU is completed.

3.2 Most Possible Depth Range (MPDR) Dynamic Prediction

For reducing the encoding complexity, the calculated DCF and the known depth information are used to guide the dynamic depth range prediction for current CTU.

Step 1. Calculating depth correlation weight
Depth correlation weight $\boldsymbol{W}_*$ is obtained by normalizing the DCF according to the formula (3).

$$W* = \frac{R*}{R_{Up} + R_{Left} + R_{Left\text{-}Up} + R_{Right\text{-}Up}} \tag{3}$$

Step 2. Predicting possible initial depth
Smallest depth $\boldsymbol{S}_*$ of four spatial adjacent CTUs in current frame and its corresponding depth correlation weight $\boldsymbol{W}_*$ are utilized to predict the initial depth $Depth_{initial}$ of current CTU.

$$Depth_{initial} = \left\lfloor \sum\nolimits_{*}^{Up,Left,Left-Up,Right-Up} S * \times W* \right\rfloor \tag{4}$$

Where $\lfloor \rfloor$ represents the down integral.

Step 3. Predicting possible terminal depth
Similarly, largest depth $\boldsymbol{L}_*$ of four spatial adjacent CTUs in current frame and its corresponding depth correlation weight $\boldsymbol{W}_*$ are utilized to predict the terminal depth $Depth_{terminal}$ of current CTU.

$$Depth_{terminal} = \left\lceil \sum\nolimits_{*}^{Up,Left,Left-Up,Right-Up} L * \times W* \right\rceil \tag{5}$$

Where $\lceil \rceil$ represents the top integral.

Finally, the dynamic MPDR is determined by the interval of $[Depth_{initial}, Depth_{terminal}]$.

So far, the CTU depth range dynamic prediction process can adaptively decide to early skip CU splitting before the encoder traversing a complete quad-tree structure. As the result, the coding complexity is reduced.

4 Experimental Results and Performance Discussion

In order to verify the performance of the proposed fast CTU depth decision method, extensive video sequences are introduced to the experiments. The resolution of the test sequences ranges from 416 × 240 to 2560 × 1600. And the specific information of the sequences is provided in Table 3.

Table 3. HEVC test sequences

Class	Resolution	Sequence	Frame rate
A	2560 × 1600	Traffic, PeopleOnStreet	30/30 fps
B	1920 × 1080	Kimono, BQTerrace, BasketballDrive	24/60/50 fps
C	832 × 480	RaceHorses, BQMall, BasketballDrill	30/60/50 fps
D	416 × 240	BQSquare, BasketballPass	60/50 fps

Gain or loss is measured with respect to HM15.0 under intra coding mode. BDBR and BDPSNR [12] are introduced to evaluate the coding performance in coding bit-rate and video quality respectively. The coding time saving was computed by formula (6). And experimental results are showed in Table 4.

$$\Delta T(\%) = \frac{EncodingTime_{Proposed} - EncodingTime_{HM15.0}}{EncodingTime_{HM15.0}} \quad (6)$$

As shown in Table 4, the average coding time reduction is 22.46%, BDPSNR decreased 0.09 dB on average with a slight BDBR increase of 1.16% in Ref. [11] compared with HM15.0. In Ref. [7], average coding time saving is 27.22% with 1.58% BDBR increase and 0.09 dB BDPSNR decrease. Our proposed method based on dynamic MPDR can achieve a coding time saving of 34.06% compared with HM15.0. It can be observed that our proposed method further reduces the coding complexity by saving more time compare with Refs. [11] and [7], with only 0.20% increase in BDBR and 0.03 dB decease in BDPSNR. Average BDBR increase of our method is even smaller than that of Ref. [7].

Table 4. Result of the proposed algorithm compared to Refs. [11] and [7]

Class	Proposed			Ref. [11]			Ref. [7]		
	ΔBR (%)	$\Delta PSNR$ (dB)	ΔT (%)	ΔBR (%)	$\Delta PSNR$ (dB)	ΔT (%)	ΔBR (%)	$\Delta PSNR$ (dB)	ΔT (%)
A	0.95	−0.14	−37.74	0.82	−0.10	−22.36	1.28	−0.21	−26.37
	1.64	−0.18	−32.61	1.53	−0.12	−20.30	3.01	−0.11	−29.01
B	1.27	−0.05	−42.39	1.03	−0.03	−33.87	2.26	−0.07	−32.35
	1.02	−0.16	−35.63	0.84	−0.13	−25.55	1.05	−0.12	−30.42
	1.54	−0.07	−42.56	1.26	−0.06	−28.08	2.33	−0.20	−38.28
C	1.35	−0.12	−31.95	1.14	−0.08	−21.37	1.53	−0.01	−24.33
	1.08	−0.15	−30.18	0.96	−0.12	−18.28	0.89	−0.05	−20.25
	1.70	−0.08	−34.71	1.53	−0.06	−22.30	1.36	−0.08	−31.12
D	1.34	−0.09	−22.82	1.03	−0.08	−15.90	0.21	−0.02	−17.75
	1.76	−0.13	−30.05	1.48	−0.08	−16.62	1.26	−0.06	−22.28
Avg	1.36	−0.12	−34.06	1.16	−0.09	−22.46	1.58	−0.09	−27.22

Performance improvement can be observed according to Table 4: (1) Coding time can be further reduced by 11.60% and 6.85% respectively compared with Refs. [11] and [7]. It is mainly due to that the unnecessary CU depths traversal procedure is skipped by early prediction mechanism according to the temporal and spatial correlation in CTU depth decision. (2) Proposed method can achieve 34.06% encoding time saving on average for different video sequences with almost unchanged BDBR and BDPSNR.

For much intuitively illustrating, examples of quad-tree partition results of HM15.0 and the proposed method are provided in Fig. 5 (Example sequences: BQSquare and BasketballDrill with QP = 22). Thereinto, differences of CTU partition results are expressed by the red regions. As shown in Fig. 5, the red regions only occupy small part of the entire image, and the decision of CTU depth still has a higher similarity. Hence, it can illustrate that the subjective video coding quality maintains at a good level in the proposed method.

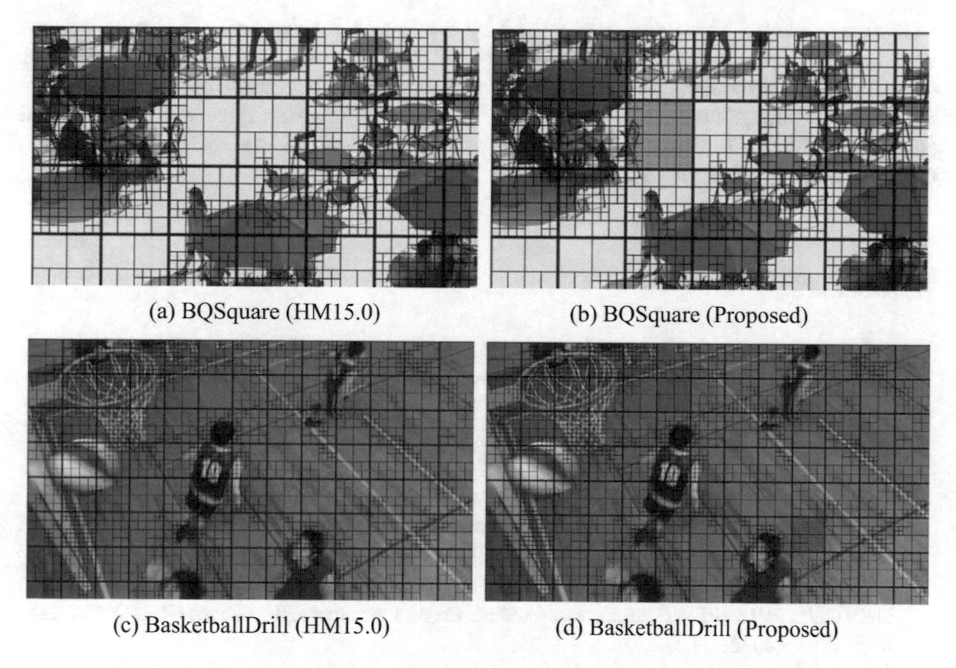

(a) BQSquare (HM15.0) (b) BQSquare (Proposed)

(c) BasketballDrill (HM15.0) (d) BasketballDrill (Proposed)

Fig. 5. Examples of quad-tree partition results

Generally speaking, the CTU depth range probability mechanism successfully reduces CTU coding complexity by early predicting the MPDR for each frame which avoid complex computing. That makes HEVC more convenient to generalize in pervasive applications.

5 Conclusions

To reduce the coding complexity of HEVC encoder, a fast CTU depth decision algorithm for intra coding based on CTU depth range possibility mechanism is proposed in this paper. First, the temporal-spatial correlation in CTU depth decision is analyzed for establishing the depth correlation factor (DCF). Second, most possible depth range (MPDR) is predicted based on the computed DCF and the known depth partition information. Therefore, some unnecessary CTU depths are excluded for the RD cost computation so as to cut down the coding complexity derived from CTU depth traversal. Experimental results show that the proposed method can achieve 34.06% encoding time saving on average with 0.12 BDPSNR decrease and 1.36 BDBR increase. The proposed method can satisfy the requirements of real-time coding application such as video conference and remote surveillance.

Acknowledgments. This paper is supported by the Project for the National Natural Science Foundation of China under Grant No. 61672064, the Beijing Natural Science Foundation under Grant No. KZ201610005007, 4172001, the China Postdoctoral Science Foundation under Grants No. 2015M580029, 2016T90022, the Beijing Postdoctoral Research Foundation under Grant No. 2015ZZ-23 and Project supported by Chaoyang District of Beijing Postdoctoral Research Foundation under Grant No. 2016ZZ-01-15.

References

1. Sullivan, G.J., Ohm, J.R., Han, W.J., et al.: Overview of the high efficiency video coding (HEVC) standard. IEEE Trans. Circuits Syst. Video Technol. **22**(12), 1649–1668 (2012)
2. Wiegand, T., Sullivan, G.J., Bjøntegaard, G., et al.: Overview of the H.264/AVC video coding standard. IEEE Trans. Circuits Syst. Video Technol. **13**(7), 560–576 (2003)
3. Li, B., Sullivan, G.J., Xu, J.: Comparison of compression performance of HEVC draft 7 with AVC high profile. In: JCTVC-J0236 r1, 10th JCT-VC Meeting (2012)
4. Kim, I.K., Min, J., Lee, T., et al.: Block partitioning structure in the HEVC standard. IEEE Trans. Circuits Syst. Video Technol. **22**(12), 1697–1706 (2012)
5. Kim, J., Choe, Y., Kim, Y.G.: Fast coding unit size decision algorithm for intra coding in HEVC. In: 2013 IEEE International Conference on Consumer Electronics (ICCE), pp. 637–638. IEEE (2013)
6. Ma, S., Wang, S., Wang, S., et al.: Low complexity rate distortion optimization for HEVC. In: Data Compression Conference, pp. 73–82 (2013)
7. Cen, Y.F., Wang, W.L., Yao, X.W.: A fast CU depth decision mechanism for HEVC. Inf. Process. Lett. **115**(9), 719–724 (2015)
8. Hou, J., Li, D., Li, Z., et al.: Fast CU size decision based on texture complexity for HEVC intra coding. In: Proceedings 2013 International Conference on Mechatronic Sciences, Electric Engineering and Computer (MEC), pp. 1096–1099. IEEE (2013)
9. Shen, Y., Zhang, S., Yang, C.: Image texture based fast CU size selection algorithm for HEVC intra coding. In: 2014 IEEE International Conference on Signal Processing, Communications and Computing (ICSPCC), pp. 363–367. IEEE (2014)

10. Cheng, Y., Teng, G., Shi, X., et al.: A fast intra prediction algorithm for HEVC. In: Advances on Digital Television and Wireless Multimedia Communications, pp. 292–298. Springer, Heidelberg (2012)
11. Shen, L., Zhang, Z., An, P.: Fast CU size decision and mode decision algorithm for HEVC intra coding. IEEE Trans. Consum. Electron. **59**(1), 207–213 (2013)
12. Bjontegaard, G.: Calcuation of average PSNR differences between RD-curves. Doc. VCEG-M33 ITU-T Q6/16, Austin, TX, USA, 2–4 April 2001

A New Approach of Shape Recognition with Affine Invariance Based on HSC

Yan Zheng, Baolong Guo([✉]), and Yunyi Yan

Institute of Intelligent Control and Image Engineering, Xidian University, Xian 710071, China
yanzheng_xidian@qq.com, blguo@xidian.edu.cn

Abstract. This paper proposes an approach for shape recognition. Shape recognition plays an important role in intelligent system and a lot of methods are produced recent years. HSC is a state-of-art method among them. But it can hardly deal with affine shapes. A novel approach based on HSC is proposed in this paper. This approach proposes an objective function with the affine state as parameters. When this objective function gets the maximum value, a shape is normalized, so the result of recognizing cannot be affected. The experimental result shows that this proposed approach is invariant to affine transformation.

Keywords: Shape recognition · Affine normalization · Affine shape

1 Introduction

Shape recognition plays an important role in intelligent system [1], such as object recognition [2] and target tracking [3]. Before recognizing shape, the interest region should be segmented by some effective algorithms of image segmentation such as Active Contour Models [4] and Level Set Method [5]. Then the recognition method recognizes what is the object in the interest region.

Shape Context [7,9] present a method to measure similarity between shapes. Shape Context is an effective approach. Then some methods based on shape context were proposed one after another, such as a visual shape descriptor using sectors and shape context of contour lines [8] and a modified shape context method for shape based object retrieval [6]. Recently HSC was proposed by Wang B.

2 Related Work

Due to the different viewing angles, the shapes of the same object may look very different, which is common in nature. To a certain extent this difference can be

This work was supported by the National Natural Science Foundation of China under Grants No. 61571346.

J.-S. Pan et al. (eds.), *Advances in Intelligent Information Hiding and Multimedia Signal Processing*, Smart Innovation, Systems and Technologies 82,
DOI 10.1007/978-3-319-63859-1_50

expressed as an affine transformation. Affine transformation will greatly reduce the accuracy of recognition. So affine invariance is so important for recognition methods. HSC [10] is a state-of-art approach in shape recognition, but it has poor robustness on affine transformation. The similarity which is computed by HSC between original shape and affine shapes is very low. In Fig. 1, (b) is an affine transformed shape of (a), nevertheless HSC cannot determine they are in the same class.

So this paper presents a novel approach of shape recognition with affine invariance based on HSC. This approach proposes an objective function with the affine state as parameters. Then a method of optimization is used to obtain the maximum of the objective function. At this time the special state of affine shape is uniquely determined. Therefore different affine states of shape can be normalized. The remainder of this paper is organized as follows: Sect. 2 introduces the new approach. Section 3 presents the experiments. The conclusions are in Sect. 4.

3 Affine Normalization Based on Optimization

The proposed approach define an objective function with the affine state as parameters.

$$\max(\mathrm{F}(\mathrm{t},\alpha)), \mathrm{t} \in [0,+\infty), \alpha \in [0,\pi] \tag{1}$$

where $\mathrm{F}\,(t,\alpha) = \mathrm{A}(\text{shape},\mathrm{t},\alpha)/\mathrm{C}^2(\text{shape},\mathrm{t},\alpha)$, $\mathrm{A}(\text{shape},\mathrm{t},\alpha)$ and $\mathrm{C}(\text{shape},\mathrm{t},\alpha)$ equal to the area and circumference of the shape in a state of t and α. A shape can be represented as $\mathbf{B} = \left[\, b_x \; b_y \,\right]$, where $\mathbf{B}$ is the points on the contour of the shape. The contour of the shape through affine transformation is

$$\mathbf{B}' = k\left[\, b_x \; b_y \,\right]\left(\begin{bmatrix} \cos(\alpha) & -\sin(\alpha) \\ \sin(\alpha) & \cos(\alpha) \end{bmatrix}\begin{bmatrix} t & 0 \\ 0 & 1 \end{bmatrix}\begin{bmatrix} \cos(\beta-\alpha) & -\sin(\beta-\alpha) \\ \sin(\beta-\alpha) & \cos(\beta-\alpha) \end{bmatrix}\right) \tag{2}$$

where k is the scaling factor, t and α are deformation factor and β is the rotation factor. As only the deformation is considered, t and α are free. To simplify the problem k is set as 1 and β is set as 0. The shape in $\mathrm{F}\,(t,\alpha)$ is surrounded by the contour points $\mathbf{B}'$.

A method of optimization should be used to obtain the maximum of F, such as an evolutionary algorithm. When the maximum of F is obtained, the values of α^* and t^* at this time are the A particular state.

$$\mathbf{B}' = \left[\, b_x \; b_y \,\right]\left(\begin{bmatrix} \cos(\alpha) & -\sin(\alpha) \\ \sin(\alpha) & \cos(\alpha) \end{bmatrix}\begin{bmatrix} t & 0 \\ 0 & 1 \end{bmatrix}\begin{bmatrix} \cos(-\alpha) & -\sin(-\alpha) \\ \sin(-\alpha) & \cos(-\alpha) \end{bmatrix}\right) \tag{3}$$

The shape surrounded by the contour points $\mathbf{B}''$ is the affine normalized shape. All shapes which are affine transformed by the same shape can be normalized to the shapes which are different from each other only on rotation and scaling. However many recognition methods have good robustness on rotation and scaling.

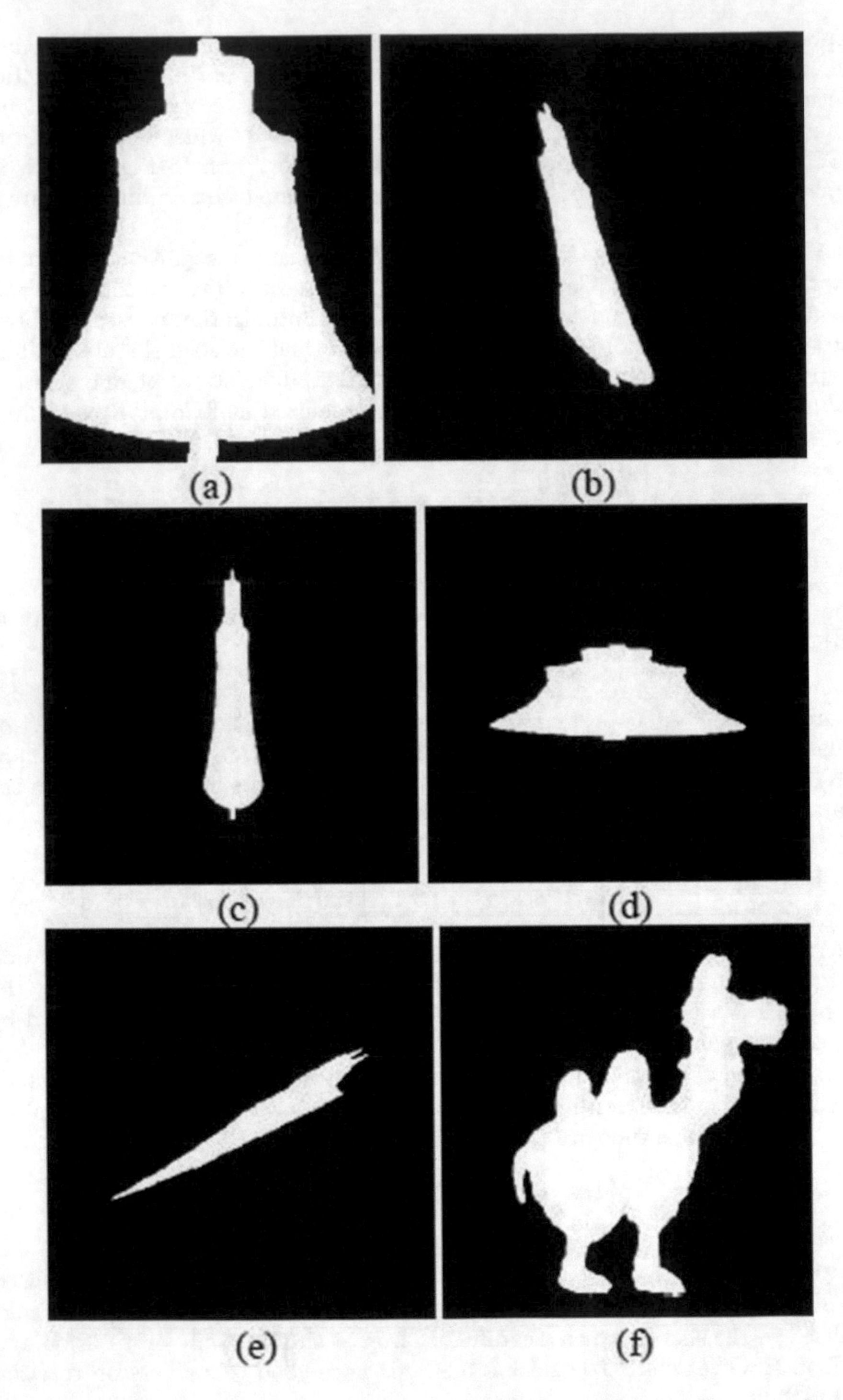

Fig. 1. (a) is the shape of an object, and (b)–(e) are randomly affine transformed shapes of (a). (f) is a shape in another class.

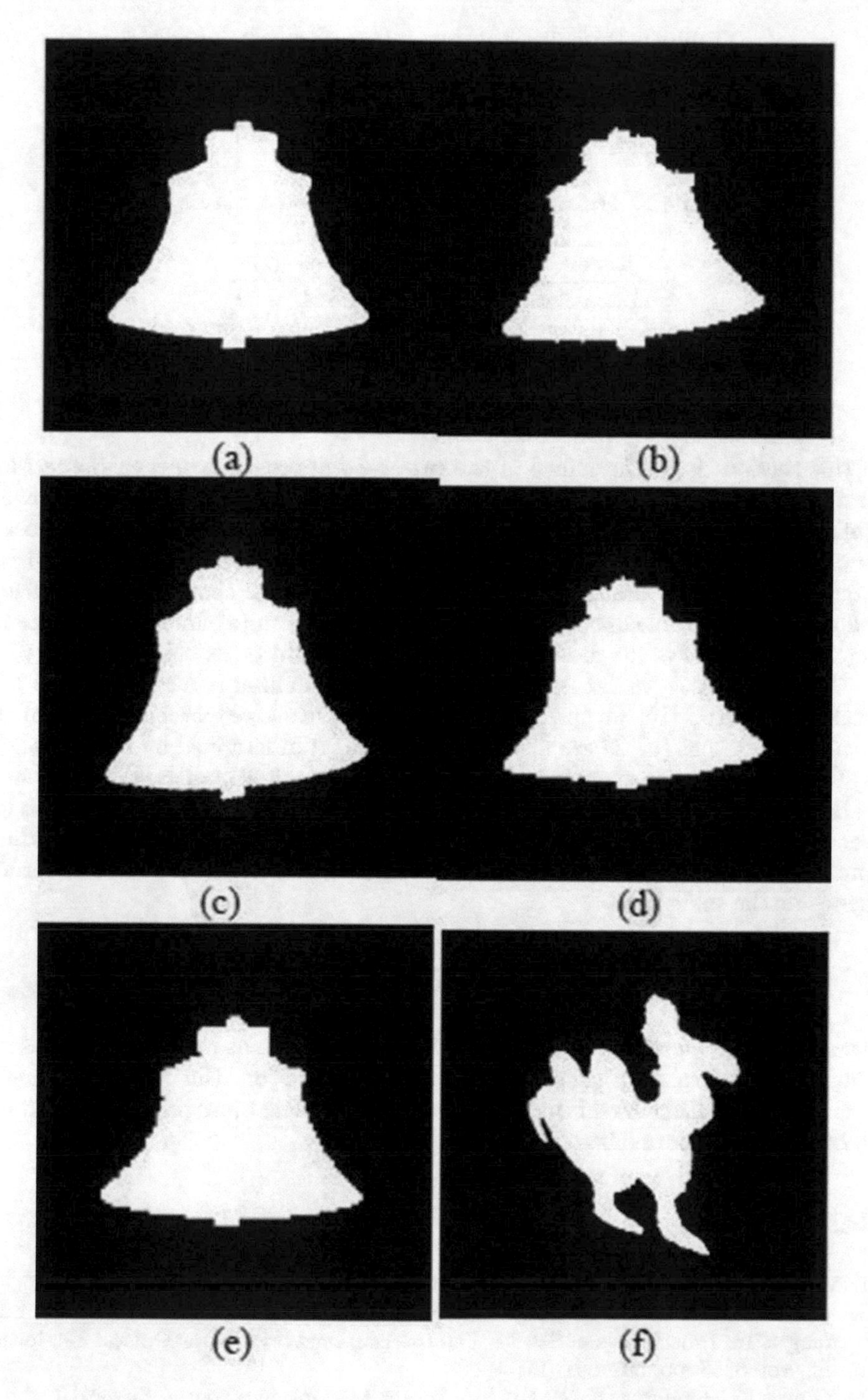

Fig. 2. (a)–(f) are the normalized shapes of Fig. 1(a)–(f) respectively.

Table 1. Dissimilarity between (a) and others in Fig. 1

Shape	(b)	(c)	(d)	(e)	(f)
Dissimilarity	0.73	0.64	0.64	1	0.67

Table 2. Dissimilarity between (a) and others in Fig. 2

Shape	(b)	(c)	(d)	(e)	(f)
Dissimilarity	0.15	0	0.16	24	0.59

4 Experimental Results

In this section the performance of the proposed approach is shown. Some randomly affine transformed shapes of the same object are processed by Affine Normalization based on Optimization. The result shows that deformation caused by affine transformation can be eliminated by the proposed approach. Figure 1 shows the randomly affine transformed shapes. Figure 2 shows the normalized shapes of Fig. 1. Table 1 shows the dissimilarity between (a) and other shapes in Fig. 1. Table 2 shows the dissimilarity between (a) and other shapes in Fig. 2.

The dissimilarity in Tables 1 and 2 which is normalized to a range of 0 to 1 is measured by HSC [10], an effective approach used in shape matching. In Table 1, it can be seen that (b) (f) are all dissimilar with (a) determined by HSC, though (a) (e) are in the same class. In Table 2, (b) (e) are all similar with (a) determined by HSC. (f) is still dissimilar with (a) as they are in different classes. So it can be seen easily that the proposed approach can recover the shapes those are affine transformed to a normalized state with keeping the dissimilarity of different classes at the same time.

5 Conclusion and Future Work

Overall, Affine Normalization based on Optimization has good properties of eliminating deformation caused by affine transformation. But it is well known that the deformation in real world is extremely complex. More parameters should be considered in normalization in the future.

References

1. Wu, H.C.: Intelligent system. J. Inf. Sci. **103**(1–4), 135–159 (1997)
2. Li, Y., Chen, H., Mei, Y., et al.: Automatic aircraft object detection in aerial images. In: Proceedings of SPIE - The International Society for Optical Engineering, vol. 5253, pp. 547–551 (2003)
3. Yilmaz, A., Shafique, K., Shah, M.: Target tracking in airborne forward looking infrared imagery. Image Vis. Comput. **21**(7), 623–635 (2003)

4. Kass, M., Witkin, A., Terzopoulos, D.: Snakes: active contour models. Int. J. Comput. Vis. **1**(4), 321–331 (1988)
5. Leventon, M.E., Grimson, W.E.L., Faugeras, O., et al.: Level set based segmentation with intensity and curvature priors. In: IEEE Workshop on Mathematical Methods in Bio-medical Image Analysis, pp. 4–11. IEEE Computer Society (2000)
6. Madireddy, R.M., Gottumukkala, P.S.V., Murthy, P.D., et al.: A modified shape context method for shape based object retrieval. Springerplus **3**(1), 1–12 (2014)
7. Belongie, S., Malik, J., Puzicha, J.: Shape matching and object recognition using shape contexts. IEEE Trans. Pattern Anal. Mach. Intell. **24**(4), 509–522 (2010)
8. Peng, S.H., Kim, D.H., Lee, S.L., et al.: A visual shape descriptor using sectors and shape context of contour lines. Inf. Sci. **180**(16), 2925–2939 (2010)
9. Belongie, S., Malik, J., Puzicha, J.: Shape context: a new descriptor for shape matching and object recognition. In: NIPS, pp. 831–837 (2000)
10. Wang, B., Gao, Y.: Hierarchical string cuts: a translation, rotation, scale and mirror invariant descriptor for fast shape retrieval. IEEE Trans. Image Process. **23**(9), 4101–4111 (2014)

Author Index

J.-S. Pan et al. (eds.), *Advances in Intelligent Information Hiding and Multimedia Signal Processing*, Smart Innovation, Systems and Technologies 82,
DOI 10.1007/978-3-319-63859-1

Printed in the United States
By Bookmasters